The Lightning Discharge

Martin A. Uman

Department of Electrical and Computer Engineering
College of Engineering
University of Florida
Gainesville, Florida

DOVER PUBLICATIONS, INC.
Mineola, New York

Copyright

Published in Canada by General Publishing Company, Ltd., 30 Lesmill
Road, Don Mills, Toronto, Ontario.

Bibliographical Note

This Dover edition, first published in 2001, is an unabridged and corrected republication of the work originally published by Academic Press,
Inc., Orlando, Florida, in 1987, as Volume 39 of the International Geophysics Series. A new Preface to the Dover Edition, written by the author,
has been added.

Library of Congress Cataloging-in-Publication Data

Uman, Martin A.
  The lightning discharge / Martin A. Uman.
    p. cm.
  Originally published: Orlando : Academic Press, 1987.
  Includes bibliographical references and index.
  ISBN 0-486-41463-9 (pbk.)
    1. Lightning. I. Title.

QC966 .U4 2001
551.56'32—dc21

00-047398

Manufactured in the United States of America
Dover Publications, Inc., 31 East 2nd Street, Mineola, N.Y. 11501

Contents

Preface to the Dover Edition

This Dover paperback edition will be the third of my hardcover books on lightning that Dover has agreed to republish after the original editions went out of print. Interestingly, the first two, *All About Lightning* (Dover, New York, 1986) and *Lightning* (Dover, New York, 1984), sell more copies each year than were sold annually in hardcover, and both paperbacks have been available for a longer period of time than were the hardcovers. Clearly, there is a very great interest in lightning. Each of my three lightning books has a different flavor. *All About Lightning* is written for the layperson, or for someone at the high school science level. *Lightning* is more technical, written from a historical point of view, and covers the basics, but since it was originally published in 1969, it does not contain the more modern information found in *The Lightning Discharge*, originally published in 1987. The differences between these two books are further discussed in the original preface to *The Lightning Discharge*, which follows. The Dover edition of *The Lightning Discharge* contains corrections to a number of errors in dates, page numbers, data, references, and spelling present in the hardcover original.

There has been one very important advance in lightning research since *The Lightning Discharge* was originally published: the continuing documentation and modeling of the wide variety of luminous phenomena that occur between the thundercloud tops and the lower ionosphere. On page 26 of *The Lightning Discharge*, four lines are devoted to reports of lightning propagating upward from the tops of the thunderclouds. There are now hundreds of published papers concerning the transient sprites, blue jets, blue starters, and elves that illuminate the rarified atmosphere between 15 and 90 km. A review paper that will get the reader started in this area is "Red Sprites, Upward Lightning, and VLF Perturbations" by C. J. Rogers in *Reviews of Geophysics,* volume 37, pp. 317–336, 1999. Additionally, there is indeed lightning on the planet Jupiter (see Chapter 14). It is described in "Galileo Images of Lightning on Jupiter" by Little *et al.* in *Icarus,* volume 141, pp. 306–323, 1999.

Preface to the 1987 Edition

In the 18 years since my technical monograph *Lightning* (McGraw-Hill, New York, 1969; Dover, New York, 1984) was first published, there have been significant advances in our understanding of lightning, but until now there has been no new monograph on the subject. A number of edited collections of papers and conference proceedings relating to lightning have been published during this period and are listed in Appendix D as well as being referenced, where appropriate, throughout the text. Besides being out-of-date, a defect in *Lightning* is its inefficient organization in that the chapters are primarily oriented toward diagnostic techniques and the material is presented in a historical manner. In the present book, the chapters are organized primarily by lightning process. Each chapter contains a reference list of essentially all literature on the subject discussed in that chapter, although all of these references may not be cited in the text.

I have attempted to make *The Lightning Discharge* as self-contained as possible by providing Appendixes on Electromagnetics, Statistics, and Experimental Techniques which discuss the background material needed to help understand most of the text. However, since my interpretation of the literature may, from time to time, contain subjective bias, there is no excuse, when doing research, for not reading and referencing the original literature. A reference to this book, except to the several original contributions of data and interpretation, should be considered an indication of less than perfect scholarship.

It has taken about 5 years to write *The Lightning Discharge*. During that time, about 30 of my students and colleagues have read and criticized various portions of the book, answered questions, and provided material for the tables and figures. In this regard, I would like to single out for special thanks, in alphabetical order, W. H. Beasley, K. Berger, A. A. Few, P. Hubert, V. Idone, E. P. Krider, L. J. Lanzerotti, M. J. Master, R. E. Orville, and E. M. Thomson. I would also like to express my appreciation to R. Crosser and J. Bartlett who, with the best of spirits, survived my seemingly endless revisions of the manuscript and the figures, respectively.

The motivation to write *The Lightning Discharge* was in large part derived from my own involvement in lightning research. That research was strongly influenced by almost 25 years of collaborative studies with E. P. Krider of the University of Arizona (21 coauthored journal articles) and recently by collaborative work at the University of Florida with W. H. Beasley and with E. M. Thomson. Most of the data for these studies has been taken at NASA Kennedy Space Center, Florida, with the much appreciated help of W. Jafferis.

If I have misinterpreted or omitted any significant work from the book, I would like to know about it, so that I can correct such errors in future review articles and perhaps in a later edition of this book.

Chapter 1 | Introduction

1.1 HISTORY

1.1.1 RELIGION AND MYTHOLOGY

Lightning and thunder have always produced fear and respect in mankind, as is evident from the significant role that they have played in the religions and mythologies of all but the most modern of civilizations.

According to Schonland (1964), who reviews 5000 years of nonscientific views on lightning and thunder, early statues of Buddha show him carrying a thunderbolt with arrowheads at each end. (The word thunderbolt is commonly used in the nonscientific literature to refer to cloud-to-ground lightning. That lightning is usually depicted as some form of arrow.) In ancient Egypt, the god Typhon (Seth) hurled the thunderbolts. The ancient Vedic books of India described how Indra, the son of Heaven and Earth, carried thunderbolts on his chariot. A Sumerian seal dating to about 2500 B.C. depicts the lightning goddess Zarpenik riding on the wind with a bundle of thunderbolts in each hand. A reproduction of that seal is found in Prinz (1977) who provides additional perspective on the role of lightning in mythology.

In ancient Greece, lightning was viewed as punishment sent by Zeus, the father of the gods, or by members of his family. The chief god of the Romans, Jupiter or Jove, was thought to use thunderbolts not only as retribution but also as a warning against undesirable behavior. The eagle emblem of Jupiter is shown on the United States one dollar bill with thunderbolts clasped in one of its talons and the olive branch of peace in the other. Interestingly, the *planet* Jupiter was observed by the Voyager 2 spacecraft to be the source of luminous impulses that are probably lightning, as discussed in Section 1.7.2 and Chapter 14. In Rome, from before 300 B.C. to as late as the fourth century A.D., the College of Augurs, composed of distinquished Roman citizens, was charged with the responsibility of determining the wishes of Jupiter relative to State affairs. This was accomplished

1

by making observations on three classes of objects in the sky: birds, meteors, and lightning. In the case of the latter, the observation was always made while looking south, and the location of the lightning relative to the direction of observation was taken as a sign of Jupiter's approval or disapproval.

Perhaps the most famous of the ancient gods associated with lightning was Thor, the fierce god of the Norsemen, who produced lightning as his hammer struck his anvil while he rode his chariot thunderously across the clouds. Thursday, the fifth day of the week, is derived from Thor's Day. In modern Danish, for example, that day is Torsday, in German Donnerstag (thunderday), and in Italian Giovedì (Jove's Day). In Scandinavia, meteorites are referred to as thunderstones, in deference to the view that the foreign material comprising such stones are broken pieces of Thor's hammer. In many other cultures meteorites are associated with thunder and lightning, and it is often believed that they have magical powers to protect against lightning (Nichols, 1965; Prinz, 1977).

Some Indian tribes of North America, as well as certain tribes in Africa (Schonland, 1984; Prinz, 1977), held the belief that lightning was due to the flashing feathers of a mystical *thunderbird* whose flapping wings produced the sounds of thunder. Drawings of the thunderbird are commonly seen in American Indian Art and are widely used commercially, for example, as the name and symbol of a modern automobile.

1.1.2   FROM THE MIDDLE AGES TO BENJAMIN FRANKLIN

Church bells in Medieval Europe often carried the Latin inscription *Fulgura Frango* (I break up the lightning flashes) since it was the practice to ring those bells in an attempt to disperse the lightning. Such activity is not itself without danger since, according to Schonland (1964) who quotes from an eighteenth century German book, in one 33-year period lightning struck 386 church steeples killing 103 bell ringers while they performed their appointed duties.

The following examples of lightning damage to churches illustrates their susceptibility to lightning during the time prior to Benjamin Franklin's invention of lightning protection. The Campanile of St. Mark in Venice, Italy was severely damaged in 1388, set on fire and destroyed in 1417, reduced to ashes again in 1489, and subsequently damaged more or less severely in 1548, 1565, 1653, and 1745. The church was protected using Franklin's grounded rods in 1766 and apparently has suffered no further damage. On April 14, 1718, 24 church towers along the Brittany coast of France were damaged by lightning during thunderstorms. In the eighteenth century, church vaults were used for storing gunpowder and the weapons that used it. In 1769, the steeple of the church of St. Nazaire in Brescia, Italy, whose vaults contained 100 tons of gunpowder, was struck by lightning. The resulting explosion killed three thousand people and destroyed one-sixth of the city.

Interestingly, many historic buildings have never suffered any lightning damage, apparently because they were accidently provided with a lightning protection system similar to that later devised by Franklin. The Temple in Jerusalem, originally built by Solomon, survived 10 centuries of lightning because its dome was covered by metal with rain drains providing a path for the lightning current to flow harmlessly to the ground. The Cathedral of Geneva, Switzerland had a wooden tower that was also covered by metallic plate connected to the ground. It suffered no damage while the nearby and lower bell tower of the Church of St. Gervois was often damaged by lightning.

In addition to nonconducting church towers, wooden ships with wooden masts were obvious targets for lightning damage. Harris (1834, 1838, 1839, 1843) (see also Bernstein and Reynolds, 1978) as part of his crusade to provide lightning protection for the wooden ships of the British navy, reported that from 1799 to 1815 there were 150 cases of lightning damage to British naval vessels. One ship in eight was set on fire, although not necessarily destroyed, about 70 sailors were killed, and more than 130 wounded. Ten ships were completely disabled and the 44-gun ship *Resistance*, its name being an unwary symbol of its electrical susceptibility, was destroyed by a lightning flash in 1798.

1.1.3 BENJAMIN FRANKLIN

The first study of lightning that could be termed scientific was carried out in the second half of the eighteenth century by Benjamin Franklin. For 150 years prior to that time, electrical science had developed to the point that positive and negative charges could be separated by electrical machines via the rubbing together of two dissimilar materials, and these charges could be stored on primitive capacitors called Leyden jars. In November 1749 Franklin wrote the following about the sparks (in his terminology, electrical fluid) he had studied (Franklin, 1774, pp. 47, 50, 331):

Electrical fluid agrees with lightning in these particulars. 1. Giving light. 2. Colour of the light. 3. Crooked direction. 4. Swift motion. 5. Being conducted by metals. 6. Crack or noise in exploding. 7. Subsisting in water or ice. 8. Rending bodies as it passes through. 9. Destroying animals. 10. Meltings metals. 11. Firing inflammable substances. 12. Sulphureous smell. The electrical fluid is attracted by points. We do not know whether this property is in lightning. But since they agree in all particulars wherein we can already compare them, is it not possible they agree likewise in this? Let the experiment be made.

Franklin was the first to design an experiment to prove that lightning was electrical, although others had previously theorized on the similarity between laboratory sparks and lightning (Prinz, 1977). In July 1750 Franklin wrote (Franklin, 1774, pp. 65–66):

> To determine the question whether the clouds that contain lightning are electrified or not, I would propose an experiment to be tried where it may be done conveniently. On the top of some high tower or steeple place a kind of sentry box ... big enough to contain a man and an electrical stand [an insulator]. From the middle of the stand let an iron rod rise and pass bending out of the door, and then upright twenty or thirty feet, pointed very sharp at the end. If the electrical stand be kept clean and dry, a man standing on it when such clouds are passing low might be electrified and afford sparks, the rod drawing fire to him from the cloud. If any danger to the man should be apprehended (though I think there would be none), let him stand on the floor of his box and now and then bring near to the rod the loop of a wire that has one end fastened to the leads, he holding it by a wax handle; so the sparks, if the rod is electrified, will strike from the rod to the wire and not affect him.

His experiment and the results he expected to achieve are illustrated in Fig. 1.1. The aim was to show that the clouds were electrically charged, for if this was the case, it followed that lightning was also electrical. Franklin did not

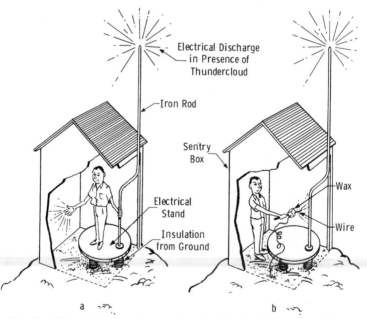

Fig. 1.1 Franklin's original experiment to show that thunderclouds are electrified. (a) Man on electrical stand holds iron rod with one hand and obtains an electrical discharge between the other hand and ground. (b) Man on ground draws sparks between iron rod and a grounded wire held by an insulating wax handle. Adapted from Uman (1971).

appreciate the danger involved in his experiment. If the iron rod were directly struck by lightning, the experimenter would likely be killed. Such was eventually to be the case as we shall see in the next paragraph.

In France in May 1752, Thomas-Francois D'Alibard successfully performed Franklin's suggested experiment. Sparks were observed to jump from the iron rod during a thunderstorm. It was proved that thunderclouds contain electrical charge. Soon after, the experiment was successfully repeated in France again, in England, and in Belgium. In July 1753, G. W. Richmann, a Swedish physicist working in Russia, put up an experimental rod and was killed by a direct lightning strike.

Before Franklin himself got around to performing the experiment, he thought of a better way of proving his theory—an electrical kite. It was to take the place of the iron rod, since it could reach a greater elevation than the rod and could be flown anywhere. During a thunderstorm in 1752 Franklin flew the most famous kite in history (Franklin, 1961a, b). Sparks jumped from a key tied to the bottom of the kite string to the knuckles of his hand as shown in Fig. 1.2. He had verified his theory and had probably done so before he knew that D'Alibard had already obtained the same proof.

In 1749 Benjamin Franklin wrote a letter that was published in *Gentlemen's Magazine*, May 1750, whose editor Edward Cave later published Franklin's book on electricity. It read, in part,

> There is something however in the experiments of points, sending off or drawing on the electrical fire, which has not been fully explained, and which I intend to supply in my next ... from what I have observed on experiments, I am of opinion that houses, ships, and even towers and churches may be eventually secured from the strokes of lightning by their means; for if instead of the round balls of wood or metal which are commonly placed on the tops of weathercocks, vanes, or spindles of churches, spires, or masts, there should be a rod of iron eight or ten feet in length, sharpened gradually to a point like a needle, and gilt to prevent rusting, or divided into a number of points, which would be better, the electrical fire would, I think, be drawn out of a cloud silently, before it could come near enough to strike.

This is Franklin's earliest recorded suggestion of the lightning rod. In the "experiments of points" he placed electrical charge on isolated conductors and then showed that the charge could be drained away (discharged) slowly and silently if a pointed and grounded conductor were introduced into the vicinity. When the pointed conductor was brought too close to the charged conductor, the discharge occurred violently via an electric spark.

In the July 1750 discussion in which he proposed the original experiment to determine if lightning were electrical (Franklin, 1774, pp. 65–66), quoted from above, Franklin repeated his suggestion for protective lightning rods, adding that they should be grounded.

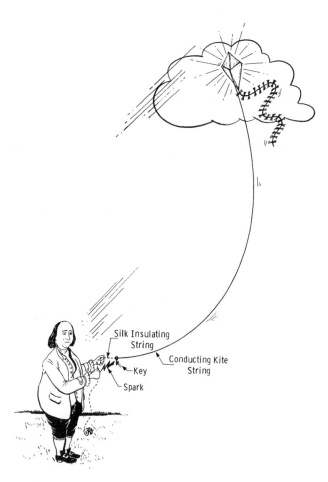

Fig. 1.2 Franklin's electrical kite experiment: sparks jump from the electrified key at the end of the electrified kite string to Franklin's hand. Adapted from Uman (1971).

Franklin originally thought—erroneously—that the lightning rod silently discharged the electric charge in a thundercloud and thereby prevented lightning. However, in 1755 he stated (Franklin, 1774, p. 169):

> I have mentioned in several of my letters, and except once, always in the alternative, viz., that pointed rods erected on buildings, and communicating with the moist earth, would either prevent a stroke, or, if not prevented, would conduct it, so that the building should suffer no damage.

It is in the latter manner that lightning rods actually work.

Lightning rods were apparently first used for protective purposes in 1752 in France and later the same year in the United States (Jernegan, 1928; Van Doren, 1938). The lightning rod was the first practical application of the study of electricity. The electric battery, for example, was not invented by Volta until 1799. Franklin's invention received widespread application and is still today the primary means of protecting structures against lightning.

In addition to showing that clouds contain electricity, Franklin, by measuring the sign of the charge delivered to rods of the type shown in Fig. 1.1 when thunderstorms were overhead, was able to infer that the lower part of the thunderstorm was generally negatively charged (Franklin, 1774, pp. 122–125), a correct observation that was not verified until the early twentieth century.

A review of Franklin's contributions to electrical science is given by Dibner (1977).

1.1.4 THE MODERN ERA

Following Benjamin Franklin there was no significant progress in understanding lightning until the late nineteenth century when photography and spectroscopy became available as diagnostic tools in lightning research. The early history of lightning spectroscopy is reviewed by Uman (1969). Among the early investigators who used time-resolved photography to identify the individual strokes that comprise a lightning discharge to ground and the leader process that precedes first strokes were Hoffert (1889) in England, Weber (1889) and Walter (1902, 1903, 1910, 1912, 1918) in Germany, and Larsen (1905) in the United States. The invention of the double-lens streak camera in 1900 by Boys (1926) in England (Section C.4) made possible the major advances in our understanding of lightning due to Schonland and co-workers in South Africa in the 1930s and thereafter. Their research is discussed throughout this book.

The first lightning current measurements were made by Pockels (1897, 1898, 1900) in Germany. He analyzed the residual magnetic field induced in basalt by nearby lightning currents and by doing so was able to estimate the values of those currents.

Modern lightning research can probably best be dated to Wilson (1916, 1920) in England, the same individual who received a Nobel Prize for his invention of the cloud chamber to track high-energy particles. Wilson was the first to use electric field measurements to estimate the charge structure in the thunderstorm and the charges involved in the lightning discharge. Contributions to our present understanding of lightning have come from researchers throughout the world and cover the time period from Wilson's work to the present. These contributions form the basis of this book. The period from about 1970 to the present has been particularly active in lightning research, as a casual inspection of the references at the ends of the chapters will attest. This activity in part is due (1) to the motivation provided by lightning damage to aircraft, spacecraft, and sensitive ground-based installations because of the vulnerability of modern solid-state electronics including computers, partly, in the case of airborne vehicles, to the decreased electromagnetic shielding afforded by new classes of lightweight structural materials being used in those vehicles (*IEEE Trans.* **EMC-24**, May 1982) and (2) to the development of new techniques of data taking involving both high-speed tape recording and direct digitization and storage under computer control of acquired analog signals.

1.2   CATEGORIZATION OF LIGHTNING FROM CUMULONIMBUS

Lightning is a transient, high-current electric discharge whose path length is measured in kilometers. The most common sources of lightning is the electric charge separated in ordinary thunderstorm clouds (cumulonimbus). The electrification and charge structure of thunderstorms are discussed in Chapter 3. Other sources of lightning are considered in Section 1.7. Well over half of all lightning discharges occur within the thunderstorm cloud and are called intracloud discharges (Section 2.4; Fig. 2.5). The usual cloud-to-ground lightning (sometimes called streaked or forked lightning) has been studied more extensively than other lightning forms because of its practical interest (e.g., as the cause of injuries and death, disturbances in power and communicating systems, and the ignition of forest fires) and because lightning channels below cloud level are more easily photographed and studied with optical instruments. Cloud-to-cloud and cloud-to-air discharges are less common than intracloud or cloud-to-ground lightning. All discharges other than cloud-to-ground are often lumped together and called cloud discharges.

Berger (1978) has categorized lightning between the cloud and earth in terms of the direction of motion, upward or downward, and the sign of charge, positive or negative, of the leader that initiates the discharge. That categorization is illustrated in Fig. 1.3. Category 1 lightning is the most common cloud-to-ground lightning. It accounts for over 90% of the worldwide cloud-to-ground flashes, accurate worldwide statistics being unavailable. It is initiated by a downward-moving negatively charged leader, as shown, and hence lowers negative charge to earth. In Section 1.3 we lay the background for the detailed discussion of this type of lightning that is found in Chapters 4 through 10. Category 3 lightning is also initiated by a

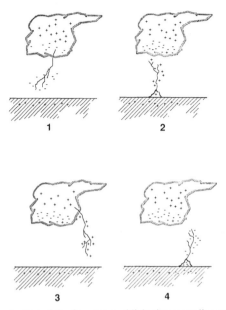

Fig. 1.3 Categorization of the four types of lightning according to Berger (1978).

downward-moving leader, but the leader is positively charged, and hence the discharge lowers positive charge. Less than 10% of the worldwide cloud-to-ground lightning is of this type. Positive cloud-to-ground discharges are discussed in Section 1.4 and Chapter 11. Categories 2 and 4 lightning are initiated by leaders that move upward from the earth and are sometimes called ground-to-cloud discharges. These upward-initiated discharges are relatively rare and generally occur from mountain tops and tall man-made structures. Category 2 lightning has a positively charged leader and may lead to the lowering of negative cloud charge; category 4 a negatively charged leader and may lead to the lowering of positive cloud charge. Upward-initiated discharges are discussed in Section 1.5 and Chapter 11.

In the previous paragraph the phrases "lowers charge" and "lowering of charge" are used. A few words of explanation are appropriate. If, for example, a positively charged upward-moving leader deposits positive charge within a volume of negative cloud charge, it is not possible to state unequivocally from remote electric field measurements that positive charge has indeed been deposited. An identical field change would occur if an equal negative charge were removed or "lowered to ground" or "neutralized." It is usual in the lightning literature to speak of the lowering to ground of cloud charge or of the neutralization of cloud charge by lightning, although this may not be what is physically occurring. Vonnegut (1983) has discussed this problem of terminology obscuring physical processes. In the case of the positive upward leader, it is likely that positive charge will be initially deposited in the cloud within the lower part of the region of cloud charge and that later some of the negative cloud charge will be drained down the existing channel, but, whatever the physical processes involved, there will be an overall "effective" lowering of negative cloud charge. Finally, it should be noted that individual charges are not lowered over the relatively large distance from cloud to ground during the relatively short time duration of the lightning discharge. Rather the charge transport is an effective one in that any flow of electrons (the primary charge carriers) into or out of, for example, the top of the lightning channel results in the flow of other electrons in other parts of the channel, much as would be the case were the channel a conducting wire. Thus coulombs of positive or negative charge can be effectively transferred to ground during the time that an individual electron in the channel moves only a few meters. In this book we will often use the word "effective" before descriptions of such charge-changing processes as lowering, neutralization, transporting, and transferring to emphasize our lack of understanding of the detailed physics of the charge transport.

1.3 NEGATIVE CLOUD-TO-GROUND LIGHTNING

A still photograph of a negative cloud-to-ground discharge is shown in Fig. 1.4. Such a discharge between cloud and ground starts in the cloud and eventually brings to earth tens of coulombs of negative cloud charge. The total discharge is termed a *flash* and has a time duration of about half a second. A flash is made up of various discharge components, among which are typically three or four high-current pulses called *strokes*. Each stroke lasts about a millisecond, the separation time between strokes being typically several tens of milliseconds. Lightning often appears to "flicker" because the human eye can just resolve the individual light pulse associated with each stroke.

In the idealized model of the cloud charges shown in Figs. 1.3 and 1.5, and discussed in Chapter 3 (see also Figs. 3.1, 3.2, 3.3, and 3.4), the main charge regions, P and N, are of the order of many tens of coulombs of positive and negative charge, respectively, and the lower p region contains a smaller positive charge. The following discussion of negative cloud-to-ground lightning is illustrated in Fig. 1.5. The *stepped leader* initiates the first *return stroke* in a flash by propagating from cloud to ground in a series of discrete steps. The stepped leader is itself initiated by a *preliminary breakdown* within the cloud, although there is disagreement about the exact form and location

Fig. 1.4   A still photograph of a typical cloud-to-ground flash. Courtesy J. Rodney Hastings.

of this process (Chapter 4). In Fig. 1.5, the preliminary breakdown is shown in the lower part of the cloud between the N and p regions. The preliminary breakdown sets the stage for negative charge to be lowered toward ground by the stepped leader. Photographically observed leader steps are typically 1 μsec in duration and tens of meters in length, with a pause time between steps of about 50 μsec. A fully developed stepped leader lowers up to 10 or more coulombs of negative cloud charge toward ground in tens of milliseconds with an average downward speed of about 2×10^5 m/sec. The average leader current is in the 100–1000 A range. The steps have pulse currents of at least 1 kA. Associated with these currents are electric and magnetic field pulses with widths of about 1 μsec or less and risetimes of about 0.1 μsec or less. The stepped leader, during its trip toward ground, branches in a downward direction producing the downward-branched geometrical structure seen in Fig. 1.4. A discussion of our present knowledge

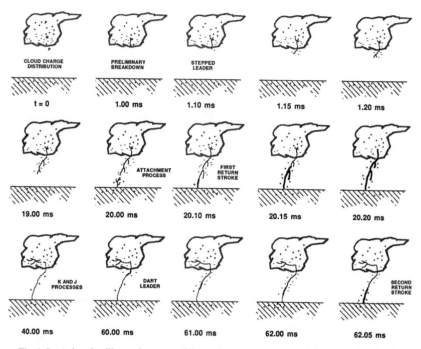

Fig. 1.5 A drawing illustrating some of the various processes comprising a negative cloud-to-ground lightning flash.

of the stepped leader is found in Chapter 5. The preliminary breakdown, the subsequent lowering of negative charge by the stepped leader, and the resultant depletion of negative charge in the cloud combine to produce a total electric field change that can be as short as a few milliseconds or as long as a few hundred milliseconds.

The electric potential of the bottom of the negatively charged leader channel with respect to ground has a magnitude in excess of 10^7 V. As the leader tip nears ground, the electric field at sharp objects on the ground or at irregularities of the ground itself exceeds the breakdown value of air and one or more upward-moving discharges are initiated from those points, thus beginning the *attachment process*, discussed in Chapter 6. When one of the upward-moving discharges from the ground contacts the downward-moving stepped leader some tens of meters above the ground, the leader tip is connected to ground potential. The leader channel is then discharged when a ground potential wave, the first *return stroke*, propagates continuously up the previously ionized and charged leader path. The upward speed of a return stroke near the ground is typically one-third or more times the speed of light, and the speed decreases with height. The total transit time from ground to the top of the channel is typically about 100 μsec. The first return stroke produces a peak current near ground of typically 30 kA, with a time from zero to a peak of a few microseconds. Currents measured at the ground fall to half of the peak value in about 50 μsec, and currents of the order of hundreds of amperes may flow for times of a few milliseconds up to several hundred milliseconds. The rapid release of return-stroke energy heats the leader channel to a temperature near 30,000 K and generates a high-pressure channel that expands and creates the shock waves that eventually become thunder (Chapter 15). The return stroke effectively lowers to ground the charge originally deposited on the stepped leader channel as well as other charges that may be available to the top of its channel, and, in so doing, produces an electric field change with time variations that range from a submicrosecond scale to many milliseconds. All aspects of the return-stroke process are considered in Chapter 7.

After the return-stroke current has ceased to flow, the flash, including charge motion in the cloud, may end. The lightning is then called a single stroke flash. On the other hand, if additional charge is made available to the top of the channel, a continuous *dart leader* may propagate down the residual first-stroke channel at a speed of about 3×10^6 m/sec. During the time between the end of the first return stroke and the initiation of a dart leader, *J*- and *K-processes* occur in the cloud (Chapter 10). There is controversy as to whether these processes are necessarily related to the initiation of the dart leader. The dart leader lowers a charge of the order of 1 C by virtue of a current of about 1 kA. The dart leader then initiates the second (or any subsequent) return stroke. Some leaders begin as dart leaders but toward the end of their trip toward ground become stepped leaders. These leaders are known as dart–stepped leaders. Dart leaders and return strokes subsequent to the first are usually not branched. Dart-leader electric field changes typically have a duration of about 1 msec. Subsequent return-stroke overall field changes are similar to, but usually a factor of two or so smaller than, first return-stroke field changes. Subsequent return-stroke currents have faster zero-to-peak rise times than do first stroke currents but similar maximum rates of change. Subsequent return-stroke characteristics are discussed in Chapter 7, and dart leader characteristics in Chapter 8.

The time between successive return strokes in a flash is usually several tens of milliseconds, as we have noted in the first paragraph of this section, but can be tenths of a second if a *continuing current* flows in the channel after a return stroke. Continuing current magnitudes are of the order of 100 A and represent a direct transfer of charge from cloud to ground. The typical electric field change produced by a continuing current is linear for roughly 0.1 sec and is consistent with the lowering of about 10 C of cloud charge to

ground. Between one-quarter and one-half of all cloud-to-ground flashes
contain a continuing current component. Continuing current is not illustrated in Fig. 1.5 but luminosity associated with it is apparent in the streak-
camera photograph in Fig. 1.7 and in the streak-camera drawing in Fig. 1.9
(see below). Also shown in Fig. 1.9 are the impulsive brightenings of the
continuing current channel known as M-components. All aspects of
continuing currents including M-components are discussed in Chapter 9.

As a way of summarizing the previous discussion and illustrating the type
of lightning data that is obtained using photographic techniques, a drawing
of both a streak photograph (obtained using a camera of the type shown in
Fig. C.8 and discussed in Section C.4) and a still photograph of a three-stroke
lightning flash similar to the cloud-to-ground discharge of Fig. 1.5 is shown
in Fig. 1.6. An actual streak photograph resolving the strokes of a flash is
shown in Fig. 1.7, and a streak photograph resolving the steps of the stepped
leader is given in Fig. 1.8. Examples of simultaneous photographic and
electric field records from two cloud-to-ground flashes are shown in Fig. 1.9.
Other examples of ground-flash electric fields on a millisecond time scale are
found in Figs. 1.12, 9.1, and 4.3. The sign convention for lightning electric
fields is discussed in Section A.1.1.

Typical histograms of the number of strokes per ground flash are given
in Fig. 1.10. A summary of almost all published measurements, 18, of this
parameter is found in Thomson (1980) who, from these data, calculated a
global mean of 3.5 strokes per flash. Schonland (1956) and Holzer (1953)
state that frontal storms generally produce more strokes per flash than do
local convective storms, an observation that may account for the fact that
Fig. 1.10a appears to have two peaks. On the other hand, the Florida

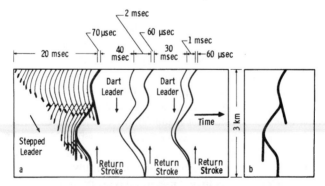

Fig. 1.6   (a) A drawing of the luminous features of a lightning flash below a 3-km cloud base
as would be recorded by a streak camera (Section C.4; Fig. C.8). Increasing time is to the right.
For clarity the time scale has been distorted. (b) The same lightning flash as would be recorded
by a camera with stationary film. Adapted from Uman (1969).

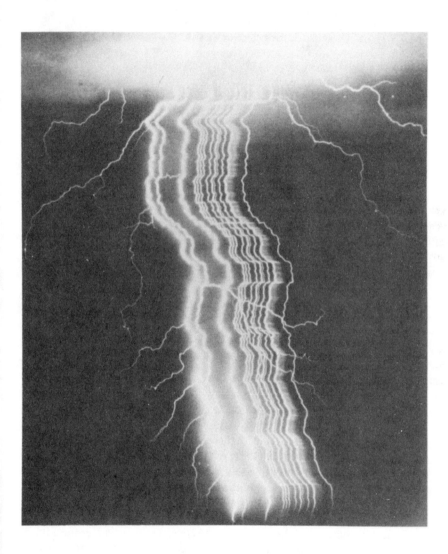

Fig. 1.7 Streak-camera photograph of a 12-stroke lightning flash. The first stroke is on the left and is the only branched stroke. Increasing time goes from left to right. Continuing current, as evidenced by continuing luminosity, flows after the eleventh stroke. Photograph is of lightning near Socorro, New Mexico. Courtesy, Marx Brook, New Mexico Institute of Mining and Technology.

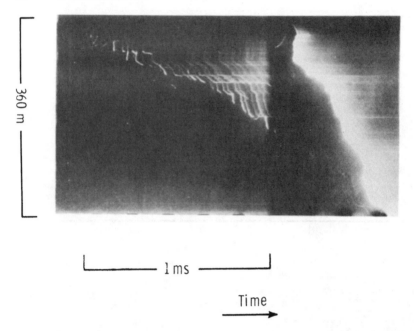

Fig. 1.8 Streak-camera photograph of a stepped leader. On the left side of the photograph the intensity of the leader is greatly enhanced. This enhancement was accomplished by varying the exposure in the reproduction process. Photograph is of lightning to Monte San Salvatore, near Lugano, Switzerland, and was originally published by Berger and Volgelsanger (1966). Courtesy, K. Berger and the Swiss High Voltage Research Committee (FKH), Zurich.

histogram for local convective storms shown in Fig. 1.10b also shows evidence of a bimodal distribution. Thomson *et al.* (1984) review and compare the available data on the number of separate channels to ground per flash and the interstroke intervals associated with these separate channels. From their own measurements Thomson *et al.* (1984) report a mean of 1.6 spatially separate channels per flash for 78 multiple-stroke flashes out of a total of 105 flashes. Clifton and Hill (1980) list the results of seven different measurements of the percentage of flashes having spatially separate channels. The percentage found in the various studies is between about 15 and 50.

Typical histograms of the time interval between strokes in a ground flash are given in Fig. 1.11. A comparison of essentially all studies, 19, of this parameter is found in Thomson (1980). From the worldwide data he calculated a geometric mean interstroke interval of 58 msec. Thompson *et al.* (1984) found that interstroke interval did not vary with stroke order, a result in agreement with Schonland (1956) but in disagreement with Kitagawa and Kobayashi (1958).

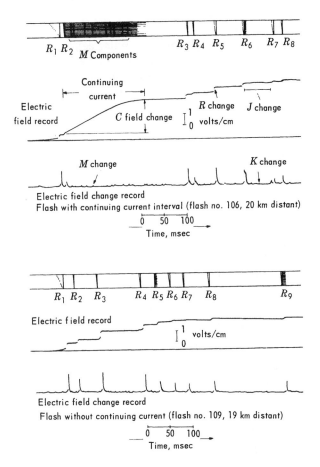

Fig. 1.9 Examples of simultaneous photographic and electric field measurements for multiple-stroke ground flashes. The flash whose records are shown at the top contained a continuing current interval while that at the bottom did not. In the records termed "electric field change" the signal is allowed to decay to zero with a 70-μsec time constant so as to emphasize and to measure accurately only the fast field changes. The electric field measuring systems used were similar to that shown in Fig. C.2 and discussed in Section C.1 with RC = 70 μsec for the "field change" system and RC = 4 sec for the "field" system. The records have been modified somewhat from the originals for illustrative purposes. Adapted from Kitagawa *et al.* (1962).

1.4   POSITIVE CLOUD-TO-GROUND LIGHTNING

Positive cloud-to-ground lightning is discussed in detail in Chapter 11 and reference to other portions of this book containing data on positive lightning is found in that chapter. A photograph of a positive flash is found in Fig. 11.1. Positive ground flashes are of considerable practical interest because their peak current and total charge transfer can be much larger than the more common negative ground flash. The largest recorded peak currents, those in

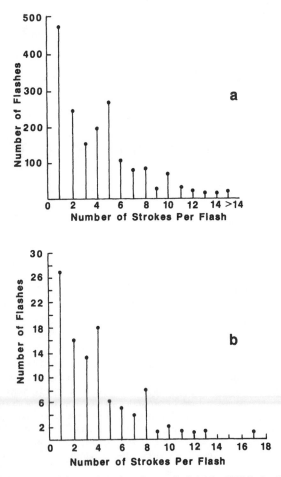

Fig. 1.10   Histograms of the number of strokes per flash (a) for 1800 flashes in South Africa studied by Schonland (1956) and (b) for 105 flashes in Florida studied by Thomson *et al.* (1984). The mean number of strokes per flash in South Africa was 4.1 and in Florida was 4.0.

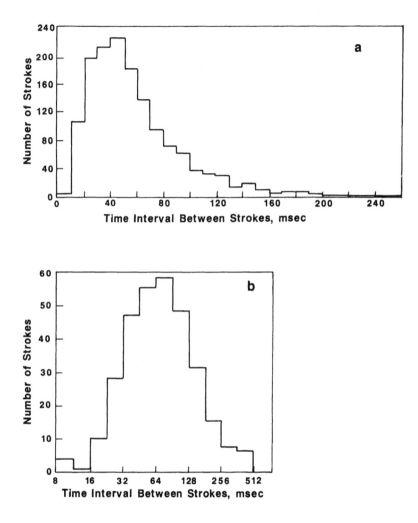

Fig. 1.11 Histograms of the interstroke time interval for (a) 1482 flashes in South Africa studied by Schonland (1956) and (b) for 105 flashes in Florida studied by Thomson *et al.* (1984). Note that the horizontal axis in b is logarithmic and that the histogram plotted with that logarithmic axis is roughly symmetrical. Thomson (1980) and Thomson *et al.* (1984) have shown that the distribution of interstroke intervals follows log normal statistics. The geometric mean interstroke interval in South Africa was 51 msec and in Florida was 69 msec. The arithmetic mean interstroke interval in South Africa was 63 msec and in Florida was 90 msec. A discussion of the log normal distribution, and of the geometric mean, which is also the median value for the case of a log normal distribution, is found in Appendix B.

the 200–300 kA range, are due to the return strokes of positive lightning. Positive flashes to ground are initiated by leaders that do not exhibit the distinct steps of their negative counterparts. Rather they show a luminosity that is more or less continuous but modulated in intensity, as shown in Fig. 11.1. Positive flashes are generally composed of a single stroke followed by a period of continuing current. Positive flashes are probably initiated from the upper positive charge in the thundercloud when that cloud charge is horizontally separated from the negative charge beneath it.

Positive flashes are relatively common in winter thunderstorms (snowstorms), which produce few flashes overall, and are relatively uncommon in summer thunderstorms. The fraction of positive lightning in summer thunderstorms apparently increases with increasing latitude and with increasing height of the ground above sea level.

1.5 ARTIFICIALLY AND UPWARD-INITIATED LIGHTNING

Lightning is initiated by upward-moving leaders propagating from tall natural geographical features and man-made structures, as noted in Section 1.2. A photograph of upward lightning from television towers is found in Fig. 12.1. Upward lightning from a tower on a mountain top is shown in Figs. 12.4 and 12.5. In addition, upward-moving leaders can be initiated when grounded conductors of the order of 100 m in length are rapidly introduced into the electric fields present beneath an active thunderstorm, as illustrated in Fig. 12.6. Considerable research has been conducted on the lightning initiated by launching small rockets trailing grounded wires toward overhead thunderstorms. We arbitrarily term upward-initiated lightning via such rockets and similar schemes and that from man-made structures as "artificially initiated." Such lightning is thought to be similar in most respects to natural upward-initiated lightning from, for example, mountain tops. The various types of upward-initiated lightning are discussed and compared in Chapter 12.

The leaders of upward-initiated lightning are generally positively charged, category 2 in Fig. 1.3. Positively charged upward leaders show a continuous modulated luminosity, as illustrated in Fig. 12.5; negative upward leaders, category 4, exhibit a stepped behavior similar to negative downward leaders, as illustrated in Fig. 12.4. Category 4 is the rarest of the types of lightning between cloud and ground. Positive upward leaders often enter the cloud and result in a more or less continuous current flow of 100–1000 A. About half the time, however, that phase of the discharge is followed by dart-leader and return-stroke sequences very similar to those following first strokes in cloud-to-ground discharges initiated by negative downward-moving leaders. Some long negative upward leaders apparently contact positive downward-moving leaders, with a resultant upward- and downward-propagating return stroke from the junction point. The result at ground level is a positive impulsive current with a longer risetime than produced by a typical negative downward leader met by a relatively short upward-moving positive connecting discharge.

1.6   CLOUD DISCHARGES

We define a cloud discharge as any lightning that does not connect to earth. Cloud discharges are discussed in Chapter 13. As noted in Section 1.2, the majority of all lightning discharges occur within the confines of the cloud. Cloud discharges can be subdivided into intracloud, intercloud, and cloud-to-air flashes, but there is no experimental data at present to distinguish between the characteristics of these three types. Indeed, based on electric field records, there is considerable similarity between these discharges. The term cloud discharge could also be applied to those portions of a flash to ground that take place within the cloud. In some cases, flashes that are primarily within the cloud, and are best characterized as cloud flashes, produce, seemingly as an unimportant by-product, a channel to ground.

Intracloud lightning discharges typically occur between positive and negative cloud charges and have total time durations about equal to those of ground discharges, about half a second. A typical cloud discharge transfers tens of coulombs of charge over a total spatial extent of 5–10 km. The discharge is thought to consist of a continuously propagating leader that generates weak return strokes called recoil streamers, with which are associated electric fields termed K-changes, when the leader contacts pockets of space charge opposite to its own. The cloud K-changes are similar, but usually of opposite polarity, to the K-changes associated with K-processes that occur in the intervals between return strokes in ground discharges.

A comparison of typical ground-flash and cloud-flash electric fields in both the time and frequency domains for discharges at about 20 km is shown in Fig. 1.12. The time-domain fields at close range exhibit a ramplike change with superimposed steps for both types of discharges, but the steps in ground discharges, which are due to the return strokes, are much larger than the cloud-flash steps due to K-changes. The outputs of narrowband receivers centered around the indicated frequencies are shown below the time-domain fields of each flash. For instance, during a ground flash the output from a 100-kHz receiver first shows a series of pulses corresponding to preliminary breakdown and leader activity and then a major pulse at the time of the first and each subsequent return stroke. The pulses above about 10 MHz are of roughly equal magnitude for both ground and cloud flashes and are probably

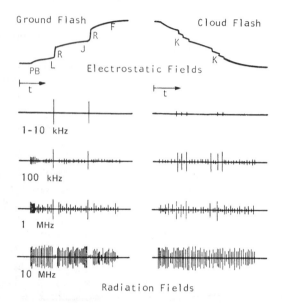

Fig. 1.12   Electric fields in the time and frequency domains for a typical cloud-to-ground flash and a typical intracloud flash both at a distance of about 20 km. Amplitude scales for different frequencies are not the same. Adapted from Malan (1958, 1963).

produced by relatively small-scale discharges that are primarily within the cloud. The return-stroke and K-change frequency spectra both have maxima in the 5-kHz range, but the return-stroke spectrum has a larger amplitude.

1.7   UNUSUAL DISCHARGES

In Section 1.7.1 we discuss unusual forms of natural lightning, and then in Section 1.7.2 we consider lightning-like discharges that are created (1) in the Earth's atmosphere by phenomena other than thunderstorms and (2) in the atmosphere of planets other than Earth.

1.7.1   NATURAL LIGHTNING

So-called heat and sheet lightning do not exist as separate physical entities. Heat lightning is lightning or lightning-induced cloud illumination not accompanied by thunder because the cloud is beyond the distance that thunder can be heard, generally about 25 km (Section 15.2.3). Sheet lightning is the name given to a similar lightning-induced cloud illumination that causes a sheetlike section of the cloud or clouds to become luminous. In both cases, if one were near the storm producing the heat or sheet lightning, normal lightning would be observed.

Rocket lightning is the name given to the long cloud-to-air discharges, generally seen best at night, which give the impression of relatively slow horizontal progression, probably because new channel sections are being illuminated at time intervals on the order of 10 msec. Rocket lightning is most likely a normal part of cloud discharges, unusual in that the discharge channels are visible by virtue of their location outside the cloud.

Ribbon lightning occurs when the cloud-to-ground discharge channel is shifted by the wind in the time between strokes so that each stroke in the flash is separated horizontally. An example is shown in Fig. 1.13.

Bead lightning is the name given to that form of lightning in which the channel to ground breaks up, or appears to break up, into luminous fragments, generally some tens of meters in length. These beads appear to persist for a longer time than does a normal cloud-to-ground channel. The literature on bead lightning is reviewed by Barry (1980) and by Uman (1969). A frame of a movie showing bead lightning occurring during a strike to a depth charge plume is reproduced in Fig. 1.14. Barry (1980) gives additional photographs and drawings. Bead lightning occurs in rocket-initiated lightning having particularly long continuing current (Section 12.1).

Ball lightning is the name given to the luminous, usually red, spheres that often appear to be associated with the location of the strike point of a cloud-to-ground flash. The orange- to basketball-size balls have a lifetime of about 1 sec during which time they generally move horizontally and maintain a roughly constant, but not bright, luminosity. In addition to appearing out of doors, ball lightning has been reliably reported to occur inside both houses and all-metal airplanes. A reproduction of a woodcut showing ball lightning that apparently entered a structure through the chimney is found in Fig. 1.15. Ball lightning, or a similar phenomenon, is also generated at electrical outlets and telephone headsets and from some types of high-power electrical equipment. The subject of ball lightning has been thoroughly reviewed by Barry (1980) in a book that contains a listing of virtually every literature reference, over 1600, to the phenomenon. Despite the relative wealth of similar ball lightning observations over a period of centuries, reports that leave little doubt as to its reality, there is still no consensus as to the physical mechanism or mechanisms responsible for ball lightning. Unfortunately, a significant fraction of the theoretical literature on ball lightning could best be described as rubbish, so the uninitiated reader should read the literature with more than the usual level of skepticism. Two other ball lightning books (Cade and Davis, 1969; Singer, 1971) have been published recently but are inferior to Barry's. Reviews of ball lightning are also found in Uman (1969, 1971).

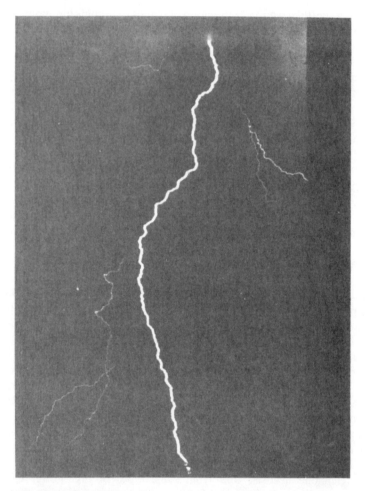

Fig. 1.13 Ribbon lightning near Tucson, Arizona photographed with a tripod-mounted camera. Courtesy, George Marcek.

Lightning has occasionally been reported to occur from a clear blue sky (McCaughan, 1926; Gisborne, 1928; Myers, 1931; Gifford, 1950; Baskin, 1952; Waldteufel *et al.*, 1980), commonly referred to as a "bolt from the blue." Some of these events refer simply to blue sky overhead and thunderstorms 10 or more km away, out of viewing range, from which the lightning originates. However, given the convincing photographs and supporting charge locations of Waldteufel *et al.* (1980), it appears that clear air can occasionally supply sufficient charge to support a lightning discharge to ground.

Fig. 1.14 Bead lightning as it appears on one frame of a motion picture showing a three-stroke flash striking the top of a depth charge plume. Courtesy, U.S. Naval Ordnance Laboratory.

Fig. 1.15 Ball lightning in a nineteenth century woodcut from "L'Atmosphere," by C. Flammarion, 2nd ed., Paris, 1873. The original title, translated from French, reads "Ball Lightning Crossing a Kitchen and a Barn." Apparently the ball lightning came down the chimney used to exhaust the cooking fires. Courtesy, Burndy Library.

Lightning has been reported to propagate upward from the top of clouds and perhaps to reach the ionosphere (Wright, 1950; Ashmore, 1950; Hoddinott, 1950; Vaughan and Vonnegut, 1982; Vonnegut, 1984; Brook *et al.*, 1985).

In almost all cases thunderstorms do not produce lightning unless their tops are at heights well above the freezing level, that is, where the temperature is well below 0°C. Hence most researchers think that the cloud electrification processes that lead to lightning must involve supercooled water and ice (Section 3.3). There are, however, a few reports of so-called "warm cloud lightning" in which the cloud tops are reported to be at an altitude lower than that of the freezing level (e.g., Foster, 1950; Moore *et al.*, 1960; Pietrowski, 1960; Michnowski, 1963; Section 3.3).

Finally, astronauts in Earth orbit have reported that there appears to be a collective or "sympathetic" organization to the lightning occurring over a wide area (Vonnegut *et al.*, 1985). Two or three lightning flashes may occur simultaneously (to the eye), and groups of 10 to 50 flashes occur in a time less than about 10 sec over an area of about 10^6 km^2. These bursts of lightning activity are separated by quiet times of a few seconds.

1.7.2 LIGHTNING-LIKE DISCHARGES

During its flyby of Jupiter in 1979, the Voyager 2 spacecraft photographed 20 luminous impulses against the disk of the planet (Fig. 14.4) and detected 167 radio frequency signals propagating in the whistler mode in its magnetosphere (Fig. 14.5), providing strong evidence for lightning on that planet. A detailed discussion of the evidence for lightning or lightning-like discharges on Jupiter as well as on Venus, Saturn, and Uranus, where the data are less convincing, is presented in Chapter 14.

On Earth, lightning-like discharges not associated with thunderstorms have been observed in the ejected material above some active volcanoes, as illustrated in Fig. 1.16. These discharges are hundreds of meters in length (e.g., Anderson *et al.*, 1965; Brook *et al.*, 1974; Pounder, 1980; Section 14.1). Shorter discharges, about 1 m in length, have been observed in New Mexico gypsum sandstorms (Kamra, 1972). Transients luminous phenomena that may be due to electrical discharges have long been observed during earthquakes and speculatively attributed to the electric fields generated by seismic strain (Finkelstein and Powell, 1970). Finally, by virtue of the detonation of thermonuclear devices (H-bombs) at ground level, negative charge is deposited in the atmosphere that results in the kilometer-length discharges shown in Fig. 1.17 (e.g., Uman *et al.*, 1972; Hill, 1973; Grover, 1981; Gardner *et al.*, 1984). The upward-going leaders associated with this

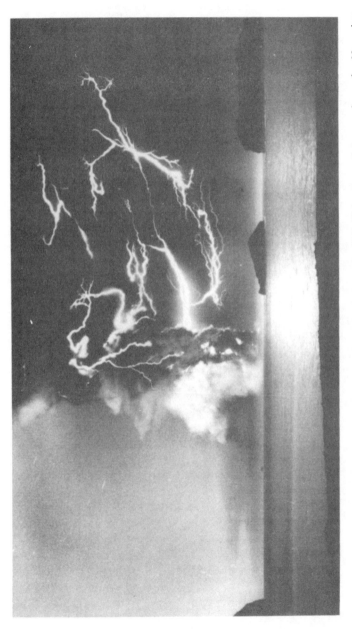

Fig. 1.16 Lightning in the volcano cloud over Surtsey, near Iceland, in December 1963. Volcanic eruptions were observed on November 14, 1963 off the southern coast of Iceland in water 130 m deep. Within 10 days an island nearly 1 km long and about 100 m above sea level was formed. The island was named Surtsey by the Icelandic government. Additional discussion is found in Anderson *et al.* (1965). The photograph originally appeared on the cover of the issue of *Science* (28 May 1965) containing that article. Copyright 1965 by the AAAS. Courtesy, Sigurgeir Jónasson, Icelandair.

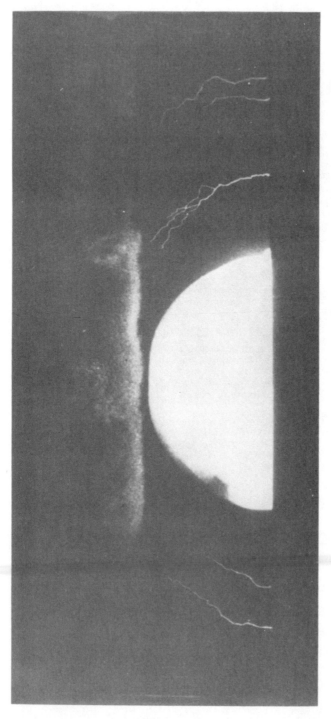

Fig. 1.17 Five lightning flashes induced by an experimental thermonuclear device exploded on October 31, 1952, at Eniwetok in the Pacific. Photograph is frame number 72 (detonation occurs in frame 1) of a 2000 frame/sec movie taken from about 30 km away. The five flashes were initiated immediately after detonation, probably from instrumentation stations projecting above sea level and propagated upward similar to typical flashes from very tall structures. All flashes lasted almost one-tenth of a second and apparently none contained subsequent strokes. The tops of the lightning channels bend toward the fireball. Scattered trade-wind cumulus clouds are visible with bases at about 600 m, roughly the radius of the fireball in frame 72. The clouds were not lightning producers. Photograph originally published in Uman *et al.* (1972). Lightning channels in the photograph are enhanced for illustrative purposes. Courtesy, U.S. Atomic Energy Commission.

so-called nuclear lightning are apparently initiated from small structures on the Earth's surface and probably are similar to the other forms of upward-initiated lightning discussed in Chapter 12.

1.8   EFFECTS OF LIGHTNING

Lightning has likely been present for the period of time during which life has evolved on earth, and, in fact, lightning has been suggested as a source for generating the necessary molecules from which life could evolve (Miller and Urey, 1959; Chameides and Walker, 1981). Evidence of 250,000-year-old lightning has been found in ancient fulgurites, glassy tubes that lightning forms in the earth (Harland and Hacker, 1966). Photographs of fulgurites are found, for example, in Uman (1971).

Lightning is a source of ignition of forest fires and hence has played a significant role in determining the composition of trees and plants in much of the world's forests. For example, according to Love (1970), until recently, frequent fires kept the California forest floor clean. These fires were small and did not damage the trees. Ironically, efforts to prevent and contain forest fires in California enabled the brush to grow more thickly, and now most fires are big ones. We may even be indebted to ancient fires for California's giant sequoias. The seedlings of these trees can germinate in ashes but are suppressed under the thick layer of needles that might cover an unburned forest floor. Further discussion of the ecological effects of lightning-caused forest fires in California is found in Kilgore and Briggs (1972). Besides being a fire producer, lightning is a predator in that it can, by direct strike, kill a tree or a group of trees. For example, lightning, by preferentially striking the taller trees in the southeastern United States, has maintained the balance between the taller pines and the smaller oaks that, without the lightning predator, would be shaded from sunlight and die out (e.g., Komarek, 1968, 1973).

Lightning produces chemicals in and around its hot discharge channel that would otherwise not be present in the atmosphere, at least in such abundance, including so-called fixed nitrogen, which plants use in the food-making process (e.g., von Liebig, 1827; Noxen, 1976; Chameides et al., 1977; Griffing, 1977; Drapcho et al., 1983; Borucki and Chameides, 1984; Hill et al., 1980, 1984; Levine et al., 1984; Ko et al., 1986).

Thunderstorms and lightning play a role in maintaining the fine-weather electric field, about 100 V/m pointing downward, that is due to the negative charge on the Earth and the positive space charge in the atmosphere (e.g., Israel, 1971; Pierce, 1974). A drawing showing how such a balance might occur is found in Fig. 1.18. The electrical conductivity of the atmosphere

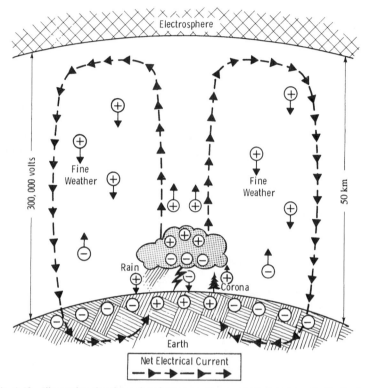

Fig. 1.18 Illustration showing a thunderstorm acting as a battery to keep the Earth charged negatively and the atmosphere charged positively. Atmospheric electrical currents flow downward in fine weather and upward in thunderstorms. Thunderstorms deliver charge to the earth by lightning, rain, and corona discharges. Adapted from Uman (1971).

increases with height. The atmosphere is a good conductor to slowly varying signals at about 50 km, a level known as the electrosphere. The value of conductivity necessary to reflect radio waves occurs somewhat higher, in the region known as the ionosphere. The voltage between the Earth and the electrosphere in regions of fair weather is about 300,000 V. To maintain this voltage the earth has about 10^6 C of negative charge on its surface, an equal positive charge being distributed throughout the atmosphere. In regions of fine weather, atmospheric currents of the order of 1000 A are continuously depleting this charge. The charge is apparently replaced by the action of thunderstorms including lightning. The thunderstorm system acts as a type of battery to keep the fine weather system charged. The process has been modeled mathematically by, for example, Chalmers (1967), Uman (1974), Hays and Roble (1979), and Tzur and Roble (1985).

Lightning is alleged to produce a variety of effects in the ionosphere and magnetosphere (e.g., Rycroft, 1973; Helliwell *et al.*, 1973; Woodman and Kudeki, 1984; Kelley *et al.*, 1984; Voss *et al.*, 1984; Carpenter *et al.*, 1984; Inan *et al.*, 1985; Goldberg *et al.*, 1986). Ground-based observations of whistlers (a part of the electromagnetic radiation generated by lightning and discussed in Sections 14.2, 14.3.3, and 14.4.2) and the correlated perturbations of electromagnetic waves propagating in the cavity between the Earth's surface and the ionosphere provided the first, albeit indirect, evidence of whistler-induced precipitation of electrons in the magnetosphere (Helliwell *et al.*, 1973; Inan *et al.*, 1985). Direct observations of such lightning-induced electron precipitation on a low-altitude satellite have been reported by Voss *et al.* (1984) in a study that involved the measurement of a sequence of bursts of 100–200 keV electrons in one-to-one association with whistlers observed at Palmer Station, Antarctica. Rocket-based observations of similar events have been reported by Rycroft (1973) and, more recently, by Goldberg *et al.* (1986) who found lightning correlated with the precipitation of electrons having energies greater than 40 keV and greater than 120 keV at altitudes near 100 km above the Earth's surface. Woodman and Kudeki (1984) and Kelly *et al.* (1984) present evidence that the occurrence of lightning causes a sudden increase in the radar signal scattered from an altitude near 250 km. This increase in radar cross section, called explosive spread F because it occurs in the F region of the ionosphere, is postulated to be due to a plasma instability triggered by a lightning electric field in the F region of the order of 10 mV/m.

Among the various so-called deleterious effects of lightning are death and injury to people and animals, damage and destruction to ground-based structures and to airborne vehicles, forests fires (although, as we have discussed above, such fires are not necessarily bad), outages of power and communication lines, and damage to sensitive solid-state electronic components such as found in modern computers. Review of our knowledge of lightning injury and death are found in Golde and Lee (1976) and in Cooper (1980). An excellent book on the lightning protection of structures as well as of trees, ships, and people is Golde (1973). Reference should also be made to the U.S. Lightning Protection Code (1980).

Finally, it is appropriate to make some comment about the present impracticality of using lightning as an energy source. Each cloud-to-ground lightning involves an energy of roughly 10^9–10^{10} J (Section A.1.6). If, worldwide, there are 100 flashes to ground per second (Section 2.4) and if all this energy could be captured, the maximum power available would be 10^{12} W, a value comparable to the peak electrical power consumed in the United States in the latter 1980s, between 0.5 and 0.6×10^{12} W. However, there are two seemingly insolvable problems in tapping the power of lightning:

(1) most of the power is converted to thunder, hot air, and radio waves which cannot, at present, be recovered, leaving only a small fraction available at the channel base for immediate use or storage and (2) it is impractical to intercept with tall towers or similar schemes any significant amount of lightning energy. For example, the total energy in a single flash, if it could be captured, would operate a single light bulb for only a few months. Since only a small fraction of the total flash is available, thousands of flashes would have to be captured during that period to power a single home. A 300-m tower in Florida is struck about 100 times per year so that many tens of towers would be needed for each home that used lightning as an energy source. Such a system of towers would, among its other negative aspects, be aesthetically unacceptable.

REFERENCES

Anderson, R., S. Bjornsson, D. C. Blanchard, S. Gathman, J. Hughes, S. Jonasson, C. B. Moore, H. J. Survilas, and B. Vonnegut: Electricity in Volcanic Clouds. *Science*, **148**:1179–1189 (1965).

Ashmore, S. E.: Unusual Lightning. *Weather*, **5**:331 (1950).

Barry, J. D.: "Ball Lightning and Bead Lightning." Plenum, New York, 1980.

Baskin, D.: Lightning without Clouds. *Bull. Am. Meteorol. Soc.*, **33**:348 (1952).

Berger, K.: Blitzstrom-Parameter von Aufwärtsblitzen. *Bull. Schweiz. Elektrotech. Ver.*, **69**:353–360 (1978).

Bernstein, T., and T. S. Reynolds: Protecting the Royal Navy from Lightning—William Snow Harris and His Struggle with the British Admiralty for Fixed Lightning Conductors. *IEEE Trans. EDUC.*, **E-21**:1–14 (1978).

Borucki, W. J., and W. L. Chameides: Lightning: Estimates of the Rates of Energy Dissipation and Nitrogen Fixation. *Rev. Geophys. Space Phys.*, **22**:363–372 (1984).

Boys, C. V.: Progressive Lightning. *Nature (London)*, **118**:749–750 (1926).

Brook, M., C. B. Moore, and J. Segurgenssen: Lightning in Volcanic Clouds. *J. Geophys. Res.*, **79**:472–475 (1974).

Brook, M., C. Rhodes, O. H. Vaughn, Jr., R. E. Orville, and B. Vonnegut: Nighttime Observations of Thunderstorm Electrical Activity from a High Altitude Airplane. *J. Geophys. Res.*, **90**:6111–6120 (1985).

Cade, C. M., and D. Davis: "The Taming of Thunderbolts." Abelard-Schuman, New York, 1969.

Carpenter, D. L., U. S. Inan, M. L. Trimpi, R. A. Helliwell, and J. P. Katsufrakis: Perturbations of Subionospheric LF and MF Signals Due to Whistler-Induced Electron Precipitation Burst. *J. Geophys. Res.*, **89**:9857 (1984).

Chalmers, J. A.: "Atmospheric Electricity," 2nd Ed. Pergamon, Oxford, 1967.

Chameides, W. L., D. H. Stedman, R. R. Dickerson, D. W. Rusch, and R. J. Cicerone: NO_x Production in Lightning. *J. Atmos. Sci.*, **34**:143–149 (1977).

Chameides, W. L., and J. C. G. Walker: Rates of Fixation by Lightning of Carbon and Nitrogen in Possible Primitive Atmospheres. *Orig. Life*, **11**:291–302 (1981).

Clifton, K. S., and C. K. Hill: Low Light Level Television Measurements of Lightning. *Bull. Am. Meteorol. Soc.*, **61**:987–992 (1980).

Cooper, M. A.: Lightning Injuries: Prognostic Signs for Death. *Ann. Emerg. Med.*, **9**:134–138 (1980).

Dibner, B.: Benjamin Franklin. *In* "Lightning, Vol. 1, Physics of Lightning" (R. H. Golde, ed.), pp. 23–49. Academic Press, New York, 1977.

Drapcho, D. L., D. Sisterson, and R. Kumar: Nitrogen Fixation by Lightning Activity in a Thunderstorm. *Atmos. Environ.*, **17**:729–734 (1983).

Finkelstein, D., and J. Powell: Earthquake Lightning. *Nature (London)*, **228**:759–760 (1970).

Foster, H.: An Unusual Observation of Lightning. *Bull. Am. Meteorol. Soc.*, **31**:40 (1950).

Franklin, Benjamin: "The Autobiography and Other Writings," p. 167. Signet Paperback, The New American Library, New York, 1961a.

Franklin, Benjamin: "The Autobiography and Other Writings," pp. 234–235. Signet Paperback, The New American Library, New York, 1961b. Gives Priestly's account of Franklin's kite flight, taken from Priestly, J.: "History of the Present State of Electricity." Dodsley, Johnson, Davenport, & Cadell, London, 1967.

Franklin, Benjamin: *In* "Experiments and Observations on Electricity Made at Philadelphia" (E. Cave, ed.). London, 1774. See also, Cohen, I. B.: "Benjamin Franklin's Experiments." Harvard Univ. Press, Cambridge, Massachusetts, 1941.

Gardner, R. L., M. H. Frese, J. L. Gilbert, and C. L. Longmire: A Physical Model of Nuclear Lightning. *Phys. Fluids*, **27**:2694–2698 (1984).

Gifford, T.: Aircraft Struck by Lightning. *Meteorol. Mag.*, **79**:121–122 (1950).

Gisborne, H. F.: Lightning from Clear Sky. *Mon. Weather Rev.*, **56**:108 (1928).

Goldberg, R. A., J. R. Barcus, L. C. Hale, and S. A. Curtis: Direct Observation of Magnetostatic Electron Precipitation Stimulated by Lightning. *J. Atmos. Terr. Phys.*, **48**:293–300 (1986).

Golde, R. H.: "Lightning Protection." Arnold, London, 1973.

Golde, R. H., and W. R. Lee: Death by Lightning. *Proc. IEEE*, **123**:1163–1179 (1976).

Griffing, G. W.: Ozone and Oxides of Nitrogen Production during Thunderstorms. *J. Geophys. Res.*, **82**:943–950 (1971).

Grover, M. K.: Some Analytical Models for Quasi-Static Source Region EMP: Application to Nuclear Lightning. *IEEE Trans. Nucl. Sci.*, **NS-28**:990–994 (1981).

Harland, W. B., and J. L. F. Hacker: "Fossil" Lightning Strikes 250 Million Years Ago. *Adv. Sci.*, **22**:663–671 (1966).

Harris, W. S.: On the Protection of Ships from Lightning. *Nautic. Mag.*, **3**:151–156, 225–233, 353–358, 402–407, 477–484, 739–744, 781–787 (1834). These separate articles have been collected by Harris under the title "A Series of Papers on the Defence of Ships and Buildings from Lightning," dated 1835, and available in the Smithsonian Library, Washington, D.C.

Harris, W. S.: Illustrations of Cases of Damage by Lightning in the British Navy. *Nautic. Mag. (Enlarged Ser.)*, **2**:590–595, 747–748 (1838). See also, **3**:113–122 (1839).

Harris, W. S.: "On the Nature of Thunderstorms." Parker, London, 1843.

Hays, P. B., and R. G. Roble: A Quasi-Static Model of Global Atmospheric Electricity, 1, The Lower Atmosphere. *J. Geophys. Res.*, **84**:3291–3305 (1979).

Helliwell, R. A., J. P. Katsufrakis, and M. L. Trimpi: Whistler-Induced Amplitude Perturbation in VLF Propagation. *J. Geophys. Res.*, **78**:4679–4688 (1973).

Hill, R. D.: Lightning Induced by Nuclear Bursts. *J. Geophys. Res.*, **78**:6355–6358 (1973).

Hill, R. D., R. G. Rinker, and H. D. Wilson: Atmospheric Nitrogen Fixation by Lightning. *J. Atmos. Sci.*, **37**:179–192 (1980).

Hill, R. D., R. G. Rinker, and A. Coucouvinos: Nitrous Oxide Production by Lightning. *J. Geophys. Res.*, **89**:1411–1421 (1984).

Hoddinott, M.: Unusual Lightning. *Weather*, **5**:331 (1950).

Hoffert, H. H.: Intermittent Lightning Flashes. *Philos. Mag.*, **28**:106-109 (1889).

Holzer, R. G.: Simultaneous Measurement of Sferics Signals and Thunderstorm Activity. *In* "Thunderstorm Electricity" (H. G. Byers, ed.), pp. 267-275. Univ. of Chicago Press, Chicago, Illinois, 1953.

Inan, U. S., D. L. Carpenter, R. A. Helliwell, and J. P. Katsufrakis: Subionospheric VLF/LF Phase Peturbations Produced by Lightning-Whistler Induced Particle Precipitation. *J. Geophys. Res.*, **90**:7457-7469 (1985).

Israel, H.: "Atmospheric Electricity," Vols. 1 and 2. U.S. Dept of Commerce, Clearinghouse for Federal Scientific and Technical Information, TT67-51394/1, 1971. Translation of 2nd Ed. of "Atmosphärische Electrizität." Israel Program for Scientific Translation, IPST Catalog No. 1995, Jerusalem, 1971.

Jernegan, M. W.: Benjamin Franklin's "Electrical Kite" and Lightning Rod. *N. Engl. Q.*, **1**:180-196 (1928).

Kamra, A. K.: Measurements of the Electrical Properties of Dust Storms. *J. Geophys. Res.*, **77**:5856-5869 (1972).

Kelley, M. C., D. T. Farley, E. Kudeki, and C. L. Siefring: A Model for Equatorial Explosive Spread F. *Geophys. Res. Lett.*, **11**:1168-1171 (1984).

Kilgore, B. M., and G. S. Briggs: Restoring Fire to High Elevation Forests in California. *J. For.*, **70**:266-270 (1972).

Kitagawa, N., and M. Kobayashi: Distribution of Negative Charge in the Cloud Taking Part in a Flash to Ground. *Pap. Meteorol. Geophys. (Tokyo)*, **9**:99-105 (1958).

Kitagawa, N., M. Brook, and E. J. Workman: Continuing Currents in Cloud-to-Ground Lightning Discharges. *J. Geophys. Res.*, **67**:637-647.

Ko, M. K. W., M. B. McElroy, D. K. Weisenstein, and N. D. Sze: Lightning: A Possible Source of Stratospheric Odd Nitrogen. *J. Geophys. Res.*, **91**:5395-5404 (1986).

Komarek, E. V.: Lightning and Lightning Fires as Ecological Forces. *Proc. Annu. Tall Timbers Fire Ecol. Conf.*, **8** (March 14-15): 169-197 (1968). Available from Tall Timbers Research Station, Tallahassee, Florida.

Komarek, E. V.: Introduction to Lightning Ecology. *Proc. Annu. Tall Timbers Fire Ecol. Conf.*, **13** (March 22-23): 421-427 (1973). Volume 13 contains many additional papers concerning lightning as both a predator and fire-starter and is available from Tall Timbers Research Station, Tallahassee, Florida.

Larsen, A.: Photographing Lightning with a Moving Camera. *Annu. Rep. Smithsonian Inst.*, 119-127 (1905).

Levine, J. S., T. R. Augustsson, I. C. Anderson, and J. M. Hoell, Jr.: Tropospheric Sources of NO_x: Lightning and Biology. *Atmos. Environ.*, **18**:1797-1804 (1984).

Love, R. M.: The Rangelands of the Western United States. *Sci. Am.*, **222**:88-96 (1970).

McCaughan, Z. A.: A Lightning Stroke far from the Thunderstorm Cloud. *Mon. Weather Rev.*, **54**:344 (1926).

Michnowski, S.: On the Observation of Lightning in Warm Clouds. *Indian J. Meteorol. Geophys.*, **14**:320-322 (1963).

Miller, S. L., and H. C. Urey: Organic Compounds Synthesis on the Primative Earth. *Science*, **130**:245-251 (1959).

Moore, C. B., B. Vonnegut, B. A. Stein, and H. J. Survilas: Observations of Electrification and Lightning in Warm Clouds. *J. Geophys. Res.*, **65**:1907-1910 (1960).

Myers, F.: Lightning from a Clear Sky. *Mon. Weather Rev.*, **59**:39-40 (1931).

Nicols, T.: "Edelgestein-Büchlein oder Beschreibung der Edelgesteine." Nauman & Wolff, Hamburg, 1675.

Noxon, J. F.: Atmospheric Nitrogen Fixation by Lightning. *Geophys. Res. Lett.*, **3**:463-465 (1976).

Pierce, E. T.: Atmospheric Electricity—Some Themes. *Bull. Am. Meteorol. Soc.*, **55**:1186-1194 (1974).

Pietrowski, E. L.: An Observation of Lightning in Warm Clouds. *J. Meteorol.*, **17**:562-563 (1960).

Pockels, F.: Über das Magnetische Verhalten Einger Basaltischer Gesteine. *Ann. Phys. Chem.*, **63**:195-201 (1897).

Pockels, F.: Bestimmung Maximaler Entladungs-Strom-Stärken aus Ihrer Magnetisirenden Wirkung. *Ann. Phys. Chem.*, **65**:458-475 (1898).

Pockels, F.: Über die Blitzentladungen Erreicht Stromstärke. *Phys. Z.*, **2**:306-307 (1900).

Pounder, C.:Volcanic Lightning. *Weather*, **35**:357-360 (1980).

Prinz, H.: Lightning in History. *In* "Lightning, Vol. 1, Physics of Lightning" (R. H. Golde, ed.), pp. 1-21. Academic Press, New York, 1977.

Rycroft, M. J.: Enhanced Energetic Electron Intensities at 100 km Altitude and a Whistler Propagating through the Plasmasphere. *Planet. Space Sci.*, **21**:239-251 (1973).

Schonland, B. F. J.: The Lightning Discharge. *Handb. Phys.*, **22**:576-628 (1956).

Schonland, B. F. J.: "The Flight of Thunderbolts," 2nd Ed. Oxford Univ. Press (Clarendon), London and New York, 1964.

Singer, S.: "The Nature of Ball Lightning." Plenum, New York, 1971.

Thomson, E. M.: The Dependence of Lightning Return Stroke Characteristics on Latitude. *J. Geophys. Res.*, **85**:1050-1056 (1980).

Thomson, E. M., M. A. Galib, M. A. Uman, W. H. Beasley, and M. J. Master: Some Features of Stroke Occurrence in Florida Lightning Flashes. *J. Geophys. Res.*, **89**:4910-4916 (1984).

Tzur, I., and R. G. Roble: The Interaction of a Dipolar Thunderstorm with Its Global Electrical Environment. *J. Geophys. Res.*, **90**:5989-5999 (1985).

Uman, M. A.: "Understanding Lightning." BEK Technical Publications, Pittsburgh, Pennsylvania, 1971. Rev. Ed. "All about Lightning." Dover, New York, 1986.

Uman, M. A.: The Earth and Its Atmosphere as a Leaky Spherical Capacitor. *Am. J. Phys.*, **42**:1033-1035 (1974).

Uman, M. A.: "Lightning," pp. 138-140, 240-242, 243-248. McGraw-Hill, New York, 1969. See also Dover, New York, 1984.

Uman, M. A., D. F. Seacord, G. H. Price, and E. T. Pierce: Lightning Induced by Thermonuclear Detonations. *J. Geophys. Res.*, **77**:1591-1596 (1972).

U.S. Lightning Protection Code 1980. National Fire Protection Association, Atlantic Avenue, Boston, Massachusetts 02210, Publication 78, 470.

Van Doren, C.: "Benjamin Franklin," pp. 156-173. The Viking Press, New York, 1938.

Vaughan, O. H., and B. Vonnegut: Lightning to the Ionosphere. *Weatherwise*, **35**:70-72 (1982).

von Liebig, J.: Une Note sur la Nitrification. *Ann. Chem. Phys.*, **35**:329-333 (1827).

Vonnegut, B.: Deductions Concerning Accumulations of Electrified Particles in Thunderclouds Based on Electric Field Changes Associated with Lightning. *J. Geophys. Res.*, **88**:3911-3912 (1983).

Vonnegut, B.: Vertical Lightning. *Weatherwise*, **37**:61 (1984).

Vonnegut, B., O. H. Vaughan, Jr., M. Brook, and P. Krehbiel: Mesoscale Observations of Lightning from Space Shuttle. *Bull. Am. Meteorol. Soc.* **66**:20-30 (1985).

Voss, H. D., W. L. Imhof, M. Walt, J. Mobilia, E. E. Gaines, J. B. Reagan, U. S. Inan, R. A. Helliwell, D. L. Carpenter, J. P. Katsufrakis, and H. C. Chang: Lightning-Induced Electron Precipitation. *Nature (London)*, **312**:740-742 (1984).

Waldteufel, P., P. Metzger, J. L. Boulay, P. Laroche, and P. Hubert: Triggered Lightning Strokes Originating in Clear Air. *J. Geophys. Res.*, **85**:2861-2868 (1980).

Walter, B.: Ein Photographischer Apparat zur Genaueren Analyse des Blitzes. *Phys. Z.*, **3**:168-172 (1902); Über die Entstehungsweise des Blitzes. *Ann. Phys.*, **10**:393-407 (1903);

Über Doppelaufnahmen von Blitzen ... *Jahrb. Hamb. Wiss. Anst.*, **27** (Beiheft 5):81–118 (1910); Steroeskopische Blitzaufnahmen. *Phys. Z.*, **13**:1082–1084 (1912); Über die Ermittelung der Zeitlichen Aufeinanderfolge Zusammengehöriger Blitze Sowie über ein Bemerkenswertes Beispiel Dieser Art von Entladungen. *Phys. Z.*, **19**:273–279 (1918).

Weber, L.: Über Blitzphotographieen. *Ber. Königl. Akad. Berlin*, 781–784 (1889).

Wilson, C. T. R.: On Some Determinations of the Sign and Magnitude of Electric Discharges in Lightning Flashes. *Proc. R. Soc. London Ser. A*, **92**:555–574 (1916).

Wilson, C. T. R.: Investigations on Lightning Discharges and on the Electric Field of Thunderstorms. *Philos. Trans. R. Soc. London Ser. A*, **221**:73–115 (1920).

Woodman, R. F., and E. Kudeki: A Causal Relationship between Lightning and Explosive Spread F. *Geophys. Res. Lett.*, **11**:1165–1167 (1984).

Wright, J. B.: A Thunderstorm in the Tropics. *Weather*, **5**:230 (1950).

Chapter 2 | Lightning Phenomenology

2.1 INTRODUCTION

We use the term "phenomenology" to refer primarily to aspects of lightning related to its frequency of occurrence, such as maximum and average flashing rates per unit area, and to the variation of those features with location and storm type. We also choose to include in phenomenology the variation with location and storm type of the characteristic of individual lightning discharges, for example, the number of strokes in a ground flash. Specifically, in this chapter we examine the available information on both ground and cloud discharge flashing rates per unit area averaged over long periods of time, that is, months or years (Section 2.2), the relation between those data and thunderstorm duration statistics (Section 2.3), the number of cloud and ground discharges as a function of location (Section 2.4), the phenomenological characteristics of the lightning within individual storms and relationships to storm meteorological parameters (Section 2.5), and the variation in the properties of ground flashes as a function of location and of storm type (Section 2.6). In this chapter we consider the phenomenological characteristics of primarily negative lightning. Information on the phenomenology of the less common positive lightning is found in Section 2.6, as well as in Sections 1.4 and 11.2.

2.2 FLASH DENSITIES AVERAGED OVER MONTHS OR YEARS

Most available data on lightning flashes per unit time per unit area, generally called "flash density," have been derived from instruments called "flash counters." Results from a flash counter are not meaningful unless the data include many storms that more or less uniformly cover an area within tens of kilometers of the counter. The reason for this is that the counter produces a registration if the time-varying electric field intensity from the lightning, after being appropriately filtered, exceeds a fixed threshold value.

Different types of counters are characterized by different filters and different trigger thresholds. Since the electric field intensity has a wide variation from stroke to stroke, any counter will fail to register small events nearby and will register large events far away. An "effective range" for the counter is defined as the distance that the "misses" within this effective range are exactly compensated for by the "hits" outside of it. An equivalent definition of effective range is the distance within which the number of flashes actually occurring equals the number registered by the counter. Thus, if the effective range can be determined from other measurements and the lightning activity is reasonably homogeneous, one can use a flash counter to determine flash density. To ensure homogeneity, flashes should be recorded for months or even years, depending on the level of lightning activity. While an accurate flash density may be obtained in, say, 1 year, the average yearly flash density itself is highly variable so that many years of measurement are required to obtain a meaningful overall average flash density. Since the effective range for ground flashes is different from that for cloud flashes, separate effective ranges must be determined for each or, alternately, calibration factors found that allow cloud flashes to be subtracted from total flash counts if only the ground flash density is desired. It is interesting to note that about 40% of the ground flashes are still registered at a distance equal to the effective range of a typical counter (Anderson *et al.*, 1979; Prentice and Mackerras, 1960; Bunn, 1968).

The two most widely used counters are endorsed by the Conference Internationale des Grands Reseaux Electriques (CIGRE) and are called the CIGRE 500-Hz counter (Pierce, 1956; Golde, 1966; Barham, 1965; Prentice and Mackerras, 1969; Prentice, 1972; Barham and Mackerras, 1972; Prentice *et al.*, 1975; Anderson *et al.*, 1979) and the CIGRE 10-kHz counter (Anderson *et al.*, 1973, 1979). The frequency designation refers to the center frequency passed by the filter that precedes the trigger circuit. Another "organization-recommended" counter, called the CCIR (International Radio Consultative Committee) or the WMO (World Meteorological Organization) counter (Horner, 1960), has apparently not attained widespread use. Its filter is centered at 10 kHz and is similar to that of the CIGRE 10 kHz (Horner, 1960; Anderson *et al.*, 1979). The 10-kHz filter allows a lower percentage of cloud flashes in the overall counts than does the 500-Hz filter (Anderson *et al.*, 1979). The effective range of the CIGRE 500-Hz counter in Australia is about 30 km for ground flashes and about 20 km for cloud flashes (Prentice and Mackerras, 1969); in South Africa it is about 37 km for ground flashes and about 17 km for cloud flashes, with about 20% of the total registrations being due to cloud flashes (Anderson, 1980). The effective range of the CIGRE 10-kHz counter in South Africa is about 20 km for ground flashes and about 6 km for cloud flashes, with about 5% of the total registration being due to cloud flashes (Anderson *et al.*, 1979; Anderson, 1980). The effective range of the CCIR counter apparently has not been properly measured, the reported values varying considerably (Horner, 1960; Muller-Hillebrand, 1963), although Cooray (1986), from theory, estimates the range to be about 140 km. Some comments on the general theory and the inherent weaknesses of flash counters are given by Brook and Kitagawa (1960) and by Bunn, and the operation of the two CIGRE and the CCIR counters is discussed in detail by Cooray (1986). If properly calibrated and operated, flash counters can provide reasonably accurate data on flash density. We consider some of those data next.

Prentice (1977) has summarized much of the published and unpublished data on average flash density that have been obtained using flash counters, visual observations, and instruments that measure electric field change. The most extensive flash counter studies come from Australia (Prentice and Mackerras, 1969; Mackerras, 1978), Scandanavia (Muller-Hillebrand, 1965), and South Africa (Anderson, 1980; Anderson and Eriksson, 1980). Southeast Queensland, Australia has a total flash density of 5 km^{-2} yr^{-1}, of which 1.2 km^{-2} yr^{-1} are ground flashes; Norway, Sweden, and Finland have measured ground flash densities between 0.2 and 3 km^{-2} yr^{-1} depending on location; and South Africa has ground flash densities from below 0.1 to about 12 km^{-2} yr^{-1}, depending on location. Flash density maps are given in the above references for these three areas of the world.

Piepgrass *et al.* (1982), using data from a network of 26 electric field mills (Section C.1; Fig. C.1) at the Kennedy Space Center, Florida, have estimated the total flash during the months of June and July, 1974 through 1980. The field mills detect the ambient electric field at ground level with a time resolution of about 0.25 sec and a frequency response that extends to d.c. Piepgrass *et al.* had to assume an effective area for the field mill network (the value was 625 km^2) in order to determine flash densities from measured electric field registrations. They state that a systematic error up to a factor of 2 is possible in this assumed area. The total flash densities ranged from 3.7 km^{-2} month^{-1} in 1977 to 21.9 km^{-2} month^{-1} in 1975. The mean for 6 years was 12 km^{-2} month^{-1} with a standard deviation of 8 km^{-2} month^{-1}. From the ratio of cloud flashes to ground flashes reported by Livingtston and Krider (1978) (Section 2.4), Pipegrass *et al.* estimate a mean ground flash density of 4.6 km^{-2} month^{-1} with a standard deviation of 3.1 km^{-2} month^{-1}.

Maier *et al.* (1979) used a lightning location system based on wideband magnetic direction finding (Sections C.2 and C.7.2) in South Florida to map highly detailed variations of flash density for periods of months and areas of up to 4×10^5 km^2. They found that flash density may vary by an order of magnitude over distances of 20–30 km, roughly the effective area of most flash counters, partly due to local meteorological effects such as occur along the Florida coastline. Similar observations have been made by Lopez and Holle (1986) for lightning in central Florida and in northeastern Colorado. For the three summer months of 1983 they found spatial maxima in the ground flash density of 7 km^{-2} in Colorado and over 8 km^{-2} in Florida. Darveniza and Uman (1984) used wideband direction finders as well as seven CIGRE 10-kHz counters and one CIGRE 500-Hz counter to measure flash density in the Tampa Bay area of Florida. The ground flash density, averaged over 2 years, from the CIGRE 10-kHz counters was between 7 and 17 flashes km^{-2} yr^{-1}, depending on the assumptions made about the effective range and counter response to cloud flashes. The CIGRE 500-Hz counter, based on the extrapolation of 4 months of summer measurement and with an effective range and cloud flash response determined in Australia, gave a yearly ground flash density of 9.5 km^{-2} yr^{-1}. The locating system was operated only during the summer months and gave an extrapolated yearly ground flash density of 12.9 ± 5.2 km^{-2} yr^{-1}.

2.3   RELATION OF GROUND FLASH DENSITY TO THUNDERDAY AND THUNDERHOUR STATISTICS

Ground flash density has not generally been considered a simple parameter to measure. The thunderday level, the number of days per month or per year on which thunder is heard, is more easily measurable, at least to weather station personnel, and hence has been recorded at most weather stations worldwide for many years. A world thunderday map is shown in Fig. 2.1; a thunderday map of the United States is shown in Fig. 2.2. There has been considerable effort to relate the thunderday level to the flash density, since it is a knowledge of the flash density that is important in the design of lightning protection systems. Prentice (1977) has reviewed 17 proposed relations between these two parameters. Most are of the form

$$N_g = aT_D^b \quad \text{km}^{-2}\,\text{yr}^{-1} \tag{2.1}$$

where N_g is the ground flash density, T_D the annual thunderdays, and a and b are empirical constants. Typically, $a = 0.1$–0.2 and $b = 1$. Apparently the only measurements of statistical significance are from South Africa and Australia. Anderson (1980) and Anderson and Eriksson (1980) review measurements made with 120 counters located throughout South Africa where the annual thunderday level varied from about 3 to about 100 and

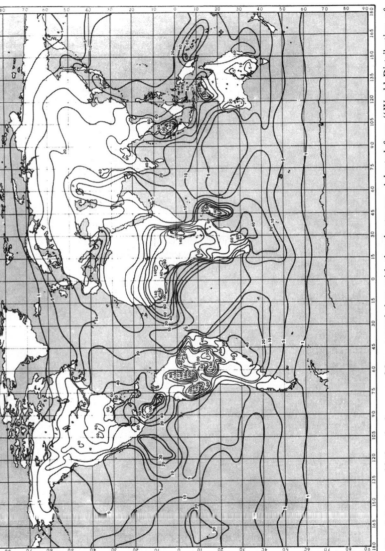

Fig. 2.1 A world thunderday map. Mean annual days with thunderstorms. Adapted from "World Distribution of Thunderstorm Days," Part 2, Tables of Marine Data and World Maps," World Meteorological Organization Publication 21, Geneva, Switzerland, 1956.

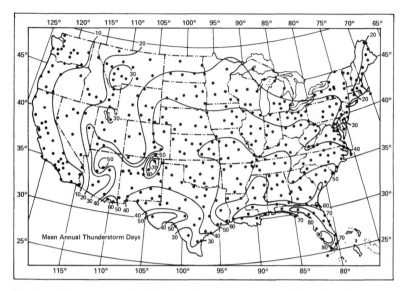

Fig. 2.2 A thunderday map of the United States compiled from data from 450 air weather stations shown as dots (Changery, 1981). Most stations had 30-year records and all at least 10-year records. Adapted from MacGorman *et al.* (1984).

where the ground flash density ranged from below 0.1 to about 12 km^{-2} yr^{-1}. The relation that best fits the South African data is

$$N_g = 0.023 T_D^{1.3} \quad km^{-2} yr^{-1} \tag{2.2}$$

although for a given T_D, N_g can vary by a factor up to about 5 from the value of N_g in Eq. (2.2), as shown in Fig. 2.3. Mackerras (1978) in Australia made measurements with 26 counters in a region where the flash density varied from about 0.2 to 3.0 km^{-2} yr^{-1} and annual thunderday level from about 10 to about 100. His data are best fit by the expression

$$N_g = 0.01 T_D^{1.4} \quad km^{-2} yr^{-1} \tag{2.3}$$

where for a given T_D, N_g varies by a factor of about 2 from the value of N_g found in Eq. (2.3).

It might be expected that the exponent b in Eq. (2.1) would exceed unity since the higher the number of thunderdays the longer one might expect the storms to last on each day. If this is true, the flash density should correlate more linearly with the number of thunderhours than with the number of thunderdays and, indeed, there is some evidence that this is the case (Prentice, 1977; MacGorman *et al.*, 1984). A thunderhour map for the United States is found in Fig. 2.4. MacGorman *et al.* (1984), using ground flash data determined from magnetic direction finding systems operated in

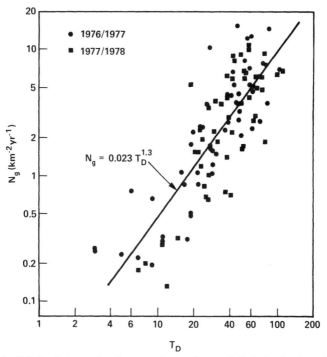

Fig. 2.3 Relation between thunderstorm days and ground flash density from 120 flash counter observations in South Africa. Adapted from Anderson and Erikisson (1980).

Florida and Oklahoma (Sections C.2 and C.7.2), have determined the following relation between ground flash density and thunderhours T_H,

$$N_g = 0.054 T_H^{1.1} \quad km^{-2} \, yr^{-1} \tag{2.4}$$

Prentice (1977) quotes a Russian study in which N_g and T_H were found to be linearly related. Maier *et al.* (1984) have investigated the correlation between the diurnal variation of lightning occurrence in Florida and the available thunder statistics and report that starting time (midafternoon) and time of peak activity (late afternoon) of both lightning and thunder statistical data are in good agreement but that the thunder data indicate a duration of activity 1 to 2 hr longer into the evening than do the lightning data, perhaps due to observer error in remembering the time of the last thunder or observer hesitance in regarding a given thunder as the last when there is distant but visable lightning out of thunder range (Section 15.2.3).

Since both the number of thunderhours or thunderdays per year and the flash density vary from year to year, both parameters must be measured over many years for the relationship of average flash density to average

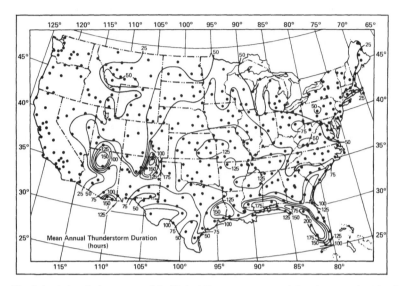

Fig. 2.4 A thunderhour map of the United States, mean annual thunderstorm duration in hours, compiled from data from 450 air weather stations shown as dots (Changery, 1981). Most stations had 30-year records and all at least 10-year records. Adapted from MacGorman *et al.* (1984).

thunderhours or thunderdays to be statistically meaningful. Anderson *et al.* (1979) recommend taking data for at least 11 years, one solar cycle, although Freier (1978) finds little evidence for coupling between solar activity and thunderstorms.

As an illustration of potential problems in relating ground flash density to thunderstorm days, Ishii *et al.* (1981) point out that near Sakata, Japan where there are about the same number of thunderstorm days, 13, in winter as in summer, the ground flash density varies an order of magnitude between the two seasons, being about 0.4 km^{-2} for the 4 summer months and 0.05 km^{-2} for the 4 winter months.

2.4 NUMBERS OF CLOUD AND GROUND FLASHES AS A FUNCTION OF LOCATION

The literature on the ratio of cloud to cloud-to-ground flashes has been reviewed by Prentice and Mackerras (1977) and by Livingston and Krider (1978). Prentice and Mackerras (1977) find the ratio to have a mean value of 5.7 for latitudes between 2° and 19°, 3.6 between 27° and 37°, 2.9 between

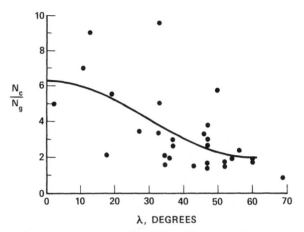

Fig. 2.5   Relation between observed values of the ratio of cloud flashes to ground flashes and latitude. Adapted from Prentice and Mackerras (1977).

43° and 50°, and 1.8 between 52° and 69°. They fit 29 observations with the curve

$$N_c/N_g = 4.16 + 2.16 \cos 3\lambda \tag{2.5}$$

where N_c is the cloud flash density and λ the latitude, as shown in Fig. 2.5. All the experimental data on N_c/N_g fall within a factor of 3 of the value given in Eq. (2.5), most within a factor of 2. More recently, Mackerras (1985) has employed an automatic device that distinguishes between the electric fields of ground and cloud discharges to determine the ratio of cloud flashes to ground flashes for the period September 1982 to May 1984 in Brisbane, Australia ($\lambda = 27.5°S$). The ratio for about 6100 total flashes was about 3 with a range from 0.9 to 24.7 for individual days having at least 100 total flashes. The ground flash density was $1.2 \, \text{km}^{-2} \, \text{yr}^{-1}$ and the cloud flash density was $3.7 \, \text{km}^{-2} \, \text{yr}^{-1}$, in excellent agreement with the flash counter data reported by Prentice (1977) (Section 2.2). Previously, Pierce (1970) had suggested using the following expression to describe the fraction of total flashes that are ground discharges

$$N_g/(N_c + N_g) = 0.1 + 0.25 \sin \lambda \tag{2.6}$$

Equation (2.6) provides a ratio $N_c/N_g = 9$ at the equator and $N_c/N_g = 1.8$ at the poles. It has often been suggested (e.g., Pierce, 1970) that the higher freezing levels associated with the lower latitudes lead to a lower probability of flashes to ground since the ground is more distant from the charge centers (see Chapter 3 for a discussion of charge locations in clouds).

Livingston and Krider (1978) found for summer thunderstorms in Florida ($\lambda = 28°N$) that between 42 and 52% of all lightning discharges were to ground during the "active storm period" (Section 2.5) and that during the final storm period only about 20% were to ground. For five storms on 3 days, 43% of 552 total flashes were to ground. Holzer (1953) has also noted that the percentage of cloud flashes increases in the latter stage of a storm. Additionally, he reports that the percentage of ground flashes is higher in frontal storms than in air mass thunderstorms (Section 1.3).

With the advent of Earth-orbiting satellites, it has become possible to chart systematically the worldwide lightning activity by detecting the light or the radio frequency radiation emitted by each discharge. The satellites used to date have recorded only a small fraction of the discharges occurring because those satellites were in relatively low Earth orbit and hence spent a relatively short time over any given storm. Further, trigger thresholds were such as to cause smaller events to be missed. The best spatial resolution has been of the order of 100 km for optical detection and about an order of magnitude greater for radio frequency detection. Nevertheless, it has been possible to estimate total flash densities and to determine ratios of activity in different geographical locations and in different seasons. It has not been possible to distinguish between cloud and cloud-to-ground discharges. Orville (1981), using photographic data from a DMSP Satellite, has studied the global distribution of lightning at midnight between 60°S and 60°N for September, October, and November 1977. He gives a series of lightning maps in which the locations of the discharges are well correlated with known features of the general atmospheric circulation. Some of these are shown in Fig. 2.6. Orville and Spencer (1979), using the same Earth coverage for midnight and dusk, estimate that the global flashing rate is about $100 \, \text{sec}^{-1}$ with a potential error of about a factor of 2. They report that there is 1.4 times more lightning during the summer in the northern hemisphere than during the southern summer. Turman and Edgar (1982) found that their satellite detector responded to about 2% of the lightning within its field of view. They were able to estimate a global flashing rate of 40 to $120 \, \text{sec}^{-1}$ with a seasonal variation of about 10%. Thirty-seven percent of the global lightning activity originated over the ocean at dawn and 15% at dusk. Turman (1978) using an array of 12 photodiodes, each having a field of view of 700 by 700 km, flown on a DMSP satellite observed 10,000 flashes from 24 storm complexes during 15 orbits in September 1974 and March 1975. The average flash density within the sensor field of view was $6 \times 10^{-8} \, \text{km}^{-2} \, \text{sec}^{-1}$ which, if extrapolated to the whole Earth, represents a global flash rate of about $30 \, \text{sec}^{-1}$.

Kotaki et al. (1981) have detected lightning from the Japanese ISS-b satellite by sensing HF radiation at 2.5, 5, 10, and 25 MHz. They present

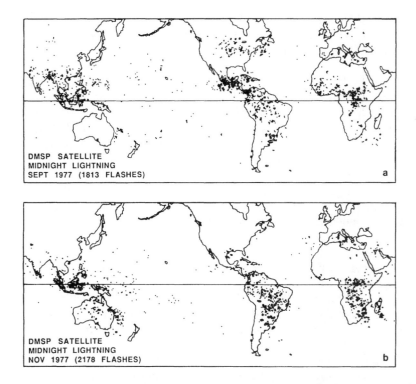

Fig. 2.6   Midnight lightning locations recorded by the DMSP Satellite for the months of September (a) and November (b) 1977. Note that the lightning shifts to the southern hemisphere as summer in that hemisphere sets in and that most of the lightning activity is over land. Adapted from Orville (1981).

worldwide lightning maps for 2 years. They find the global lightning frequency to be $64 \, \text{sec}^{-1}$ for the northern spring, $55 \, \text{sec}^{-1}$ for the summer, $80 \, \text{sec}^{-1}$ for the fall, and $54 \, \text{sec}^{-1}$ for the winter.

It is interesting to note that Brooks (1925), on the basis of worldwide thunderday data, estimated that the global flash rate was about $100 \, \text{sec}^{-1}$. The satellite data are in remarkably good agreement with this value, considering the assumptions that must necessarily be made to analyze the satellite data. A global flash rate of 100 sec^{-1} represents a global flash density of about $6 \, \text{km}^{-2} \, \text{yr}^{-1}$, which would not appear unreasonable considering the data on total and ground flash density over land discussed in Sections 2.2 and 2.3, the fact that in general there is less lightning over the oceans (e.g., Figs. 2.1 and 2.6), and the fact that in the Tropics there are many cloud flashes for each ground flash, as noted earlier in this section.

2.5 PHENOMENOLOGICAL PROPERTIES OF THE LIGHTNING IN INDIVIDUAL STORMS AND RELATIONSHIPS TO METEOROLOGICAL PARAMETERS

We consider now lightning flashing rates and other properties of individual storms. Livingston and Krider (1978) give statistical data on lightning for 22 storm days at the Kennedy Space Center, Florida during 1974 and 1975. The data were derived from the electric fields recorded by the KSC field mill network (Sections 2.2 and C.1). The number of discharges per storm ranged from 8 to 1987. About 70% of all lightning occurred during what was termed the "active storm period" which represented only about 30% of the total storm duration. A strong correlation was found between the total number of discharges during the active period, n, and the duration of that period, D_A, in minutes. The curve fitting the 23 data points is

$$\log_{10} n = 0.01D_A + 1.90 \tag{2.7}$$

The average flashing rate per storm varied between 0.3 and 9.3 min^{-1}. On 5 of the storm days, the flashing rate averaged over 5-min intervals exceeded 20 min^{-1}, the maximum rate being 26 min^{-1}. Small storms, which perhaps can be regarded as single storm cells, produced a maximum flashing rate of typically 1 to 4 min^{-1}. Piepgrass *et al.* (1982) have further added to the Florida results by presenting statistics on lightning and rainfall for 79 additional storms which occurred at the Kennedy Space Center during the summers of 1976 through 1980. In addition to showing a correlation between lightning and rainfall (there were about 10^4 m^3 of rain per lightning) and reviewing the previous literature on that subject, they present the following data on the frequency of lightning occurrence. The number of discharges per storm varied from 1 to 3687. Storm averaged flashing rates were between 0.1 and 12.2 min^{-1}. Maximum flashing rates, averaged over 5-min intervals, were between 0.2 and 30.6 min^{-1}.

Lhermitte and Krehbiel (1979) report a maximum flashing rate of about 60 min^{-1} over a several minute interval for a normal-size storm at the Kennedy Space Center. The discharges were identified using a VHF time-of-arrival source location technique (Section C.7.3). The primary contributors to the high flashing rate were small intracloud discharges associated with the development of a strong updraft through precipitation at the $-10°C$ level.

Maier and Krider (1982) used a lightning location system based on wide-band magnetic direction finding (Sections 2.2, C.2, and C.7.2) to measure cloud-to-ground lightning rates in three severe storms in the midwest and 268 air-mass storms in south Florida. The severe thunderstorms lasted from 2 to 8 hr, and produced 200 to 2800 cloud-to-ground flashes over areas of 2000 to 5000 km^2. The Florida air-mass storms rarely lasted longer than 1 to 2 hr, producing about 50 discharges over an area of about $450 \, km^2$. However, there were usually only 1 or 2 severe storms per storm day in the midwest while in Florida there were 10 to 50 storms per storm day. The air-mass storms produced cloud-to-ground lightning at a mean rate of about $1 \, min^{-1}$, while the severe storm rate varied from about 2 to $6 \, min^{-1}$. The peak flashing rate in severe storms approached $20 \, min^{-1}$, averaged over 5-min intervals, while the peak rate in the 268 air-mass storms was about $12 \, min^{-1}$ with a average peak rate for all the air mass storms of about $7 \, min^{-1}$.

Peckham *et al.* (1984) used the same type of ground flash locating system as Maier and Krider (1982) to study cloud-to-ground lightning in the Tampa Bay area of Florida. For 111 storms on 8 days the following parameters were determined: duration, area, number of ground flashes, average ground flash density, average ground flashing rate, and maximum ground flashing rate averaged over a 5-min period. The means and standard deviations of these data are presented in Table 2.1 where the storms are divided into three categories: (1) single-peak storms, those with a single peak in their flashing rate vs time curves; (2) multiple-peak storms, those with multiple-peak flashing rates; and (3) storm systems composed of groups of the first two types of storms which would be detected as one overall storm by a nondirectional lightning detector like a flash counter or perhaps a network of electric field mills. Peckham *et al.* (1984) define the storm duration as the period of the

Table 2.1

Storm Parameters[a]

	Single-peak storms	Multiple-peak storms	Storm systems
Duration (min)	41	77	130
	(16)	(26)	(51)
Area (km^2)	103	256	900
	(63)	(154)	(841)
Number of ground flashes	73	270	887
	(69)	(216)	(720)
Ground flash density ($km^{-2} \, min^{-1}$)	0.018	0.015	0.010
	(0.011)	(0.008)	(0.006)
Average ground flashing rate (min^{-1})	1.7	3.4	6.8
	(1.2)	(2.3)	(4.1)
Maximum ground flashing rate (min^{-1})	3.7	7.3	14
	(2.6)	(4.4)	(9)

[a] Means and standard deviations (in parentheses) of storm parameters for 111 storms on 8 days in the Tampa Bay area of Florida as a function of storm category. Adapted from Peckham *et al.* (1984).

ground flashing rate curve during which there is no more than one consecutive 5-min interval without lightning. A map of one day's ground flash activity is shown in Fig. 2.7. The letters indicate the chronology of storm starting time for single- and multiple-peak storms. Two storm systems are composed of storms CFJ and storms ABDEGHIKL. As can be seen in Table 2.1, Peckham *et al.* (1984) found an average ground flash density for the storm area containing lightning of about 0.01 $km^{-2} min^{-1}$. This value can be compared with the satellite data of Turman (1978) who found a mean of $2 \times 10^{-3} km^{-2} min^{-1}$ with a range from 3×10^{-4} to $1 \times 10^{-2} km^{-2} min^{-1}$ using satellite photographs to determine the thunderstorm areas that,

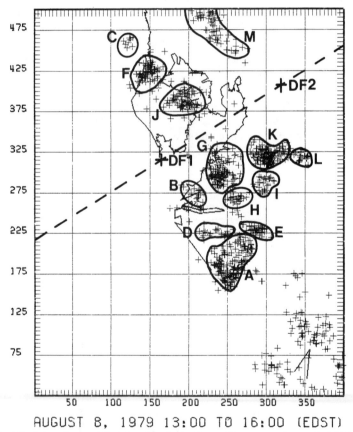

AUGUST 8, 1979 13:00 TO 16:00 (EDST)

Fig. 2.7   Map of ground flash locations in the Tampa Bay area on August 8, 1979 between 13:00 and 16:00 PM EDT. The two direction finders comprising the location system are labeled DF1 and DF2. The map scales are in units of thousands of feet. Adapted from Peckham *et al.* (1984).

according to Turman (1978), must considerably overestimate the active storm areas because of the spread of the thunderstorm anvil and the presence of decayed cells. Peckham *et al.* (1984) found that the number of ground flashes, n_g, in a storm and storm duration, D_S, in minutes was related by

$$\log_{10} n_g = 0.014 D_S + 1.2 \tag{2.8}$$

an expression similar to that found from the Kennedy Space Center field mill network and given in Eq. (2.7). The Peckham *et al.* (1984) data are given in Fig. 2.8.

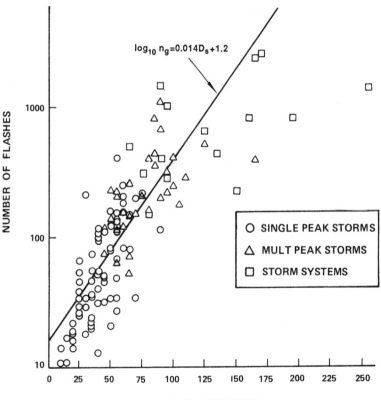

Fig. 2.8 Number of ground flashes vs storm duration. The regression line (correlation coefficient $r = 0.78$) is determined excluding the storm system data, which are nevertheless plotted, because the storm systems are a combination of single- and multiple-peak storms and because they have a subjective definition. Interestingly, the storm systems lie near the regression line. Adapted from Peckham *et al.* (1984).

Reap (1986) analyzed lightning location data for summers 1983 and 1984 from the network of 33 wideband magnetic direction finders operated by the U.S. Bureau of Land Management in the 11 westernmost states of the continental United States. The lightning locations and occurrence rates were compared with radar and satellite data and were examined in relation to the topological features of the generally mountainous western terrain. Over two million lightning ground flashes were detected in the 149-day study, and each was assigned to a grid block of size 47 × 47 km. Reap (1986) gives the percentage of days on which at least two ground flashes occurred in each individual grid block, and this map is shown to be in good general agreement with the thunderstorm day map of Fig. 2.2, with the exception that the lightning map is considerably more detailed, especially in the higher elevations. A map is also given of the total number of ground flashes per grid block for the two summers, the maximum value being about 5000, a moderate underestimate because the network fails to detect between 30 and 50% of the ground flashes. Reap (1986) found a high correlation between terrain elevation, which ranged from sea level to near 3 km, and the time of maximum lightning activity: at the higher elevations, maximum lightning activity occurred in the early to midafternoon whereas at lower elevations it occurred later. The lightning activity as measured by the number of flashes per grid block was also found to increase with increasing terrain elevation. Reap (1986) reported that about 40% of the lightning flashes occurred without the presence of a radar echo, an observation attributed to the inadequate radar coverage in the West by National Weather Service and Federal Aviation Agency radars. Further, 87% of the flashes were associated with radar intensity levels less than the threshold used in the East for defining thunderstorms, according to Reap (1986) because of the relatively low moisture content of the clouds over the elevated western terrain. In general, however, the lightning frequency was observed to increase with radar echo intensity. A correlation was also observed between the cloud top temperature as determined from GOES satellite data and the lightning activity. For example, there was a linear relation between lightning frequency and decreasing cloud top temperature in the temperature range from -35 to $-63°C$.

Lopez and Holle (1986) used networks of wideband magnetic direction finders in northeastern Colorado and in central Florida to study the diurnal and the spatial variation of the lightning ground flash density (Section 2.2) and to relate those results to the geographic and climatic characteristics of the two regions. The geographic feature of primary importance in Colorado is the mountains and in Florida is the interface between the waters of the Gulf of Mexico and the Atlantic Ocean and the land peninsula. For the summer of 1983, detailed maps of the ground flash density as a function of the time of day are given for both locations. Lopez and Holle (1986) show from these maps that the temporal and spatial distributions of lightning are clearly related to the local topographic features. Further, for northeast Colorado, radar and surface wind studies from earlier summers were found to be entirely consistent with the lightning data for summer 1983 indicating that the flow and convective patterns are similar from summer to summer.

Prentice (1977) presents unpublished data from Fuquay and Baugham derived from electric field measurements showing the flashing rate for an intense, a moderate, and a small storm in Montana. The maximum flashing rate averaged over a 5-min interval occurred for the intense storm and was about 10 flashes $\min^{-1}$, equally divided between ground and cloud discharges. Prentice (1977) quotes a variety of sources that indicate a maximum flashing rate of greater than $100 \min^{-1}$, but it is not clear in these cases whether there were separate storms occurring simultaneously or over what area they occurred.

Shackford (1960), Larsen and Stansbury (1974), Jacobson and Krider (1976), Carte and Kidder (1977), Livingston and Krider (1978), Marshall and Radhakant (1978), Lhermitte and Krehbiel (1979), Holle and Maier (1982), and Cherna and Stansbury (1986) have related flashing rates to cloud properties as measured by radar. Jacobson and Krider (1976) and Livingston and Krider (1987) in Florida show that total flashing rates greater than $10 \min^{-1}$ are associated only with storms having maximum radar echo tops above 14 km. At the Kennedy Space Center, a total flashing rate of $10 \min^{-1}$ is probably indicative of a ground flash rate of roughly $4 \min^{-1}$ (Livingston and Krider, 1978). Holle and Maier (1982) have reviewed the worldwide literature regarding the cloud top temperature at the time of the first lightning. Larsen and Stansbury (1974), Marshall and Radhakant (1978), and Cherna and Stansbury (1986) in Canada have shown that flash rate is related to the area of the storm as measured by radar, but the dominant parameter determining the rate of electrical activity, according to Cherna and Stansbury (1986), is the thunderstorm height. Lhermitte and Krehbiel (1979), in addition to the results discussed above, were apparently the first to demonstrate conclusively that lightning flash rate is correlated with in-cloud updraft velocity.

2.6   PROPERTIES OF GROUND FLASHES AS A FUNCTION OF LATITUDE AND STORM TYPE

The fraction of flashes that lower positive charge to ground and the characteristic of those positive flashes are considered in Chapter 11. Information from Chapter 11 on the occurrence of positive discharges can be summarized as follows. Positive lightning is apparently more frequent where the freezing level is closer to the ground and where there is strong wind shear separating horizontally the upper positive and lower negative cloud charge centers. Conditions favoring the occurrence of positive flashes are apparently high elevations, cold weather, high latitudes, and severe storms. The overall fraction of positive discharges to normal ground appears to be of the order of 1% or less and in mountainous regions of the order of 10%, although in specific storms the majority of flashes can be positive.

Thomson (1980) has reviewed the literature on interstroke time interval and number of strokes per flash, as noted in Section 1.3, and finds no statistically significant correlation with latitude. This does not, however, preclude a latitude dependence which some investigators (Pierce, 1970; Takeuti *et al.*, 1975) have inferred should exist while others (Harris and Salman, 1972) have not found to exist. Rather, Thomson (1980) concludes that "the distributions of both interstroke intervals and the number of strokes per flash are more sensitive to local influences, including measurement and sampling techniques, than they are to latitude effects."

The literature on the height and magnitude of the negative charge N (Section 3.2) and the electric moment change per flash (Section A.1.2) has been reviewed by Jacobson and Krider (1976), and the literature on flash duration and on frequency of occurrence of continuing current has been reviewed by Livingston and Krider (1978). The locations of negative charge centers appear always to occur at about the same cloud temperature level, typically -10 to $-34°C$, although these temperatures occur at different heights above the local terrain in different locations.

REFERENCES

Anderson, R. B.: Lightning Research in Southern Africa. *Trans. South Afr. Inst. Electr. Eng.*, **71** (Part 4):3–27 (1980).

Anderson, R. B., and A. J. Eriksson: Lightning Parameters for Engineering Application. *Electra*, **69**:65–102 (1980).

Anderson, R. B., H. R. Van Niekerk, and J. J. Gertenbach: Improved Lightning—Earth Flash Counter. *Electron. Lett.*, **9**:394–395 (1973).

Anderson, R. B., H. R. Van Niekerk, H. Kroninger, and D. V. Meal: Development and Field Evaluation of a Lightning Earth-Flash Counter. *IEE Proc. A*, **131**:118–124 (1984).

Anderson, R. B., H. R. Van Niekerk, S. A. Prentice, and D. Mackerras: Improved Lightning Flash Counters. *Electra*, **66**:85–98 (1979).

Barham, R. A.: Transistorized Lightning-Flash Counter. *Electron. Lett.*, **1**:173 (1965).

Barham, R. A., and D. Mackerras: Vertical Aerial CIGRE-Type Lightning-Flash Counter. *Electron. Lett.*, **8**:480–482 (1972).

Brook, M., and N. Kitagawa: Electric-Field Changes and the Design of Lightning-Flash Counters. *J. Geophys. Res.*, **65**:1927–1931 (1960).

Brooks, C. E. P.: The Distribution of Thunderstorms over the Globe. *Geophys. Mem.*, **13**:147-164 (1925). (Air Ministry, Meteorological Office, London.)

Bunn, C. C.: Application of Electric Field Change Measurements to the Calibration of a Lightning Flash Counter. *J. Geophys. Res.*, **73**:1907-1912 (1968).

Carte, A. E., and R. E. Kidder: Lightning in Relation to Precipitation. *J. Atmos. Terr. Phys.*, **39**:139-148 (1977).

Changery, M. J.: "National Thunderstorm Frequencies for the Contiguous United States," 57 pp. USNRC Report NUREG/CR-2252, November, 1981.

Cherna, E. V., and E. J. Stansbury: Sferics Rate in Relation to Thunderstorm Dimensions. *J. Geophys. Res.*, **91**:B701-B707 (1986).

Cooray, V.: Response of CIGRE and CCIR Lightning Flash Counters to the Electric Field Changes from Lightning: A Theoretical Study. *J. Geophys. Res.*, **91**:2835-2842 (1986).

Davis, M. H., M. Brook, H. Christian, B. G. Heikes, R. E. Orville, C. G. Park, R. G. Roble, and B. Vonnegut: Some Scientific Objectives of a Satellite-Borne Lightning Mapper. *Bull. Am. Meteorol. Soc.*, **64**:114-119 (1983).

Darveniza, M., and M. A. Uman: Research into Lightning Protection of Distribution Systems II—Results from Florida Field Work 1978 and 1979. *IEEE Trans. PAS*, **PAS-103**:673-682 (1984).

Eriksson, A. J.: The Lightning Ground Flash—An Engineering Study. Ph.D. thesis, University of Natal, Pretoria, South Africa, 1979. Available as CSIR Special Report ELEK 189 from National Electrical Engineering Research Institute, CSIR, P.O. Box 395, Pretoria 0001, South Africa.

Freier, G. D.: A 10-Year Study of Thunderstorm Electric Fields. *J. Geophys. Res.*, **83**:1373-1376 (1978).

Golde, R. H.: A Lightning Flash Counter. *Electron. Eng.*, **38**:164-166 (1966).

Harris, D. J., and Y. E. Salman: The Measurement of Lightning Characteristics in Northern Nigeria. *J. Atmos. Terr. Phys.*, **34**:775-786 (1972).

Holle, R. L., and M. W. Maier: "Radar Echo Height Related to Cloud-Ground Lightning in South Florida," pp. 330-333. Preprints, 12th Conference on Severe Local Storms, San Antonio, Texas, January 12-15, 1982. Am. Meteorol. Soc., Boston, Massachusetts.

Holzer, R. E.: Simultaneous Measurement of Sferics Signals and Thunderstorm Activity. *In* "Thunderstorm Electricity" (H. R. Byers, ed.), pp. 267-275. Univ. of Chicago Press, Chicago, Illinois, 1953.

Horner, F.: The Design and Use of Instruments for Counting Local Lightning Flashes. *Proc. IEE*, **107**:321-330 (1960).

Ishii, M., T. Kawamura, J. Hojyo, and T. Iwaizumi: Ground Flash Density in Winter Thunderstorm. *Res. Lett. Atmos. Electr.*, **1**:105-108 (1981).

Jacobson, E. A., and E. P. Krider: Electrostatic Field Changes Produced by Florida Lightning. *J. Atmos. Sci.*, **33**:103-117 (1976).

Kitterman, C. G.: Characteristics of Lightning from Frontal System Thunderstorms. *J. Geophys. Res.*, **85**:5503-5505 (1980).

Kotaki, M., I. Kurika, C. Kotoh, and H. Sugiuchi: Global Distribution of Thunderstorm Activity Observed with ISS-b. *J. Radio Res. Lab. Tokyo*, **28**:49-71 (1981).

Kurihara, Y.: World Distribution of Thunderstorm Activity Obtained from Ionosphere Sounding Satellite-b Observations June 1978 to May 1980. Radio Research Laboratories, Ministry of Posts and Telecommunications, Japan, October 1981.

Larsen, H. R., and E. J. Stansbury: Association of Lightning Flashes with Precipitation Cores Extending to Height 7 km. *J. Atmos. Terr. Phys.*, **36**:1547-1553 (1974).

Lhermitte, R., and P. R. Krehbiel: Doppler Radar and Radio Observations of Thunderstorms. *IEEE Trans. Geosci. Electron.*, **GE-17**:162-171 (1979).

Livingston, J. M., and E. P. Krider: Electric Fields Produced by Florida Thunderstorms. *J. Geophys. Res.*, **83**:385-401 (1978).

Lopez, R. E., and R. L. Holle: Diurnal and Spatial Variability of Lightning Activity in Northeastern Colorado and Central Florida during the Summer. *Mon. Weather Rev.*, **114**:1288-1312 (1986).

MacGorman, D. R., M. W. Maier, and W. D. Rust: "Lightning Strike Density for the Contiguous United States from Thunderstorm Duration Records," 44 pp. NUREG/CR-3759, Office of Nuclear Regulatory Research, U.S. Nuclear Regulatory Commission, Washington, D.C., May, 1984.

Mackerras, D.: Prediction of Lightning Incidence and Effects in Electrical Systems. *Electr. Eng. Trans., Inst. Eng. Aust.* **EE-14**:73-77 (1978).

Mackerras, D.: Automatic Short-Range Measurement of the Cloud Flash to Ground Flash Ratio in Thunderstorms. *J. Geophys. Res.*, **90**:6195-6201 (1985).

Maier, L. M., E. P. Krider, and M. W. Maier: Average Diurnal Variation of Summer Lightning over the Florida Peninsula. *Mon. Weather Rev.*, **112**:1134-1140 (1984).

Maier, M. W., A. G. Boulanger, and R. I. Sax: "An Initial Assessment of Flash Density and Peak Current Characteristics of Lightning Flashes to Ground in South Florida," 43 pp. NUREG/CR-1024, Office of Standards Development, U.S. Nuclear Regulatory Commission, Washington, D.C., September, 1979.

Maier, M. W., and E. P. Krider: "A Comparative Study of the Cloud-to-Ground Lightning Characteristics in Florida and Oklahoma Thunderstorms," pp. 334-337. Preprints, 12th Conference on Severe Local Storms, San Antonio, Texas, January 12-15, 1982. Am. Meteorol. Soc., Boston, Massachusetts.

Marshall, J. S., and S. Radhakant: Radar Precipitation Maps as Lightning Indicators. *J. Appl. Meteorol.*, **17**:206-212 (1978).

Muller-Hillebrand, D.: Lightning Counters. II—The Effect of Changes of Electric Field on Counter Circuits. *Ark. Geofys.*, **4**:271-292 (1963).

Muller-Hillebrand, D.: Lightning-Counter Measurements in Scandinavia. *Proc. IEE (London)*, **112**:203-210 (1965).

Orville, R. E.: Global Distribution of Midnight Lightning—September to November 1977. *Mon. Weather Rev.*, **109**:391-395 (1981).

Orville, R. E., and D. W. Spencer: Global Lightning Flash Frequency. *Mon. Weather Rev.*, **107**:934-943 (1979).

Peckham, D. W., M. A. Uman, and C. E. Wilcox, Jr.: Lightning Phenomenology in the Tampa Bay Area. *J. Geophys. Res.*, **89**:11,789-11,805 (1984).

Piepgrass, M. V., E. P. Krider, and C. B. Moore: Lightning and Surface Rainfall during Florida Thunderstorms. *J. Geophys. Res.*, **87**:11,193-11,201 (1982).

Pierce, E. T.: The Influence of Individual Variations in the Field Changes Due to Lightning Discharges upon the Design and Performance of Lightning Flash Counters. *Arch. Meteorol. Geophys. Biokl., Ser.*, **9**:78-86 (1956).

Pierce, E. T.: Latitudinal Variation of Lightning Parameters. *J. Appl. Meteorol.*, **9**:194-195 (1970).

Prentice, S.A.: CIGRE Lightning Flash Counter. *Electra*, **22**:149-171 (1972).

Prentice, S. A.: Frequency of Lightning Discharges. *In* "Lightning, Vol. 1, Physics Of Lightning (R. H. Golde, ed.), pp. 465-495. Academic Press, New York, 1977.

Prentice, S. A., and D. Mackerras: Recording Range of a Lightning-Flash Counter. *Proc. IEE (London)*, **116**:294-302 (1969).

Prentice, S. A., and D. Mackerras: The Ratio of Cloud to Cloud-Ground Lightning Flashes in Thunderstorms. *J. Appl. Meteorol.*, **16**:545-550 (1977).

Prentice, S. A., D. Mackerras, and R. P. Tolmie: Development and Testing of a Vertical Aerial Lightning Flash Counter. *Proc. IEE (London)*, **122**:487–491 (1975).

Proctor, D. E.: Lightning and Precipitation in a Small Multicellular Thunderstorm. *J. Geophys. Res.*, **88**:5421–5440 (1983).

Proctor, D. E.: Correction to "Lightning and Precipitation in a Small Multicellular Thunderstorm." *J. Geophys. Res.*, **89** (D):11,826 (1984).

Reap, R. M.: Evaluation of Cloud-to-Ground Lightning Data from the Western United States for the 1983–1984 Summer Seasons. *J. Climate Appl. Meteorol.*, **25**:785–799 (1986).

Schonland, B. F. J.: The Lightning Discharge. *Handb. Phys.*, **22**:576–638 (1956).

Shackford, C. R.: Radar Indications of a Precipitation–Lightning Relationship in New England Thunderstorms. *J. Meteorol.*, **17**:15–19 (1960).

Sparrow, J. G., and F. E. Ney: Lightning Observations by Satellite. *Nature (London)*, **232**:540–541 (1971).

Takeuti, T., M. Nakano, and M. Nagatani: Lightning Discharges in Guam and Philippine Islands. *J. Meteorol. Soc. Japan*, **53**:360–361 (1975).

Thomson, E. M.: The Dependence of Lightning Return Stroke Characteristics on Latitude. *J. Geophys. Res.*, **85**:1050–1056 (1980).

Turman, B. N.: Analysis of Lightning Data from the DMSP Satellite. *J. Geophys. Res.*, **83**:5019–5024 (1978).

Turman, B. N., and B. C. Edgar: Global Lightning Distributions at Dawn and Dusk. *J. Geophys. Res.*, **87**:1191–1206 (1982).

Vonnegut, B., O. H. Vaughan, M. Brook, and P. Krehbiel: Mesoscale Observations of Lightning from Space Shuttle. *Bull. Am. Meteorol. Soc.*, **66**:20–29 (1985).

Vorpahl, J. A., J. G. Sparrow, and E. P. Ney: Satellite Observations of Lightning. *Science*, **169**:860–862 (1970).

Chapter 3 | Cloud and Lightning Charges

3.1 INTRODUCTION

Most studies of cloud fields and charges have been concerned with the cloud type called cumulonimbus, also referred to as the thundercloud or thunderstorm. There have, however, been some measurements of the electrical properties of stratus, stratocumulus, cumulus, nimbostratus, altocumulus, and altostratus clouds (e.g., Imyanitov *et al.*, 1972) and of the electric fields and the charge on precipitation at ground level from non-thunderstorm rain, sleet, and snow (e.g., Simpson, 1949). In this chapter we will consider only the electrical properties of the primary generator of lightning, the thunderstorm. In Section 1.7.2 and in Chapter 14 some comments are given on the conditions that might produce lightning on planets other than Earth and on the occurrence of lightning-like long electrical sparks on Earth in nonthunderstorm situations.

3.2 CUMULONIMBUS ELECTRIC FIELDS AND CHARGES

As discussed in the historical review by Uman (1969), a model for the charge structure of a thundercloud was developed by the early 1930s. The model was derived from ground-based measurements of both the electric fields associated with the static cloud charges and the electric field changes associated with the effective neutralization of a portion of those cloud charges by lightning (e.g., Wilson, 1916, 1920; Appleton *et al.*, 1920; Schonland and Craib, 1927). In that model the primary thundercloud charges form a positive electric dipole, that is, a positive charge region P containing charge Q_P located above a negative charge region N containing charge Q_N, as shown in Figs. 1.3, 1.5, 3.1, and 3.2. Later in the 1930s, Simpson and Scrase (1937) verified the existence of this fundamental dipole structure from in-cloud measurements made with instrumented balloons, and, additionally, identified a localized region of positive charge Q_p at the base of the cloud, as is also illustrated in Figs. 1.3, 1.5, 3.1, and 3.2.

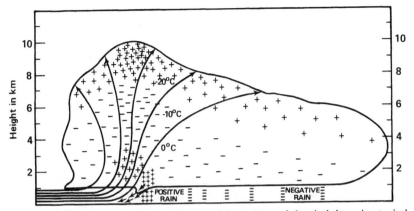

Fig. 3.1 A diagram showing the distribution of air currents and electrical charge in a typical convective thunderstorm in England inferred by Simpson and Scrase (1937) from instrumented balloon flights. A summary of the charge locations found in this study and in a later study reported by Simpson and Robinson (1941) is given in Simpson and Robinson's Table 3.

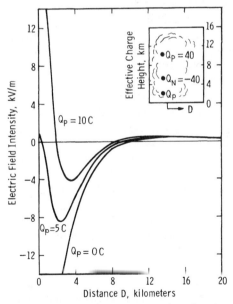

Fig. 3.2 Calculated electric field intensity at the ground as a function of distance for model thunderstorm charges and charge locations: $Q_P = +40$ C at 10 km height, $Q_N = -40$ C at 5 km height, and three values of Q_P at 2 km height. Expressions from which these fields are calculated are found in Sections A.1.2 and 13.2. The field sign convention is discussed in Section A.1.1. Drawing adapted from Malan (1952, 1963) and Uman (1969).

Subsequent measurements of cloud electric fields made both inside and outside the cloud have confirmed the general validity of the double-dipole charge structure (e.g., Simpson and Robinson, 1941; Kuettner, 1950; Malan, 1952; Huzita and Ogawa, 1976; Winn et al., 1981; Marshall and Winn, 1982; Taniguchi et al., 1982; Weber et al., 1982, 1983; Byrne et al., 1983). A fundamental problem, however, in developing an essentially static cloud charge model is that neither the fields nor the charges that produce them are really steady in time. Among other effects, there are substantial rapid field changes due to lightning and subsequent slower field recoveries. Illustrating these effects are the records of the electric field vs time at ground near thunderstorms found, for example, in Jacobson and Krider (1976).

From field measurements made *outside* the cloud, the following charges Q_P, Q_N, and Q_p and the heights of these charge centers above ground level have been derived: in South Africa, $+10\,C$ at 2 km, $-40\,C$ at 5 km, and $+40\,C$ at 10 km, ground level being about 1.8 km above sea level, as illustrated in Fig. 3.2 (Malan, 1952, 1963), and in Japan, $+24\,C$ at 3 km, $-120\,C$ at 6 km, and $+120\,C$ at 8.5 km, ground level being about 1 km above sea level (Huzita and Ogawa, 1976). Calculated electric fields at ground level due to the thunderstorm charge distribution determined by Malan (1952, 1963) are presented in Fig. 3.2. Expressions from which the fields in Fig. 3.2 are derived are found in Sections 13.2 and A.1.2. The field sign convention is discussed in Section A.1.1. The reversal in the direction of the vertical electric field in the 8- to 10-km distance range is perhaps the most characteristic feature of the primary positive dipole structure. A mathematical expression for the distance at which this field reversal occurs is given in Eq. (13.3). The electric field magnitude beneath the cloud is limited by corona processes occurring at the ground (Standler and Winn, 1979), an effect not included in the calculations shown in Fig. 3.2. Because of the difficulty of interpreting electric field measurements made outside the cloud when, in addition to the effects of lightning, there are spatially and time-varying conductivities in and around the cloud, the magnitudes and heights of the cloud charges obtained from remote measurements are, to some extent, uncertain (Kasemir, 1965; Moore and Vonnegut, 1977). For example, space-charge screening layers on the surface of the cloud, that is, layers of charge opposite in sign to the main charge inside (Brown et al., 1971; Hoppel and Phillips, 1971; Klett, 1972; Winn et al., 1978; Weber et al., 1982), can lead to an underestimation of remotely measured main charge magnitudes.

From measurements of cloud electric fields made *inside* the cloud with rockets, aircraft, and balloons, a relatively accurate charge height can be determined, but the charge magnitude cannot be found unless assumptions are made about the size and shape of the individual charge centers, since the cloud probe making the measurements can sense the field only along a more-or-less straight path. Simpson and Robinson (1941) launched several instrumented balloons through each of eight thunderstorms in England and summarized their results in a model in which $+4$ C is present at a height of 1.5 km, -20 C at 3 km, and $+24$ C at 6 km, the stated heights being above a ground that is about 1 km above sea level. Winn *et al.* (1981) in New Mexico found about $+1$ C on rain at about 4 km above sea level, a negative charge density of 5×10^{-9} C/m^3 between 4.8 and 5.8 km, the -2 to $-5°$C clear-air temperature range, and positive charge above 10 km. Marshall and Winn (1982) in New Mexico describe the detection of two lower positive charge regions in one storm, estimated as having 0.4 and 2 C, with a vertical extent of about 1 km and located below the main negative charge center. Taniguchi *et al.* (1982) found that Japanese winter thunderstorms have qualitatively similar charge distributions to summer convective storms but that the charges are lower in altitude, the main negative charge being between 1 and 3 km, a clear-air temperature of 0 to $-10°$C. The measurements of Krehbiel *et al.* (1983), illustrated in Fig. 3.3, are supportive of this observation although Krehbiel *et al.* (1983) found the negative charge to be at a slightly colder temperature. Weber *et al.* (1982) for one thunderstorm in New Mexico found negative charge extending from 5.5 to 8.0 km above sea level, at clear-air temperatures between -5 and $-20°$C, and positive charge above 8 km.

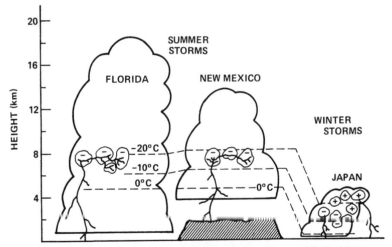

Fig. 3.3 Drawing illustrating the altitude and distribution of ground-flash charge sources observed in summer thunderstorms in Florida and New Mexico and winter thunderstorms in Japan as determined from simultaneous measurement of electric field at a number of ground stations (Section C.7.1) by Krehbiel *et al.* (1983). Note the positive lightning shown in the winter thunderstorm, a common feature of that type of storm (Chapter 11).

Similar results are reported by Weber *et al.* (1983) for another storm in which a 200-m-thick layer of negative screening charge was found at the cloud's upper surface. Four thunderstorms in New Mexico studied by Byrne *et al.* (1983) had positive dipoles with the negative charge in an average vertical height of 1 km between the 0 and $-10°C$ clear-air isotherms. Charge densities in both the main negative and main positive regions were roughly 1×10^{-9} C/m^3.

Thus, despite the various uncertainties, both internal and external measurements of cloud electric fields lead to a similar view of the cloud charge distribution. Supporting this model are both the measurements of the location and value of the negative charge effectively neutralized by lightning as derived from measurements of the change in electric field due to that lightning, illustrated in Figs. 3.3 and 3.4, and the measurements of the location of the first VHF sources in flashes that subsequently produce negatively charged channels. Jacobson and Krider (1976) have summarized much of the available data on the location and size of the negative charge region that is effectively neutralized by lightning. The bulk of these data are similar to their results in Florida obtained using the NASA Kennedy Space Center field mill network (Sections C.1 and B.5). Jacobson and Krider (1976) found a total flash charge of -10 to -40 C being lowered to ground from a height of 6–9.5 km above sea level where the clear-air temperature is between -10 and $-34°C$. Using the same field mill network, Maier and Krider (1986) show that the altitude from which the negative flash charge is lowered varies surprisingly little from flash to flash throughout a given day, but does vary from day to day. Different days produced mean altitudes from 6.9 km ($-14°C$) to 8.8 km ($-26°C$), with a standard deviation on each day usually less than 20% of the mean. Krehbiel *et al.* (1979) in New Mexico report on the heights of individual strokes in ground discharges, as shown in Figs. 3.3 and 3.4. They located stroke charges between a clear-air temperature of -9 and $-17°C$. Proctor (1983) reported that the initial VHF sources for each of 26 consecutive flashes in one South African thunderstorm, all of which produced negatively charged channels, was located in a narrow range of altitudes where the clear-air temperature was between -5 and $-16°C$. The average altitude of the initial radiation was about 4 km above ground ($-10°C$) with a standard deviation of only 440 m. In four additional storms, D. E. Proctor (personal communication, 1984) found flashes originating between 3.1 and 3.9 km above ground, between 0 and $-10°C$ clear-air temperature. In two of these storms, first sources also appeared at 6.5 km ($-15°C$) and at 7.6 km ($-31°C$). Rustan *et al.* (1980) found that the first VHF sources from one ground flash in Florida were between 5.1 and 7.2 km above sea level, at temperatures between about 0 and $-20°C$.

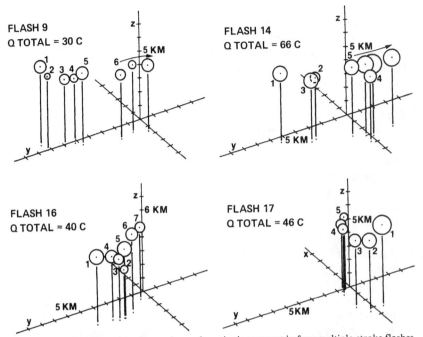

Fig. 3.4 Charge locations for strokes and continuing current in four multiple-stroke flashes in New Mexico. Each charge location is numbered sequentially. The numbers that contain two or more locations (i.e., flash 9, charge 6; flash 14, charge 5; flash 17, charge 4) represent sequential continuing-current charge locations. Continuing currents are discussed in Section 1.3 and Chapter 9. Krehbiel *et al.* (1979) determined the points representing the centers of the assumed spherical charge distributions and the value of the charges involved. The size of the spheres shown surrounding the measured charge location points was arbitrarily determined by assuming the charge density to be 20 C/km³. The electric field vs time at eight stations for flash 14 is given in Fig. 4.3 and a map showing the location of those stations is given in Fig. 4.2. Adapted from Krehbiel *et al.* (1979).

The overall charge associated with the negative charge region is thought not to be uniformly distributed, but rather localized in pockets of relatively high space-charge concentration. Evidence for this localized charge concentration is found in the fact that individual return strokes in a multiple-stroke ground flash tap different negative charge regions, Krehbiel *et al.* (1979) found that the negative charge regions discharged by individual strokes to ground were displaced primarily horizontally from one another. Detailed charge locations for the strokes and continuing current of four flashes in New Mexico are shown in Fig. 3.4. The charge locations found in New Mexico, Florida, and Japan by the same research group are illustrated in Fig. 3.3. Apparently, the negative charge region contributing to ground discharges is located in the same relatively narrow temperature range, roughly -10 to $-25°C$, for storms in these very different environments, an observation consistent with most of the data presented previously in this section. Some additional data from the Florida measurements are reproduced by Williams (1985). These show that over the initial 8-min period of one storm, during which time 15 discharges occurred, the height of the negative charge involved in the discharges remained roughly constant at about 6.8 km while the height of the upper positive charge increased with time from a height of about 10 to about 13 km. Thirteen of the 15 discharges were intracloud flashes so that, for these, both positive and negative charge heights could be obtained from the multiple-station electric field measurements.

It should be noted, as discussed by Vonnegut (1983), that the location of the change in charge due to lightning does not necessarily coincide precisely with the location of the overall charge region responsible for initiating the lightning. Some additional comments on cloud and lightning charges are given in Section 1.2.

The value of electric field intensity for electrical breakdown between plane, parallel electrodes at sea level in dry air is about 3×10^6 V/m. Griffiths and Phelps (1976) argue that corona discharge from solid and liquid precipitation will limit the ambient cloud fields to $2.5-9.5 \times 10^5$ V/m, these values being derived from a variety of laboratory measurements. Winn $et\ al.$ (1974) report having measured a peak horizontal field of 1×10^6 V/m with an instrumented rocket, although they express reservations about the validity of that measurement, and Fitzgerald (1976), using an instrumented aircraft, reports measuring a peak field of 1.2×10^6 V/m. Gunn (1948) found an electric field of 3.4×10^5 V/m on the underbelly of an aircraft just before it was struck. He considered this value to be an underestimate of the overall field. Generally, however, measured fields are $1-2 \times 10^5$ V/m, as is evident from Table 3.1.

Table 3.1

Thunderstorm Electric Fields Measured in Airborne Experiments

Investigation	Typical (V/m)	High values occasionally observed (V/m)	Measurement type
Winn $et\ al.$ (1974)	$5-8 \times 10^4$	2×10^5	Rockets
Winn $et\ al.$ (1981)	—	1.4×10^5	Balloons
Kasemir and Perkins (1978)	1×10^5	2.8×10^5	Aircraft
W. D. Rust and H. W. Kasemir (personal communication)	1.5×10^5	3.0×10^5	Aircraft
Imyanitov $et\ al.$ (1972)	1×10^5	2.5×10^5	Aircraft
Evans (1969)	—	2×10^5	Parachuted sonde
Fitzgerald (1976)	$2-4 \times 10^5$	8×10^5	Aircraft

Calculations of the atmospheric electric fields and currents resulting from the thunderstorm model dipole structure imbedded in an atmosphere whose conductivity is a function of position have been performed, for example, by Holzer and Saxon (1952), Kasemir (1965), Anderson and Frier (1969), Park and Dejnakarintra (1973), Hays and Roble (1979), Nisbet (1983, 1985a, b), Tzur and Roble (1985), and Ogawa (1985).

3.3 ELECTRIFICATION PROCESSES

Recent reviews of the charge generation and separation processes postulated to occur in cumulonimbus clouds have been given by Magono (1980), Latham (1981), Lhermitte and Williams (1983), Illingworth (1985), and Williams (1985). According to Mason (1953), any adequate theory of thunderstorm electrification must account for the fact that the mature stage of a thunderstorm of moderate intensity is characterized by lightning, strong vertical air motion, and the presence of precipitation, that it lasts for about 30 min during which time the average current (negative charge downward) is about 1 A, that the first lightning usually occurs within 20 min of the formation of precipitation, that the basic electrical structure is a positive dipole, that the location of negative charge is near $-5°C$, and that there is an association of field growth with the development of soft hail. Latham (1981) finds this list still "largely acceptable" in the light of recent research, which he reviews, and adds the condition that electric fields of magnitude near 4×10^5 V/m (see Table 3.1) be generated within the central regions of the thunderstorm within about 20 min of the formation of precipitation. The presence of soft hail, as specified by Mason (1953), plays a prominent role in the electrification processes favored by Latham (1981) but is not necessary to some of the other theories.

Basically, there are two types of theories for the generation of the main cloud charge dipole: (1) precipitation theories and (2) convection theories. The former are viewed in the literature as the more significant, but both types could well play some part in cloud electrification. We now consider each of these types of theories. (1) In the precipitation theories, heavy, falling precipitation particles interact with lighter particles carried in updrafts. The interaction process serves to charge the heavy particles negatively and the light particles positively, after which gravity and updrafts separate the opposite charges to form a positive cloud dipole. Charge transfer can be by collision in which two initially uncharged precipitation particles become, after collision, oppositely charged, as, for example, in collisions between hail and ice crystals, the presently preferred mechanism for thunderstorm electrification (e.g., Reynolds *et al.*, 1957; Caranti and Illingworth, 1980;

Illingworth, 1985), or by induction in which two uncharged but electrically polarized (one sign of charge on the top, the other sign on the bottom due to the ambient field) precipitation particles collide in such a way that the small light hydrometeor absorbs charge from the bottom of the larger heavy precipitation particle, as the lighter particle moves upward (e.g., Sartor, 1967). The induction process serves to enhance the initial field in which it operates. Kuettner *et al.* (1981) discuss the relative importance of inductive and noninductive interaction mechanisms in thunderstorm electrification. Individual hydrometeors may also acquire charges in the process of melting or freezing or by the capture or release of free ions or charged aerosol particles (e.g., Illingworth, 1985). (2) In the convective electrification theories, charge that has been accumulated near the Earth's surface or across regions of varying air and cloud conductivity, including the screening layer at the cloud boundary, is moved in bulk to the observed locations by the air flow associated with the thunderstorm (e.g., Grenet, 1947, 1959; Vonnegut, 1953, 1955; Wagner and Telford, 1981). Magono (1980) provides an interesting discussion of the arguments for and against convective theories. Perhaps related to the argument for the existence of convective electrification is Magono's review of the observations of lightning in clouds whose tops do not reach the freezing level, so-called warm cloud lightning (see also Section 1.7.1). If such observations are accepted as true, precipitation charging mechanisms involving collisions between different forms of ice or between ice and supercooled water are not necessary for electrification, although this certainly does not rule out the possibility that they occur and even that they are the dominant charging mechanism in the usual thunderstorm. Williams (1985) reviews a variety of reasons why a precipitation mechanism alone is insufficient to account for the available observations of thunderstorm electrification and hence why a convective mechanism could play an important role in charge separation.

The small positive charge region at the base of the cloud may not be present in all thunderstorms or it may simply be well localized and hence relatively difficult to detect (Simpson and Robinson, 1941). The positive charge could be produced by one or more of the following proposed mechanisms: (1) a mechanism similar to those postulated for producing the primary dipole charges, for example, by hailstones falling through a cloud of supercooled droplets and ice crystals at temperatures between 0 and $-20°C$ as suggested by Jayaratne and Saunders (1984) on the basis of laboratory experiments (see also Marshall and Winn, 1985; and Jayaratne and Saunders, 1985a); (2) the release of positive corona at the ground and its subsequent upward motion to the cloud base, as suggested by Malan and Schonland (1951) and Malan (1952); or (3) the deposition by lightning, as suggested by Marshall and Winn (1982).

REFERENCES

Anderson, F. J., and G. D. Frier: Interaction of the Thunderstorm with a Conducting Atmosphere. *J. Geophys. Res.*, **74**:5390-5396 (1969).

Appleton, E. V., R. A. Watson-Watt, and J. F. Herd: Investigations on Lightning Discharges and on the Electric Fields of Thunderstorms. *Proc. R. Soc. London Ser. A*, **221**:73-115 (1920).

Arabadzhi, V. I.: The Measurement of Electric Field Intensity in Thunderclouds by Means of Radiosonde. *Dokl. Akad. Nauk. SSSR*, **111**:85-88 (1956).

Barnard, V.: The Approximate Mean Height of the Thudercloud Charge Taking Part in a Flash to Ground. *J. Geophys. Res.*, **56**:33-35 (1951).

Brown, K. A., P. R. Krehbiel, C. B. Moore, and G. N. Sargent: Electrical Screening Layers around Charged Clouds. *J. Geophys. Res.*, **76**:2825-2836 (1971).

Byrne, G. J., A. A. Few, and M. E. Weber: Altitude, Thickness and Charge Concentration of Charged Regions of Four Thunderstorms during Trip 1981 Based Upon *in Situ* Balloon Electric Field Measurements. *Geophys. Res. Lett.*, **10**:39-42 (1983).

Byrne, G. J., A. A. Few, M. F. Stewart, A. C. Conrad, and R. L. Torczon: *In Situ* Measurements and Radar Observations of a Severe Storm: Electricity, Kinematics, and Precipitation. *J. Geophys. Res.*, **92**:1017-1031 (1987).

Caranti, J. M., and A. J. Illingworth: Surface Potentials of Ice and Thunderstorm Charge Separation. *Nature (London)*, **284**:44-46 (1980).

Chalmers, J. A.: "Atmospheric Electricity," 2nd Ed. Pergamon, Oxford, 1967.

Dye, J. E., J. J. Jones, W. P. Winn, T. A. Cerni, B. Gardiner, D. Lamb, R. L. Pitter, J. Hallett, and C. P. R. Saunders: Early Electrification and Precipitation Development in a Small, Isolated Montana Cumulonimbus. *J. Geophys. Res.*, **91**:1231-1327, 6747-6750 (1986).

Evans, W. H.: The Measurement of Electric Fields in Clouds. *Rev. Pure Appl. Geophys.*, **62**:191-197 (1965).

Evans, W. H.: Electric Fields and Conductivity in Thunderclouds. *J. Geophys. Res.*, **74**:939-948 (1969).

Fitzgerald, D. R.: Experimental Studies of Thunderstorm Electrification. Air Force Geophysics Laboratory, AFGL-TR-76-0128, AD-A0322374, June 1976.

Gish, O. P., and G. R. Wait: Thunderstorms and the Earth's General Electrification. *J. Geophys. Res.*, **55**:473 (1950).

Grenet, G.: Essai d'Explication de la Charge Électrique des Nuages d'Orages. *Ann. Geophys.*, **3**:306-307 (1947).

Grenet, G.: Le Nuage d'Orage: Machine Électrostatique. *Meteorologie*, **I-53**:45-47 (1959).

Griffiths, R. F., and C. T. Phelps: The Effects of Air Pressure and Water Vapor Content on the Propagation of Positive Corona Streamers, and Their Implications to Lightning Initiation. *Q. J. R. Meteorol. Soc.*, **102**:419-426 (1976).

Gunn, R.: Electric Field Intensity Inside of Natural Clouds. *J. Appl. Phys.*, **19**:481-484 (1948).

Gunn, R.: Electric Field Intensity at the Ground under Active Thunderstorms and Tornadoes. *J. Meteorol.*, **13**:269-273 (1956)

Gunn, R.: The Electrification of Precipitation and Thunderstorms. *Proc. IRE*, **45**:1331-1358 (1957).

Gunn, R.: The Electric Field Intensity and Its Systematic Changes under an Active Thunderstorm. *J. Atmos. Sci.*, **22**:498-500 (1965).

Handel, P. H.: Polarization Catastrophe Theory of Cloud Electricity—Speculation on a New Mechanism for Thunderstorm Electrification. *J. Geophys. Res.*, **90**:5857-5863 (1985).

Hays, P. B., and R. G. Roble: A Quasi-Static Model of Global Atmospheric Electricity, 1, The Lower Atmosphere. *J. Geophys. Res.*, **84**:3291-3305 (1979).

Holzer, R. E., and D. S. Saxon: Distribution of Electrical Conduction Current in the Vicinity of Thunderstorms. *J. Geophys. Res.*, 57:207–216 (1952).

Hoppel, W. A., and B. B. Phillips: The Electrical Shielding Layer around Charged Clouds and Its Roll in Thunderstorm Electricity. *J. Atmos. Sci.*, 28:1258–1271 (1971).

Huzita, A., and T. Ogawa: Charge Distribution in the Average Thunderstorm Cloud. *J. Meteorol. Soc. Japan*, 54:284–288 (1976).

Illingworth, A. J.: Charge Separation in Thunderstorms: Small Scale Processes. *J. Geophys. Res.*, 90:6026–6032 (1985).

Imyanitov, I. M., Y. V. Chubarian, and Y. M. Shvarts: Electricity in Clouds. NASA Technical Translation from Russia, NASA TT-F-718, 1972.

Jacobson, E. A., and E. P. Krider: Electrostatic Field Changes Produced by Florida Lightning. *J. Atmos. Sci.*, 33:113–117 (1976).

Jayaratne, E. R., and C. P. R. Saunders: The Rain Gush, Lightning, and the Lower Positive Charge Center in Thunderstorms. *J. Geophys. Res.*, 89:11,816–11,818 (1984).

Jayaratne, E. R., and C. P. R. Saunders: Reply. *J. Geophys. Res.*, 90:10,755 (1985a).

Jayaratne, E. R., and C. P. R. Saunders: Thunderstorm Electrification: The Effect of Cloud Droplets. *J. Geophys. Res.*, 90:13,063–13,066 (1985b).

Jayaratne, E. R., and C. P. R. Saunders: Reply. *J. Geophys. Res.*, 91:10,950 (1986).

Kasemir, H. W.: The Thundercloud. *In* "Problems of Atmospheric and Space Electricity" (S. C. Coroniti, ed.), pp. 215–235. American Elsevier, New York, 1965.

Kasemir, H. W., and F. Perkins: Lightning Trigger Field of the Orbiter. Final Report, Kennedy Space Center Contract CC 69694A, 1978.

Klett, J. D.: Charge Screening Layers around Electrified Clouds. *J. Geophys. Res.*, 77:3187–3195 (1972).

Krehbiel, P. R., M. Brook, and R. A. McCrory: An Analysis of the Charge Structure of Lightning Discharges to the Ground. *J. Geophys. Res.*, 84:2432–2456 (1979).

Krehbiel, P. R., M. Brook, R. L. Lhermitte, and C. L. Lennon: Lightning Charge Structure in Thunderstorms. *In* "Proceedings in Atmospheric Electricity" (L. H. Ruhnke and J. Latham, eds.), pp. 408–411. Deepak Hampton, Virginia, 1983.

Krider, E. P., and J. A. Musser: Maxwell Currents under Thunderstorms. *J. Geophys. Res.*, 87:11,171–11,176 (1982).

Kuettner, J.: The Electrical and Meteorological Conditions Inside Thunderclouds. *J. Meteorol.*, 7:322–332 (1950).

Kuettner, J.: The Development and Masking of Charges in Thunderstorms. *J. Meteorol.*, 13:456–470 (1956).

Kuettner, J. P., Z. Levin, and J. D. Sartor: Thunderstorm Electrification—Inductive or Non-Inductive. *J. Atmos. Sci.*, 38:2470–2484 (1981).

Latham, J.: The Electrification of Thunderstorms. *Q. J. R. Meteorol. Soc.*, 107:277–298 (1981).

Lhermitte, R., and E. Williams: Cloud Electrification. *Rev. Geophys. Space Sci.*, 21:984–992 (1983).

Lhermitte, R., and E. Williams: Thunderstorm Electrification: A Case Study. *J. Geophys. Res.*, 90:6071–6078 (1985).

Livingston, J. M., and E. P. Krider: Electric Fields Produced by Florida Thunderstorms. *J. Geophys. Res.*, 83:385–401 (1978).

Magono, C.: "Thunderstorms." Elsevier, Amsterdam, 1980.

Maier, L. M., and E. P. Krider: The Charges That Are Deposited by Cloud-to-Ground Lightning in Florida. *J. Geophys. Res.*, 91:13,279–13,289 (1986).

Malan, D. J.: Les Descharges dans l'Air et la Charge Inférieure Positive d'un Nuage Orageux. *Ann. Geophys.*, 8:385–401 (1952).

Malan, D. J.: "Physics of Lightning." English Univ. Press, London, 1963.

Malan, D. J., and B. F. J. Schonland: The Distribution of Electricity in Thunderclouds. *Proc. R. Soc. London Ser. A*, 209:158–177 (1951).

Marshall, T. C., and W. P. Winn: Measurements of Charged Precipitation in a New Mexico Thunderstorm: Lower Positive Charge Centers. *J. Geophys. Res.*, 87:7141–7157 (1982).

Marshall, T. C., and W. P. Winn: Comments on "The 'Rain Gush,' Lightning, and the Lower Positive Center in Thunderstorms" by E. R. Jayaratne and C. P. R. Saunders. *J. Geophys. Res.*, 90:10,753–10,754 (1985).

Mason, B. J.: A Critical Examination of Theories of Charge Generation in Thunderstorms. *Tellus*, 5:446–498 (1953).

Moore, C. B., and B. Vonnegut: The Thundercloud. *In* "Lightning, Vol. 1, Physics of Lightning" (R. H. Golde, ed.), pp. 51–98. Academic Press, New York, 1977.

Nisbet, J. S.: A Dynamic Model of Thundercloud Electric Fields. *J. Atmos. Sci.*, 40:2855–2873 (1983).

Nisbet, J. S.: Thundercloud Current Determination from Measurements at the Earth's Surface. *J. Geophys. Res.*, 90:5840–5856 (1985a).

Nisbet, J. S.: Currents to the Ionosphere from Thunderstorm Generators: A Model Study. *J. Geophys. Res.*, 90:9831–9844 (1985b).

Ogawa, T.: Fair-Weather Electricity. *J. Geophys. Res.*, 90:5951–5960 (1985).

Ogawa, T., and M. Brook: Charge Distribution in Thunderstorm Clouds. *Q. J. R. Meteorol. Soc.*, 95:513–525 (1969).

Park, C. G., and M. Dejnakarintra: Penetration of Thundercloud Electric Fields into the Ionosphere and Magnetosphere, 1, Middle and Subauroral Latitudes. *J. Geophys. Res.*, 78:6623–6633 (1973).

Phillips, B. B.: Charge Distribution in a Quasi-Static Thundercloud Model. *Mon. Weather Rev.*, 95:847–853 (1967a).

Phillips, B. B.: Convected Cloud Charge in Thunderstorms. *Mon. Weather Rev.*, 95:863–870 (1967b).

Proctor, D. E.: Lightning and Precipitation in a Small Multicellular Thunderstorm. *J. Geophys. Res.*, 88:5421–5440 (1983).

Proctor, D. E.: Correction to "Lightning and Precipitation in a Small Multicellular Thunderstorm." *J. Geophys. Res.*, 89:11,826 (1984).

Reynolds, S. E., M. Brook, and M. F. Gourley: Thunderstorm Charge Separation. *J. Meteorol.*, 14:426–436 (1957).

Rustan, P. L., M. A. Uman, D. G. Childers, W. H. Beasley, and C. L. Lennon: Lightning Source Locations from VHF Radiation Data for a Flash at Kennedy Space Center. *J. Geophys. Res.*, 85:4893–4903 (1980).

Sartor, J. D.: The Role of Particle Interactions in the Distribution of Electricity in Thunderstorms. *J. Atmos. Sci.*, 24:601–613 (1967).

Schonland, B. F. J., and J. Craib: The Electric Fields of South African Thunderstorms. *Proc. R. Soc. London Ser. A*, 114:229–243 (1927).

Simpson, G. C.: Atmospheric Electricity during Disturbed Weather. *Geophys. Mem. London*, 84:1–51 (1949).

Simpson, G. C., and G. D. Robinson: The Distribution of Electricity in the Thunderclouds, Pt. 11. *Proc. R. Soc. London Ser. A*, 117:281–329 (1941).

Simpson, Sir George, and F. J. Scrase: The Distribution of Electricity in Thunderclouds. *Proc. R. Soc. London Ser. A*, 161:309–352 (1937).

Standler, R. B., and W. P. Winn: Effects of Corona on Electric Fields beneath Thunderclouds. *Q. J. R. Meteorol. Soc.*, 105:285–302 (1979).

Taniguchi, T., C. Magona, and T. Endoh: Charge Distribution in Active Winter Clouds. *Res. Lett. Atmos. Electr.*, 2:35–38 (1982).

Tzur, I., and R. G. Roble: The Interaction of a Dipolar Thunderstorm with Its Global Electrical Environment. *J. Geophys. Res.*, **90**:5989-5999 (1985).

Uman, M. A.: "Lightning," pp. 48-59. McGraw-Hill, New York, 1969. See also, Dover, New York, 1984.

Vonnegut, B.: Possible Mechanism for the Formation of Thunderstorm Electricity. *Bull. Am. Meteorol. Soc.*, **34**:378 (1953).

Vonnegut, B.: Possible Mechanism for the Formation of Thunderstorm Electricity. *Proc. Int. Conf. Atmos. Electr., Portsmouth*, 169-181. See also *Geophys. Res. Pap.*, 42. AFCRC-TR-55-222, Air Force Cambridge Research Laboratories, Bedford, Massachusetts, 1955.

Vonnegut, B.: Deductions Concerning Accumulations of Electrified Particles in Thunderclouds Based on Electric Field Changes Associated with Lightning. *J. Geophys. Res.*, **88**:3911-3912 (1983).

Vonnegut, B.: Reduction of Thunderstorm Electric Field Intensity Produced by Corona from a Nearby Object. *J. Geophys. Res.*, **89** (D1):1468-1470 (1984).

Vonnegut, B., and C. B. Moore: Comments on "The 'Rain Gush,' Lightning, and the Lower Positive Charge Center in Thunderstorms" by E. R. Jayaratne and C. P. R. Saunders. *J. Geophys. Res.*, **91**:10,949 (1986).

Wagner, P. B., and J. V. Telford: Charge Dynamics and a Electric Charge Separation Mechanism in Convective Clouds. *J. Rech. Atmos.*, **15**:97-120 (1981).

Weber, M. E., H. J. Christian, A. A. Few, and M. F. Stewart: A Thundercloud Electric Field Sounding: Charge Distribution and Lightning. *J. Geophys. Res.*, **87**:7158-7169 (1982).

Weber, M. E., M. F. Stewart, and A. A. Few: Corona Point Measurements in a Thundercloud at Langmuir Laboratory. *J. Geophys. Res.*, **88**:3907-3910 (1983).

Weinheimer, A. J., and A. A. Few, Jr.: Comment on "Contributions of Cloud and Precipitation Particles to the Electrical Conductivity and the Relaxation Time of the Air in Thunderstorms" by A. K. Kamra. *J. Geophys. Res.*, **86**:4302-4304 (1981).

Williams, E. R: Large Scale Charge Separation in Thunderclouds. *J. Geophys. Res.*, **90**:6013-6025 (1985).

Williams, E. R., and R. M. Lhermitte: Radar Tests of the Precipitation Hypothesis for Thunderstorm Electrification. *J. Geophys. Res.*, **88**:10,984-10,992 (1983).

Wilson, C. T. R.: On Some Determinations of the Sign and Magnitude of Electric Discharges in Lightning Flashes. *Proc. R. Soc. London Ser. A*, **92**:555-574 (1916).

Wilson, C. T. R.: Investigation on Lightning Discharges and on the Lightning Field of Thunderstorms. *Philos. Trans. R. Soc. London Ser. A*, **221**:73-115 (1920).

Winn, W. P., and L. G. Byerley: Electric Field Growth in Thunderclouds. *Q. J. R. Meteorol. Soc.*, **101**:979-994 (1975).

Winn, W. P., and C. B. Moore: Electric Field Measurements in Thunderclouds Using Instrumented Rockets. *J. Geophys. Res.*, **76**:5003-5018 (1971).

Winn, W. P., G. W. Schwede, and C. B. Moore: Measurements of Electric Fields in Thunderclouds. *J. Geophys. Res.*, **79**:1761-1767 (1974).

Winn, W. P., C. B. Moore, C. R. Holmes, and L. G. Byerly III: Thunderstorm on July 16, 1975, Over Langmuir Laboratory: A Case Study. *J. Geophys. Res.*, **83**:3079-3091 (1978).

Winn, W. P., C. B. Moore, and C. R. Holmes: Electric Field Structure in an Active Part of a Small, Isolated Thundercloud. *J. Geophys. Res.*, **86**:1187-1193 (1981).

Chapter 4 | Preliminary Breakdown

4.1 EXISTENCE AND STATISTICS

The electric field change prior to the first return stroke in a negative cloud-to-ground flash has a duration from a few milliseconds to a few hundred milliseconds, with a typical value of some tens of milliseconds (e.g., Clarence and Malan, 1957; Kitagawa and Brook, 1960; Takeuti et al., 1960; Harris and Salman, 1972; Thomson, 1980; Beasley et al., 1982). Clarence and Malan (1957) in South Africa report pre-first-stroke field changes with durations up to 200 msec, 50% of the measured values exceeding 30 msec and 10% exceeding 120 msec. Kitagawa and Brook (1960) in New Mexico found durations between 10 and 200 msec, the most frequent value being about 30 msec. Thomson (1980) in Papua New Guinea found a mean duration of 240 msec, with 68% of pre-first-stroke field changes exceeding 100 msec. Beasley et al. (1982) in Florida report a mean duration for pre-first-stroke field changes of 118 msec with a median of 65 msec. Photographic evidence for discharges occurring in the cloud which precede the stepped leader is given by Malan (1952, 1955) who, using a special streak camera, showed that clouds often produce luminosity for a hundred or more milliseconds before the emergence of the stepped leader from the cloud base. Since most evidence indicates that the stepped leader duration is typically 10 to 30 msec, as we shall see in Chapter 5, it follows that some of the observed pre-first-stroke field change and luminosity must be attributed to a "preliminary breakdown" within the cloud.

Although, as indicated above, there is considerable literature on the duration of the electric field change prior to the first return stroke, the only published statistical data on the duration of the electric field change prior to the stepped leader would appear to be due to Beasley et al. (1982). They have examined the electric field change prior to 79 first return strokes and found that the mean duration of what they term "preliminary variations," field changes prior to the initiation of the stepped leader, was 90 msec, with a median value of 42 msec. The data for individual storms are shown in

Table 4.1
Duration of Preliminary Variations[a]

Storm year–day	Longest (msec)	Shortest (msec)	Mean (msec)	Standard deviation (msec)	Median (msec)	Number of events	Most probable (msec)
1976–181	258	18	79	101	42	5	
1976–195	373	0	61	98	30	13	
1976–201	34	11	24	10	~26	4	
1976–203	52	0	24	22	~22	4	
1977–203	90	0	45	64	—	2	
1977–211	240	0	61	59	39.5	26	
1977–212a	484	123	349	137	400	7	
1977–212b	254	38	126	77	114	12	
1977–220	53	0	12	20	~5	6	
Overall	484	0	90	115	42	79	0–20

[a] From Beasley *et al.* (1982).

Table 4.1. It is a matter of dispute whether relatively long pre-first-stroke field changes, say over 100 msec, should be associated with processes that initiate stepped leaders, as argued by Clarence and Malan (1957), or whether these long pre-first-stroke field changes should be treated as relatively independent cloud discharges, as argued by Kitagawa and Brook (1960) and by Thomson (1980).

4.2 LOCATION

The location of the preliminary breakdown in the cloud has been determined in three different ways: (1) from the measurement at a single station of the variation of the preliminary-breakdown electric field change as a function of the distance to various flashes (Clarence and Malan, 1957), (2) from the simultaneous measurement of the electric field change of the preliminary breakdown at eight different ground stations (Krehbiel *et al.*, 1979) (Section C.7.1), and (3) from a determination of the location of the sources of the initial VHF radiation produced within the cloud (Rustan *et al.*, 1980; Hayenga and Warwick, 1981; Beasley *et al.*, 1982; Proctor, 1983) (Section C.7.3).

1. Clarence and Malan (1957), assuming that the preliminary breakdown channels were vertical, concluded that the initial breakdown began between the main negative charge center N and the lower positive charge center p, shown in Figs. 1.3, 1.5, 3.1, and 3.2, and that the discharge was just as likely to start from the positive center and go upward as to start from the negative

center and go downward. They arrived at this conclusion from an examination of their reported typical electric fields shown in Fig. 4.1. A field waveshape of this type is also shown in Fig. 8.3, flash 208. Specifically, from the variation of the polarity of the initial field change, labeled B, with distance, the charge height reponsible for that field change was determined (Section A.1.2). Clarence and Malan (1957) found the location of the preliminary breakdown in South African thunderstorms to be between 1.4 and 3.6 km above the ground, ground being about 1.8 km above sea level. In Fig. 4.1 B, I, and L represent breakdown, intermediate, and leader, respectively, and precede the return-stroke field change R (see Chapter 7), which in turn precedes the interstroke process J (see Chapter 10). The B field change was reported by Clarence and Malan (1957) to be characterized by a significant field change in a period of a few milliseconds, the I portion by a slow or irregular field change, and the L portion by a more rapid field change. Clarence and Malan (1957) suggested that the B stage was due to electrical breakdown between the p and N charge centers in the cloud, that the I stage was due to negative charging of the breakdown channel by subsidiary discharges, and that the L stage occurred after the discharge channel at the cloud base was sufficiently charged to initiate a stepped leader. Thus only the L portion of the pre-first-stroke field change was attributed to the stepped leader. Clarence and Malan (1957) reported that during the B stage large VLF pulses occurred, their detailed shapes being unresolved.

Investigators following Clarence and Malan (1957) have generally not been able to verify the validity of the B, I, L scheme (Kitagawa and Brook, 1960; Thomson, 1980; Krehbiel *et al.*, 1979; Beasley *et al.*, 1982). Beasley *et al.* (1982) reported that only 6% of their measured field changes fits the B, I, L scheme well, with another 15% fitting if the B change is allowed to start without a discernible discontinuity in slope. Beasley *et al.* (1982) review additional literature in which the B, I, L categorization is attempted. It is interesting to note that Schonland (1956), who had worked closely with

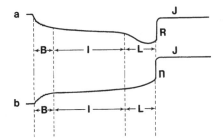

Fig. 4.1 Diagram of typical electric field changes of discharges to ground according to Clarence and Malan (1957). (a) Fields at 2 km. (b) Fields at 5 km.

Clarence and Malan, uses B to label all field changes preceding L, although earlier Schonland *et al.* (1938) distinguished between B and I in describing the postulated fast and slow part of the so-called β-stepped leader (Section 5.2). Krehbiel *et al.* (1979) and Proctor (1983) have observed that the B, I, L waveshape arises from charge motion along a nonvertical path. Krehbiel *et al.* (1979) found that, for two flashes with long-duration preliminary field changes, the motion of negative charge was initially vertically downward and later became horizontal, and they suggested that the I field change observed by Clarence and Malan (1957) could be interpreted as due to horizontal charge motion and thus that the B, I characterization was not justified. In both flashes, the actual leader field change began with a discontinuity 30–40 msec prior to the first stroke. A map of the eight field stations used by Krehbiel *et al.* (1979) is given in Fig. 4.2 and the field change data recorded at each station for their flash 14 are shown in Fig. 4.3. Proctor (1983), from VHF source location measurements, concluded that the waveshape of the pre-first-stroke field change was due to the shape of the path followed by the stepped leader, and ascribed the total duration of the field change to the stepped leader because the VHF radiation pattern remained the same during the field change.

2. Krehbiel *et al.* (1979) found that the preliminary breakdown takes place at roughly the height from which negative charge is eventually lowered to earth, 4 to 6 km above ground where the clear-air temperature is the range -10 to $-20°C$ (see also discussion in Section 3.2), ground in New Mexico being 1.8 km above sea level. These heights, roughly 6–8 km above sea level, are almost twice as high as those determined from the single-station field change measurements of B by Clarence and Malan (1957), roughly 3.2 to 5.4 km above sea level. Thus the locations reported by Krehbiel *et al.* (1979) would not appear to be associated with the lower p charge.

3. Proctor (1983) and Rustan *et al.* (1980), by measuring the difference in the time-of-arrival of VHF impulses at an array of radio receivers (Section C.7.3), located the initial sources of individual VHF impulses produced by lightning. Proctor (1983) in South Africa found that the initial VHF sources for 19 cloud flashes and 7 ground flashes in one storm had a mean height of about 4.0 km above ground or about 5.8 km above sea level with a standard deviation of 440 m. All flashes began between 3 and 5 km above ground (clear-air temperature -5 to $-16°C$) and had negatively charged channels. Rustan *et al.* (1980) found that the initial sources of VHF for one lightning in Florida (ground level equals sea level) were roughly vertical between a height of 5.1 and 7.2 km (between 0 and $-20°C$) and that the appearance of these sources preceded any significant change in the electrostatic field. The stepped leader emerged from the bottom of this preliminary breakdown region and was characterized by a significant change in the

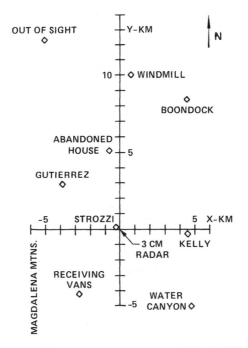

Fig. 4.2   Map of the eight stations used for measurement of electric field change by Krehbiel *et al.* (1979).

electrostatic field and by VHF pulses of much lower amplitude and shorter duration than those of the preceding preliminary breakdown. Both the preliminary breakdown and the stepped leader lowered negative charge toward ground. Figure 4.4 shows VHF source locations for the preliminary breakdown and for the stepped leader as determined by Rustan *et al.* (1980).

4.3   ELECTRIC FIELDS

Many investigators (e.g., Kitagawa, 1957; Clarence and Malan, 1957; Kitagawa and Kobayashi, 1959; Kitagawa and Brook, 1960; Krider and Radda, 1975; Weidman and Krider, 1979; Beasley *et al.*, 1982) have presented data that indicate that the end of the preliminary breakdown or the beginning of the stepped leader is often characterized by a train of relatively large bipolar pulses that can be observed with systems operating at frequencies from VLF to VHF. In much of the literature (e.g., Clarence and Malan, 1957) these pulses are identified as the beginning of a β-leader,

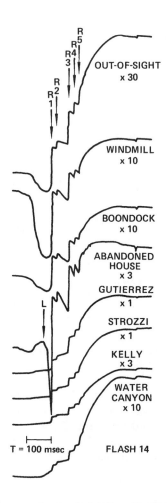

Fig. 4.3 Electric field change vs time for flash 14 from Krehbiel *et al.* (1979). Arrows at top indicate the time of the five return-stroke field changes. Arrow above gutierrez waveform indicates the beginning of the stepped leader. The field change records are ordered according to the north–south position of the stations shown in Fig. 4.2, with the northernmost station on the top of the figure, the distance from out-of-sight to water canyon being about 17 km. The largest total field change was at gutierrez, 28.7 kV/m; the smallest was at out-of-sight, 2.43 kV/m. The stepped leader descended over abandoned house and contacted ground near it. The preliminary breakdown apparently began at about 5-km altitude, a few kilometers NNW of abandoned house. A continuing current (Chapter 9) flowed to earth following the fifth return stroke, as is evident from the positive field change at all stations following that stroke. The charge locations associated with the individual strokes and the continuing current are shown in Fig. 3.4.

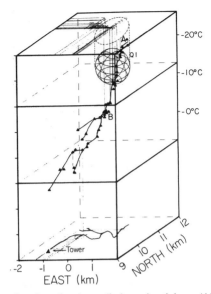

Fig. 4.4 VHF source locations for the preliminary breakdown (AB) and stepped leader (below B) for a lightning to ground which struck the 150-m weather tower at the Kennedy Space Center. Locations are 94-μsec averages determined by finding time delays between VHF signal arrival at four stations from cross-correlating the recorded signals for that time period. The channel sketched connecting the average VHF locations and the projection of that channel on the ground are a best estimate determined from both the location and time of occurrence of the sources. The first stroke charge Q_1 is reported in Uman *et al.* (1978) and was found using the technique of Krehbiel *et al.* (1979). Adapted from Rustan *et al.* (1980).

as discussed in Section 5.2. Examples of these pulses are shown in Fig. 4.5. A frequency spectrum for the pulses is given by Weidman *et al.* (1981). Beasley *et al.* (1982) have presented evidence that these pulses are produced at the location at which the preliminary breakdown observed by Rustan *et al.* (1980) ends and the stepped leader begins. Uman and McLain (1970) have presented theory to relate the currents and electromagnetic fields of similar pulses (Sections 5.5 and A.4). Optical radiation apparently associated with one of these pulses had been published by Brook and Kitagawa (1960). Weidman and Krider (1979) have characterized the pulse shape as follows. The initial polarity is the same as the ground-stroke electric field change that follows; the overall shape is bipolar with a 10-μsec rise to peak upon which is superimposed 2- or 3-μsec-width pulses followed by a smooth negative half-cycle and an overall duration of about 50 μsec; the amplitude approaches that of the return stroke, and several pulses separated by about 100 μsec make up a characteristic sequence. Since the detailed shapes of the

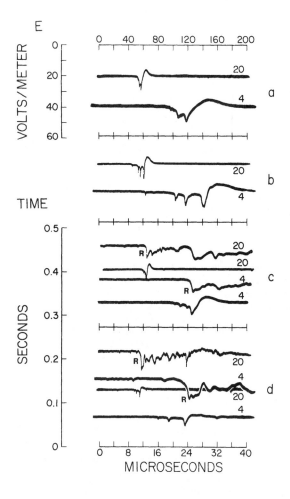

Fig. 4.5 Large bipolar electric fields radiated by lightning discharges at the end of preliminary breakdown or beginning of stepped leader as observed at distances of 30–50 km. Each waveform is shown on both a slow (20 μsec/div) and fast (4 μsec/div) time scale, the former indented with respect to the latter. The polarity of the waveforms is consistent with the lowering of negative charge or the raising of positive. The time interval between the return stroke R and the discharge preceding it in c and d is shown on the scale at the left. Adapted from Weidman and Krider (1979).

β-pulses of Clarence and Malan (1957) were not resolved by their experimental apparatus, it is not known whether their pulses were similar to those shown in Fig. 4.5, but it is likely they were, particularly in view of the fact that Weidman and Krider (1979) found that the shapes of the pulses in cloud discharges that did not precede return strokes (Figs. 13.6 and 13.7) were similar to the shapes of the pulses that did precede return strokes. The polarities of the two types of pulses, however, were opposite. Since the cloud pulses and the preliminary breakdown pulses are apparently generated in part by an intermittent or stepping process, it is natural to expect that the literature regarding the differences between these pulses and pulses associated with the photographed stepped leader would be confused, as is apparently the case.

Prior to the occurrence of pulses of the type shown in Fig. 4.5, there is appreciable VHF radiation from the cloud (e.g., Malan, 1958; Brook and Kitagawa, 1964) presumably due to localized breakdown processes involving relatively small charge transfers.

4.4 PHYSICS

The detailed physics of the preliminary breakdown is not understood, if indeed there is a unique event called preliminary breakdown, as opposed to a variety of types of stepped-leader-initiating cloud discharges or stepped leaders alone without preliminary discharges. There have been, however, a number of suggestions as to how breakdown could start in the cloud, continue to grow, and eventually produce a stepped leader (e.g., Pierce, 1957; Loeb, 1966, 1970; Dawson and Duff, 1970; Phelps, 1974; Griffiths and Phelps, 1976a, b; Diachuk and Muchnik, 1979). A good review of the recent attempts at explaining how the stepped leader is initiated in the presence of ambient electric fields that are too small for direct spark breakdown is given by Griffiths and Phelps (1976a). The model developed by Griffiths and Phelps (1976a) involves the enhancement of the local electric field via positive streamer systems that are able to propagate upward in electric fields greater than 1.5×10^5 V/m. These positive corona streamers are initiated by corona from hydrometeors which occurs at fields above 2.5×10^5 V/m. The streamers travel upward about 100 m and enhance the field at their lowest point by an order of magnitude in a time of 2 to 10 msec. The ambient fields in which the positive streamer systems are initiated and propagate are near the values categorized in Table 3.1 as occasionally observed. This type of discharge, however, would occur between the lower p region and the primary N region above it, a location inconsistent with the bulk of the available data on the position of the preliminary breakdown.

REFERENCES

Beasley, W. H., M. A. Uman, and P. L. Rustan: Electric Fields Preceding Cloud-to-Ground Lightning Flashes. *J. Geophys. Res.*, **87**:4883–4902 (1982).

Brook, M., and N. Kitagawa: Electric-Field Changes and the Design of Lightning-Flash Counters. *J. Geophys. Res.*, **65**:1927–1931 (1960).

Brook, M., and N. Kitagawa: Radiation from Lightning Discharges in the Frequency Range 400 to 1,000 Mc/s. *J. Geophys. Res.*, **69**:2431–2434 (1964).

Clarence, N. D., and D. J. Malan: Preliminary Discharge Processes in Lightning Flashes to Ground. *Q. J. R. Meteorol. Soc.*, **83**:161–172 (1957).

Crabb, J. A., and J. Latham: Corona from Colliding Drops as a Possible Mechanism for the Triggering of Lightning. *Q. J. R. Meteorol. Soc.*, **100**:191–200 (1974).

Dawson, G. A., and D. G. Duff: Initiation of Cloud-to-Ground Lightning Strokes. *J. Geophys. Res.*, **75**:5858–5867 (1970).

Dawson, G. A., and W. P. Winn: A Model for Streamer Propagation. *Z. Phys.*, **183**:159–171 (1965).

Diachuk, V. A., and V. M. Muchnik: Corona Discharge of Watered Hailstones as a Basic Mechanism of Lightning Initiation. *Dokl. Akad. Nauk SSSR*, **248**:63–70 (1979).

Gladshteyn, N. D.: Statistical Properties of Noise from the Predischarge Part of a Lightning Discharge. *Geomagn. Aeron.*, **5**:741–742 (1967).

Griffiths, R. F., and C. T. Phelps: A Model of Lightning Initiation Arising from Positive Corona Streamer Development. *J. Geophys. Res.*, **31**:3671–3676 (1976a).

Griffiths, R. F., and C. T. Phelps: The Effects of Air Pressure and Water Vapor Content on the Propagation of Positive Corona Streamers, and Their Implications to Lightning Initiation. *Q. J. R. Meteorol. Soc.*, **102**:419–426 (1976b).

Harris, D. J., and Y. E. Salman: The Measurement of Lightning Characteristics in Northern Nigeria. *J. Atmos. Terr. Phys.*, **34**:775–786 (1972).

Hayenga, C. O.: Characteristics of Lightning VHF Radiation near the Time of Return Strokes. *J. Geophys. Res.*, **89**:1403–1410 (1984).

Hayenga, C. O., and J. W. Warwick: Two-Dimensional Interferometric Positions of VHF Lightning Sources. *J. Geophys. Res.*, **86**:7451–7462 (1981).

Kitagawa, N.: On the Electric Field Change Due to the Leader Processes and Some of Their Discharge Mechanism. *Pap. Meteorol. Geophys. Tokyo*, **7**:400–414 (1957).

Kitagawa, N., and M. Brook: A Comparison of Intracloud and Cloud-to-Ground Lightning Discharges. *J. Geophys. Res.*, **65**:1189–1201 (1960).

Kitagawa, N., and M. Kobayashi: Field Changes and Variations in Luminosity Due to Lightning Flashes. *In* "Recent Advances in Atmospheric Electricity" (L. G. Smith, ed.), pp. 485–501. Pergamon, Oxford, 1959.

Krehbiel, P. R., M. Brook, and R. A. McCrory: An Analysis of the Charge Structure of Lightning Discharges to Ground. *J. Geophys. Res.*, **84**:2432–2456 (1979).

Krider, E. P., and G. J. Radda: Radiation Waveforms Produced by Lightning Stepped Leaders. *J. Geophys. Res.*, **80**:2653–2657 (1975).

Krider, E. P., C. D. Weidman, and D. M. LeVine: The Temporal Structure of the HF and VHF Radiation Produced by Intracloud Lightning Discharges. *J. Geophys. Res.*, **74**:5750–5762 (1979).

Latham, J., and J. M. Stromberg: Point Discharge. *In* "Lightning, Vol. 1, Physics of Lightning" (R. H. Golde, ed.), pp. 99–117. Academic Press, New York, 1977.

Loeb, L. B.: The Mechanism of Stepped and Dart Leaders in Cloud-to-Ground Lightning Strokes. *J. Geophys. Res.*, **71**:4711–4721 (1966).

Loeb, L. B.: Mechanism of Charge Drainage from Thunderstorm Clouds. *J. Geophys. Res.*, **75**:5882–5889 (1970).

Malan, D. J.: Les Decharges dans l'Air et la Charge Inférieure Positive d'un Nuage Oraguex. *Ann. Geophys.*, **8**:385–401 (1952).

Malan, D. J.: Les Decharges Lumineuses dans Les Nuages Oraguex. *Ann. Geophys.*, **11**:427–434 (1955).

Malan, D. J.: Radiation from Lightning Discharges and Its Relation to Discharge Processes. *In* "Recent Advances in Atmospheric Electricity" (L. G. Smith, ed.), pp. 557–563. Pergamon, Oxford, 1958.

Malan, D. J., and B. F. J. Schonland: The Distribution of Electricity in Thunderclouds. *Proc. R. Soc. London Ser. A*, **209**:158–177 (1951).

Phelps, C. T.: Positive Streamer System Identification and Its Possible Role in Lightning Initiation. *J. Atmos. Terr. Phys.*, **26**:103–111 (1974).

Pierce, E. T.: Recent Advances in Meteorology: Lightning. *Sci. Prog. (London)*, **45**:62–75 (1957).

Proctor, D. E.: Lightning and Precipitation in a Small Multicellular Thunderstorm. *J. Geophys. Res.*, **88**:5421–5440 (1983).

Proctor, D. E.: Correction to "Lightning and Precipitation in a Small Multicellular Thunderstorm." *J. Geophys. Res.*, **89**:11,826 (1984).

Rustan, P. L., M. A. Uman, D. G. Childers, W. H. Beasley, and C. L. Lennon: Lightning Source Locations from VHF Radiation Data for a Flash at Kennedy Space Center. *J. Geophys. Res.*, **85**:4893–4903 (1980).

Schonland, B. F. J.: The Lightning Discharge. *Handb. Phys.*, **22**:576–628 (1956).

Schonland, B. F. J., D. B. Hodges, and H. Collens: Progressive Lightning, Part 5, A Comparison of Photographic and Electrical Studies of the Discharge Process. *Proc. R. Soc. London Ser. A*, **166**:56–75 (1938).

Takeuti, T., H. Ishikawa, and M. Takagi: On the Cloud Discharge Preceding the First Ground Stroke. *Proc. Res. Inst. Atmos., Nagoya Univ.*, **7**:1–6 (1960).

Thomson, E. M.: Characteristics of Port Moresby Ground Flashes. *J. Geophys. Res.*, **85**:1027–1036 (1980).

Uman, M. A., and D. K. McLain: Radiation Field and Current of the Lightning Stepped Leader. *J. Geophys. Res.*, **75**:1058–1066 (1970).

Uman, M. A., W. H. Beasley, J. A. Tiller, Y. T. Lin, E. P. Krider, C. D. Weidman, P. R. Krehbiel, M. Brook, A. A. Few, J. L. Bohannon, C. L. Lennon, H. A. Poehler, W. Jafferis, J. R. Gulick, and J. R. Nicholson: An Unusual Lightning Flash at Kennedy Space Center. *Science*, **201**:9–16 (1978).

Weidman, C. D., and E. P. Krider: The Radiation Field Waveforms Produced by Intracloud Lightning Discharge Processes. *J. Geophys. Res.*, **84**:3157–3164 (1979).

Weidman, C. D., E. P. Krider, and M. A. Uman: Lightning Amplitude Spectra in the Interval from 100 kHz to 20 MHz. *Geophys. Res. Lett.*, **8**:931–934 (1981).

Zonge, K. L., and W. H. Evans: Prestroke Radiation from Thunderclouds. *J. Geophys. Res.*, **71**:1519–1523 (1966).

Chapter 5 | Stepped Leader

5.1 INTRODUCTION

A significant fraction of what is known about stepped leaders was determined in the 1930s by Schonland and co-workers in South Africa using streak-photographic measurements (e.g., Schonland, 1938; Schonland *et al.*, 1938a,b). The South African photographic studies were sometimes supplemented by millisecond-scale electric field measurements at close range (e.g., Schonland *et al.*, 1938a; Malan and Schonland, 1947). Recently, the characteristics of the overall stepped-leader electric field change has been reexamined by Beasley *et al.* (1982) and by Thomson (1985), measurements of the electromagnetic fields due to individual leader steps have been made with submicrosecond resolution (e.g., Krider and Radda, 1975; Krider *et al.*, 1977; Weidman and Krider, 1980; Baum *et al.*, 1980; Weidman *et al.*, 1981; Cooray and Lundquist, 1982, 1985), and photoelectric measurements of leader-step light have been obtained (Krider, 1974; Beasley *et al.*, 1983).

5.2 TYPES OF STEPPED LEADERS

A streak photograph of a stepped leader is shown in Fig. 1.8. On the basis of measured step length and average earthward speed obtained from streak photographs, Schonland (1938) and Schonland *et al.* (1938a,b) divided stepped leaders into two categories, α and β. Type-α leaders were reported to have a uniform earthward speed of the order of 10^5 m/sec and to have steps that do not vary appreciably in length or brightness and that are shorter and much less luminous than β-steps. Type-β leaders were reported to begin with long, bright steps and high average earthward speed, of the order of 10^6 m/sec, to exhibit extensive branching near the cloud base, and, as they approach the earth, to assume the characteristics of α leaders. Type-β leaders are apparently indistinguishable from heavily branched air discharges, both of which Schonland (1956) includes in the β-category in Table 5.1.

Table 5.1

Average Downward Speeds of Type-α Stepped Leaders and Type-β Stepped Leaders and Air Discharges in South Africa[a]

	Number of type-α leaders	Number of type-β leaders and air discharges
$0.8-2.0 \times 10^5$	27	
2.0–3.0	9	
3.0–4.0	13	
4.0–5.0	4	
5.0–6.0	3	1
6.0–7.0	2	1
7.0–8.0	2	1
8.0–9.0		3
9.0–10.0		2
10.0–11.0		2
11.0–12.0		2
12.0–13.0		1
17.0–18.0		1
20.0–21.0		1
23.0–24.0		1
25.0–26.0		1

[a] Adapted from Schonland (1956).

Schonland (1956) states that the majority of photographed leaders are type-α (photography generally views only that part of the leader beneath the cloud base although occasionally leaders are photographed emerging from the side of the cloud), whereas the majority of electric field measurements indicate type-β. Type-β leaders are identified on electric field records by the presence of large pulses at the beginning of the stepped-leader field change, followed by smaller pulses. According to Malan and Schonland (1951), a β-leader has roughly 1 msec of large electric field pulses followed by about 9 msec of α-activity. Beasley *et al.* (1982) suggest that these initial large pulses are the bipolar pulses shown in Fig. 4.5, and point out that there have been no simultaneous electric field and streak-photographic measurements of stepped leaders so that the identification of the large electric field pulses with photographed long, bright steps is speculative. Beasley *et al.* (1982) argue that there is only *one* type of stepped leader, not the two types-α and β, and that *every* stepped leader begins, or every preliminary breakdown ends, with electric field pulses of the type shown in Fig. 4.5. As we shall discuss in Section 5.3, electric field pulses from leader steps near ground are relatively small and unipolar and hence of a different character than those shown in Fig. 4.5.

Schonland *et al.* (1938b) report that there are two variations of the type of leader they identify photographically as β. These are labeled β_1 and β_2. Type-β_1 differs from the normal β-leader in that it exhibits a marked discontinuity in downward speed at some point in its earthward trip. Before the discontinuity the β_1 is heavily branched and exhibits an average speed of about 0.6–2.6×10^6 m/sec. After the discontinuity the leader becomes essentially a type-α leader with few branches, weak luminosity, and average speed of about 0.7–3.2×10^5 m/sec. Type-β_2 leaders differ from normal β-leaders in that during the slow second stage, near earth, the β_2-leader channel is illuminated by a continuously moving or dart leader passing from cloud base to leader tip. Profuse branching is observed at the earthward end of the dart. A number of these dart leaders may overtake the relatively slow-moving stepped leader. The time interval between the darts is of the order of 10 msec, as is the time interval between the appearance of the first stage of the type-β_2 leader and the first dart. Schonland *et al.* (1938b) reported that the second stage of the β_2-leader is the slowest and least luminous form of leader process. There is not much difference in the leader speed before or after the appearance of a dart. Workman *et al.* (1936) have apparently photographed a type-β_2 leader that appeared to travel to ground in four successive large darts. The leader steps were unresolved on the photograph.

5.3 PROPERTIES OF LEADER STEPS

The step lengths of type-α leaders reported by Schonland (1956) range between 10 and 200 m, with a pause time between steps from 37 to 124 μsec. Longer pause times are generally followed by longer step lengths. Berger and Vogelsanger (1966) report pause times between 29 and 52 μsec with step lengths between 3 and 50 m. Apparently the shorter step lengths occur near the ground. From electric field records, Kitagawa (1957) observed a mean time between steps of 50 μsec far above the ground, decreasing to 13 μsec as the leader approached the ground. Recent work has verified that the electric field pulses produced by leader steps a few hundred microseconds prior to the return stroke occur at about 5- to 20-μsec intervals with a mean value in the 15-μsec range (Krider and Radda, 1975; Krider *et al.*, 1977; Beasley *et al.*, 1983; Cooray and Lundquist, 1982, 1985). Some of the decrease in time between pulses as a stepped leader approaches ground may be due to the fact that near ground there can be several leader branches stepping downward together, making the apparent time between leader electric field pulses shorter than the time between steps in any one branch. Examples of stepped-leader electric field pulses just prior to the return stroke are shown in Fig. 5.1, and a drawing of these pulses is found in Fig. 7.4c. The final

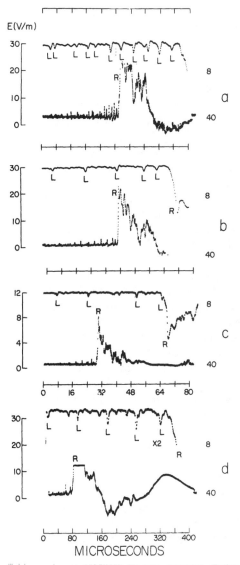

Fig. 5.1 Electric field waveforms produced by 1001 lightning discharges in Florida at distances of 20–100 km. Each record contains an abrupt return-stroke transition R preceded by small pulses characteristic of leader steps. The polarities of all waveforms reproduced are consistent with negative charge being lowered to ground. The same waveform is shown on both a slow (40 μsec/div) and an inverted fast (8 μsec/div) time scale. The vertical gain of the fast trace for d has been magnified by a factor of 2 in relation to the lower trace. Adapted from Krider *et al.* (1977).

stepped-leader pulse is generally the largest and, on average, has about 0.1 of the peak value of the following return stroke field (Krider *et al.*, 1977; Cooray and Lundquist, 1985). Cooray and Lundquist (1985) show that, for three distant storms, the amplitudes of the last leader pulse and the following return stroke peak are correlated.

Beasley *et al.* (1983) have published simultaneous leader-step light and electric field pulses thus proving unequivocally that the electric field pulses that were previously associated with leader steps are indeed due to leader steps.

As can be seen in Fig. 5.1, the stepped-leader electric field pulses occurring a few hundred microseconds before the return stroke are essentially unipolar and hence differ considerably from the large bipolar pulses (Fig. 4.5) that occur near the beginning of the stepped leader, typically 10–20 msec before the return stroke. It is significant that the large bipolar pulses, the stepped-leader pulses, and the return-stroke field change all have the same initial polarity indicating that they all begin with the motion of negative charge downward.

The stepped-leader electric field pulses just prior to the return stroke for two storms for which propagation was partly over salt water were found by Krider *et al.* (1977) to have mean 10–90% risetime of 0.2 and 0.3 μsec, respectively. Weidman and Krider (1980) show examples of stepped-leader pulses with 10–90% risetimes as short as 50 nsec when the propagation paths were entirely over salt water. Beasley *et al.* (1982) found pulse risetime of 0.5–0.8 μsec for stepped leaders 2 km distant over land. Beasley *et al.* (1983) found full widths of electric field pulses at half-maximum in the 0.8- to 1.5-μsec range, whereas Krider *et al.* (1977) found a mean of 0.4 and 0.5 μsec for the two storms they studied. Apparently the propagation over poorly conducting Florida soil preferentially degrades the higher frequencies, as discussed in more detail in relation to return strokes in Sections 7.2.1 and 7.3. A frequency spectrum from about 200 kHz to 20 MHz for electric field pulses from individual leader steps is given by Weidmen *et al.* (1981).

From measurements of the electric fields radiated by leader steps (Krider *et al.*, 1977; Weidman and Krider, 1980; Baum *et al.*, 1980; Cooray and Lundquist, 1985) it can be inferred (e.g., Krider *et al.*, 1977; Cooray and Lundquist, 1985) that step currents are in the kiloampere range or larger with risetimes of the order of 0.1 μsec or less.

From both photographic and photoelectric measurements, the luminosity of the step has been found to rise to peak in about 1 μsec and to fall to half this value in roughly the same time (Schonland *et al.*, 1935; Schonland, 1956; Orville, 1968; Krider, 1974; Beasley *et al.*, 1983). Perhaps scattering of the step light by aerosols increases the risetime and width from the values that would be observed close to the lightning. Optical scattering is discussed in relation to return strokes in Section 7.2.4. Since a step becomes luminous in 1 μsec or less and has a typical length of perhaps 50 m, the velocity with which this luminosity fills the step channel is in excess of 5×10^7 m/sec (Schonland et al., 1935; Orville and Idone, 1982). For several individual steps in dart-stepped leaders (Section 8.2), Orville and Idone (1982) reported a photographed structure consisting of a distinct bright tip at the bottom that fanned out and became more diffuse toward the upper portion of the step. Negatively charged stepped leaders are photographically dark between steps, but positively charged leaders (Sections 11.3 and 12.2) emit some light between steps and hence have less distinct steps (Schonland, 1956; Berger and Vogelsanger, 1966). Photographs of both downward-moving stepped leaders (Schonland et al., 1935) and upward-moving stepped leaders (Berger and Vogelsanger, 1966; Section 12.2) show a faint corona discharge extending for about one step length in front of the bright leader step. The luminosity of this advance corona does not appear to develop between steps but rather occurs simultaneously with the creation of the bright step behind it.

5.4 OVERALL LEADER CHARACTERISTICS

Schonland (1956) reports average two-dimensional stepped-leader speeds between 0.8 and 26×10^5 m/sec for the bottom 2–3 km of the channel and estimates the minimum three-dimensional value to be about 1×10^5 m/sec. The most often measured two-dimensional speed is between 1 and 2×10^5 m/sec. Two-dimensional speed data from Schonland (1956) are given in Table 5.1. Berger and Vogelsanger (1966) in Switzerland find average speeds in the final 1.3 km above ground between 0.9 and 4.4×10^5 m/sec for 14 photographed stepped leaders. For 2 photographed stepped leaders Orville and Idone (1982) found average two-dimensional speeds over the last few steps of 6×10^5 and 15×10^5 m/sec. Thomson et al. (1985) used measurements of 10 stepped-leader electric field changes to estimate stepped-leader speeds between 1.3 and 19×10^5 m/sec with a mean of 5.7×10^5 for the bottom 0.6–2.0 km of the channel.

Luminous stepped-leader diameters have been measured photographically to be between 1 and 10 m (Schonland, 1953). The leader charge is probably contained within a channel of similar size because such diameters are required if the surface electric field is to be below the breakdown value (Section A.1.5; Schonland, 1953; Uman, 1969). The leader current, on the other hand, must be carried in a central core some centimeters in diameter, as in any arc of hundreds or more amperes in air, the magnitude of current needed to lower the coulombs of charge consistent with the measured electric field change in the tens of milliseconds of leader duration. Evidence of a stepped-leader central arc core is provided by the spectral measurements of Orville (1968) which show a step optical radiation characteristic of a high-current spark flowing in a centimeter-size channel. Why this central core has not been reported in photographs is unclear.

Average stepped-leader currents of 50 and 63 A have been measured for two leaders using remote magnetic field measurements (Williams and Brook, 1963). Sixty-two stepped-leader currents derived from electric field measurements in Florida for stepped leaders near ground have a mean of 1.3 kA and a standard deviation of 1 kA, with range from 100 A to 5 kA (Thomson *et al.*, 1985). Krehbiel (1981) from multiple-station electric field measurements in Florida found final leader currents averaged over milliseconds in the range 200 A to 3.3 kA with a mean of 1.3 kA for seven stepped leaders.

The millisecond-scale electric fields of the stepped leader are reasonably well understood (Malan and Schonland, 1947; Beasley *et al.*, 1982; Thomson, 1985) and adequately modeled for many purposes by the lowering of negative charge density along a vertical line. The electric field change for such a model with uniform charge density is given in Eq. (A.11) and derived in Sections A.1.2 and A.1.3. Plotted results are shown in Fig. 5.2 and are discussed in Section A.1.3. Thomson (1985) has calculated the stepped-leader electrostatic field and its time derivative for vertical one-dimensional channels, inclined channels, and three-dimensional charge distributions with arbitrary height dependence. Thomson (1985) argues that the derivative of the electric field is preferable to the field itself for the characterization and interpretation of leader field waveshapes from close lightning. The derivative is a function of how the leader tip is propagating, whereas the total electric field at the same time depends on the overall behavior of the leader up to that time. Thomson (1985) shows that final leader current and speed can be derived from the time-derivative records, results using this technique having been given above (Thomson *et al.*, 1985). Further, he points out that the time derivative of the electric field is zero when the model leader begins (see Fig. 5.2) and is positive when the model leader touches ground (a fact that is not obvious for close leaders on a slow time scale such as shown, for example, in Fig. 5.3). The implication of the former observation is that an initial increase in electric field sometimes attributed to the beginning of the stepped leader is either due to a leader different from the accepted model or to another process in the cloud, and hence electric field measurements of the time duration of the leader, as listed in Table 5.2, are, to some extent, suspect. Typical stepped-leader electric fields measured on the ground at various distances by Beasley *et al.* (1982) are given in Fig. 5.3 (see also Figs. 4.1 and 4.3). The distance categories adopted by Beasley *et al.* (1983) are arbitrary. Overall, the waveshapes are in good qualitative agreement with the model illustrated in Fig. 5.2.

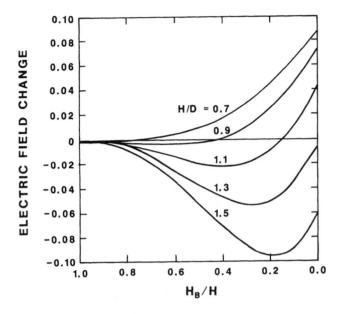

Fig. 5.2 Electric field change for a model negative leader moving downward from a negative charge center at height H. The assumed uniform linear charge density ρ_L on the leader is derived from the negative charge at H. The factor H_B represents the height of the leader tip above the ground. Multiply the field change given in the graph by $\rho_L/2\pi\varepsilon_0 D$ to obtain SI units of volts per meter where D is the horizontal distance to the leader. See Section A.1.3 for derivation. Adapted from Uman (1969).

Table 5.2

Durations of Electric Field Change of Stepped Leaders[a]

Reference	Number of flashes	Distance range (km)	Duration (msec)		
			Minimum	Maximum	Mode
Schonland et al. (1938a)	69	0–24	0–3	66	9–12
Pierce (1955)	240	40–100	0–20	525–550	20–40
Clarence and Malan (1957)	234	0–80	6	442	—
Kitagawa (1957)	41	0–15	8	89	20–30
Kitagawa and Brook (1960)	290	—	0–10	210	10–30
Thomson (1980)	53	6–40	~4	~36	—
Beasley et al. (1982)	79	0–20	2.8	120	6–20

[a] Adapted from Beasley et al. (1982).

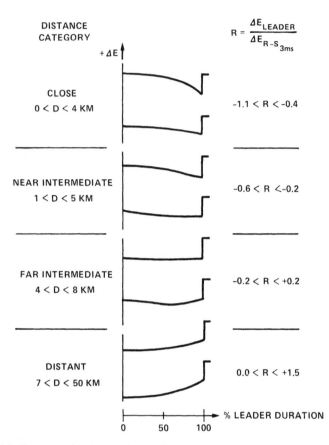

Fig. 5.3 Representative shapes of stepped-leader electric field changes in four distance categories. Note that the categories overlap. Also given is the ratio of the stepped-leader to return-stroke field change for which detailed data are found in Fig. 5.5. Adapted from Beasley *et al.* (1982).

Total charge on stepped leaders ranges from a few coloumbs to 10 to 20 C (Brook *et al.*, 1962), with a resulting average charge lowered per unit length of the order of 10^{-3} C/m (Schonland, 1953). Durations of stepped-leader electric field changes as reported by various investigators are given in Table 5.2. As noted above, these should be viewed as subject to considerable error. The measured leader field change has been reported to be relatively smooth which implies that the leader lowers charge continuously between steps and that the step process itself does not lower appreciable charge (Chapman, 1939; Malan and Schonland, 1947; Schonland, 1953). On the other hand, as pointed out by Uman and McLain (1970), if the frequency response of the instruments used to measure the field was sufficient to resolve electrostatic field changes of a few microseconds duration, such measurements should be obscured by the pulsed radiation fields. If the frequency response was not sufficient to see the pulsed fields, any existing step electrostatic fields would be instrumentally smeared into the time between steps. Measured leader total electric field change vs distance for Florida storms is given in Fig. 5.4, and the measured ratio of the leader to return-stroke field vs distance is given in Fig. 5.5. Additional information on the ratio is found in Fig. 5.3. The solid

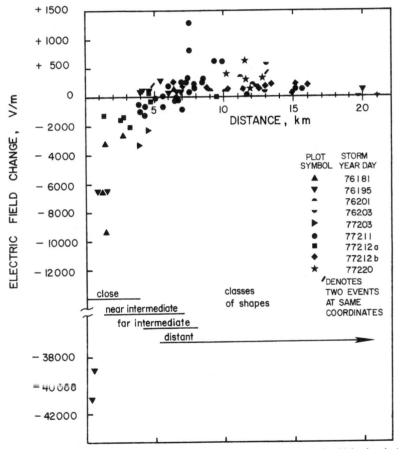

Fig. 5.4 Stepped-leader electric field change as a function of distance for 80 leaders in 9 Florida storms. The distance categories listed are the same as defined in Fig. 5.3. Adapted from Beasley *et al*. (1982).

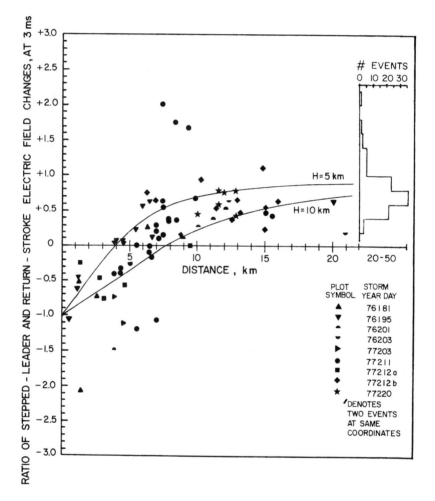

Fig. 5.5 Ratio of stepped-leader to return-stroke electric field change, with return-stroke field change measured 3 msec after the start of return stroke, for the same individual events shown in Fig. 5.4. Additional data are given for the range 20–50 km. Solid curves represent predicted ratio for an uniformly charged leader of height 5–10 km assuming the return stroke removes all of leader charge but no additional charge from the cloud. Ratios for return-stroke field change periods less than 3 msec are given by Beasley *et al.* (1982). Adapted from Beasley *et al.* (1982).

curves in Fig. 5.5 are calculated for model uniformly charged leaders of height 5 and 10 km assuming that all leader charge but no additional charge is brought to earth by the return stroke (Section A.1.4, Fig. A.5). Beasley *et al.* (1982) point out that, while the fact that, for the distant data, the ratio of leader to return-stroke field change is near unity can be interpreted to indicate that stepped leaders are uniformly charged (Section A.1.4), there are also other interpretations of that result. For example, a distant ratio of unity could result from a leader field change due to charge being preferentially distributed on the lower portion of the channel and a return-stroke field change that includes the effects of charge flow from cloud to ground following the return-stroke transit time. Thomson (1985) has investigated theoretically the effects of such nonuniform leader charging on the observed electric field. It might be expected that there is more charge per unit length of channel closer to the ground because of the increased capacitance between the lower channel sections and ground (Little, 1978) and because there are more branches closer to the ground. In support of this view, Lin *et al.* (1980) find that they cannot adequately explain measured return-stroke field changes unless the leader charge density decreases from the ground upward so that it is a fraction of its value near ground at a height of 1–2 km (Section 7.3).

5.5 THEORY

Originally Schonland (1938, 1953, 1956) viewed the stepped leader as a channel of uniform cross-sectional properties with a radius of the order of 1 m. Later, Schonland (1962) stated that it was the general concensus that the stepped-leader channel carries at its center a thin conducting core similar to an electric arc and that the leader current flows in the core, the core being surrounded by a corona sheath several meters in radius that carries the leader charge. This latter view is similar to that first suggested by Bruce (1941, 1944). Schonland (1938) considered the observed minimum average speed of the stepped leader, about 1×10^5 msec, and the relatively constant ratio of the step length to pause time sufficient grounds for postulating the existence of a *pilot* leader. The pilot leader was supposed to travel continuously downward with the step process intermittently catching up. According to Schonland (1962), the pilot could not move forward unless the channel behind it was capable of carrying the current to sustain it, and thus one function of the step process was to render that channel suitably conducting The pilot was presumably not bright enough to be easily photographed. The pilot speed was supposed to be no slower than the electron drift velocity at the pilot tip, which set the minimum pilot speed and overall leader average speed. The fact that the measured stepped-leader field change is relatively smooth and that at close range no appreciable electrostatic field changes associated with steps are observed, as discussed in the previous section, has been interpreted to imply that there is a continuous lowering of negative charge by the pilot. Support for the view that the steps themselves do not transfer appreciable charge is provided by the measurement of relatively smooth currents in upward-moving stepped leaders that are observed photographically to exhibit stepping (Section 12.2). On the other hand, analysis of the pulsed electric fields associated with the steps indicate that the pulses can supply the same order of magnitude of charge to the leader step as inferred from the electrostatic field change during the time between steps (Krider *et al.*, 1977; Cooray and Lundquist, 1985).

Why does the stepped leader step? There is no satisfactory quantitative explanation, but a number of qualitative views have been advanced. Two theories of the mechanism of stepping were proposed by Schonland in 1938 and a third was proposed by Schonland in 1953 and 1956. These invoke verbal combinations of space charge, recombination, electron capture, and ionization processes. An explanation of stepping more directly based on laboratory experiments has been proposed by Bruce (1941, 1944). According to Bruce, the region ahead of the conducting leader channel formed by the steps is in a state of brush or glow discharge because of the high potential at the conducting leader tip. This glow discharge is identified with Schonland's pilot leader. The current required to maintain the glow discharge increases as the glow discharge lengthens. When the current in the glow exceeds a value of the order of 1 A, a sudden transition from glow to arc discharge occurs. The resultant elongation of the arc channel appears as a luminous step. In Bruce's view, radial current flow from the high-voltage arc to the surrounding air plays an important part in the overall physics of the leader. Pierce (1955) has discussed the role of space charge in inhibiting this radial current flow.

Wagner and Hileman (1958, 1961) propose a detailed leader model based on laboratory experiments in which "the actual transition from the glow to the arc is of the nature proposed by Bruce but on a filamentary basis." That is, the corona region ahead of the last leader step is composed of many filamentary channels. The current at the base of each filamentary channel increases as its length increases. When one filament develops sufficient current, it transforms into an arc, short circuits the other filaments, and becomes the leader step.

The observation of a pulsed corona occurring simultaneously with and ahead of the last leader step (Section 5.3) does not appear to fit easily into the views of step formation discussed above.

While the above theories give some physical feelings for how steps could occur and provide reasonable numbers for voltages, electric fields, currents, and current densities, they do not address the basic questions of the detailed physics of the corona and arc formation. Some idea of the complexity of the breakdown process can be obtained by reading the stepped-leader theory of Loeb (1966) which is based on more sensitive optical measurements of laboratory breakdown than were available to either Schonland or Bruce. Computer modeling of the stepped-leader process using a fluid dynamic analysis has been attempted by Klingbeil and Tidman (1974a) (see also Phelps, 1974; Klingbeil and Tidman, 1974b).

Uman and McClain (1970) provided theory from which to find the step current given the observed electromagnetic fields (Section A.4). In their analysis they used an electric field waveform that Pierce (1955) and Arnold and Pierce (1964) claimed to be typical of the stepped leader. The waveform analyzed is more similar to those shown in Fig. 4.5 than those shown in Fig. 5.1 and hence is to be associated with preliminary breakdown pulses (or pulses from the early stage of the stepped leader) rather than with the typical stepped-leader pulses near ground.

REFERENCES

Arnold, H. R., and E. T. Pierce: Leader and Junction Process in the Lightning Discharge as a Source of VLF Atmospherics. *Radio Sci.*, **68D**:711–776 (1964).

Baum, C. E., E. L. Breen, J. P. O'Neill, C. B. Moore, and D. L. Hall: Measurement of Electromagnetic Properties of Lightning with 10 Nanosecond Resolution. "Lightning Technology," pp. 39–84. NASA Conference Publication 2128, FAA-RD-80-30, April 1980.

Beasley, W. H., M. A. Uman, and P. L. Rustan: Electric Fields Preceding Cloud-to-Ground Lightning Flashes. *J. Geophys. Res.*, **87**:4883–4902 (1982).

Beasley, W. H., M. A. Uman, D. M. Jordan, and C. Ganesh: Simultaneous Pulses in Light and Electric Field from Stepped Leaders near Ground Level. *J. Geophys. Res.*, **88**:8617–8619 (1983).

Bensema, W. D.: Pulse Spacing in High-Frequency Atmospheric Noise Burst. *J. Geophys. Res.*, **74**:2780–2782 (1969).

Berger, K., and E. Vogelsanger: Messungen und Resultate der Blitzforschung der Jahre 1951–1963 auf dem Monte San Salvatore. *Bull. Schweiz. Elektrotech. Ver.*, **56**:2–22 (1965).

Berger, K., and E. Vogelsanger: Photographische Blitzuntersuchunger der Jahre 1955–1965 auf dem Monte San Salvatore. *Bull. Schweiz. Elektrotech. Ver.*, **57**:599–620 (1966).

Brook, M., and N. Kitagawa: Electric Field Changes and the Design of Lightning-Flash Counters. *J. Geophys. Res.*, **65**:1927–1931 (1960).

Brook, M., N. Kitagawa, and E. J. Workman: Quantitive Study of Strokes and Continuing Currents in Lightning Discharges to Ground. *J. Geophys. Res.*, **67**:649–659 (1962).

Bruce, C. E. R.: The Lightning and Spark Discharges. *Nature (London)*, **147**:805–806 (1941).

Bruce, C. E. R.: The Initiation of Long Electrical Discharges. *Proc. R. Soc. London Ser. A*, **183**:228–242 (1944).

Chapman, F. W.: Atmospheric Disturbances Due to Thundercloud Discharges, Pt. 1. *Proc. Phys. Soc. (London)*, **51**:876–894 (1939).

Clarence, N. D., and D. J. Malan: Preliminary Discharge Processes in Lightning Flashes to Ground. *Q. J. R. Meteorol. Soc.*, **83**:161–172 (1957).

Cooray, V., and S. Lundquist: On the Characteristics of Some Radiation Fields from Lightning and Their Possible Origin in Positive Ground Flashes. *J. Geophys. Res.*, **87**:11,203-11,214 (1982).

Cooray, V., and S. Lundquist: Characteristics of the Radiation Fields from Lightning in Sri Lanka in the Tropics. *J. Geophys. Res.*, **90**:6099-6109 (1985).

Gupta, S. P., M. Rao, and B. A. P. Tantry: VLF Spectra Radiated by Stepped Leaders. *J. Geophys. Res.*, **77**:3924-3927 (1972).

Hagenguth, J. H., and J. G. Anderson: Lightning to the Empire State Building, Pt. 111. *AIEE Trans.*, **71**:641-749 (1952).

Hayenga, C. O.: Characteristics of Lightning VHF Radiation near the Time of Return Strokes. *J. Geophys. Res.*, **89**:1403-1410 (1984).

Hodges, D. B.: A Comparison of the Rates of Change of Current in the Step and Return Processes of Lightning Flashes. *Proc. Phys. Soc. (London)*, **B67**:582-587 (1954).

Ishikawa, H., M. Takgai, and T. Takeuti: On the Leader Waveforms of Atmospherics near the Origin. *Proc. Res. Inst. Atmos., Nagoya Univ.*, **5**:1-11 (1958).

Khastgir, S. R.: Leader Stroke Current in a Lightning Discharge According to the Streamer Theory. *Phys. Rev.*, **106**:616-617 (1957).

Khastgir, S. R., and D. Ghosh: Theory of Stepped-Leader in Cloud-to-Ground Electrical Discharges. *J. Atmos. Terr. Phys.*, **34**:109-113 (1972).

Kitagawa, N.: On the Electric Field Change Due to the Leader Process and Some of Their Discharge Mechanism. *Pap. Metereol. Geophys. (Tokyo)*, **7**:400-414 (1957).

Kitagawa, N., and M. Brook: A Comparison of Intracloud and Cloud-to-Ground Lightning Discharges. *J. Geophys. Res.*, **65**:1189-1201 (1960).

Kitagawa, N., and M. Kobayashi: Distribution of Negative Charge in the Cloud Taking Part in a Flash to Ground. *Pap. Metereol. Geophys. (Tokyo)*, **9**:99-105 (1958).

Klingbeil, R., and D. A. Tidman: Theory and Computer Model of the Lightning Stepped Leader. *J. Geophys. Res.*, **79**:865-869 (1974a).

Klingbeil, R., and D. A. Tidman: Reply to Comments on the Brief Report Theory and Computer Model of the Lightning Stepped Leader by C. T. Phelps. *J. Geophys. Res.*, **79**:5669-5670 (1974b).

Krehbiel, P. R.: An Analysis of the Electric Field Change Produced by Lightning. Ph.D thesis, University of Manchester Institute of Science and Technology, Manchester, England, 1981.

Krehbiel, P. R., M. Brook, and R. A. McCrory: An Analysis of the Charge Structure of Lightning Discharges to Ground. *J. Geophys. Res.*, **84**:2432-2456 (1979).

Krider, E. P.: The Relative Light Intensity Produced by a Lightning Stepped Leader. *J. Geophys. Res.*, **79**:4542-4544 (1974).

Krider, E. P., and G. J. Radda: Radiation Field Wave Forms Produced by Lightning Stepped Leaders. *J. Geophys. Res.*, **80**:2635-2657 (1975).

Krider, E. P., C. D. Weidman, and R. C. Noggle: The Electric Fields Produced by Lightning Stepped Leaders. *J. Geophys. Res.*, **82**:951-960 (1977).

Larigaldie, S.: Linear Gliding Discharge Over Dielectric Surfaces. *J. Phys.*, **40** (C7):429-430 (1979).

Larigaldie, S., G. Labaune, and J. P. Moreau: Lightning Leader Laboratory Simulation by Means of Rectilinear Surface Discharges. *J. Appl. Phys.*, **52**:7114-7120 (1981).

LeVine, D. M.: The Effect of Pulse Internal Statistics on the Spectrum of Radiation from Lightning. *J. Geophys. Res.*, **82**:1773-1777 (1977).

Lin, Y. T., M. A. Uman, and R. B. Standler: Lightning Return Stroke Models. *J. Geophys. Res.*, **85**:1571-1583 (1980).

Little, P. F.: Transmission Line Representation of a Lightning Return Stroke. *J. Phys. D: Appl. Phys.*, **11**:1893-1910 (1978).

Loeb, L. J.: The Mechanism of Stepped and Dart Leaders in Cloud-to-Ground Lightning Strokes. *J. Geophys. Res.*, **71**:4711–4721 (1966).

Loeb, L. J.: Confirmation and Extension of a Proposed Mechanism of the Stepped Leader Lightning Stroke. *J. Geophys. Res.*, **73**:5813–5817 (1968).

Malan, D. J., and B. F. J. Schonland: Progressive Lightning, Pt. 7, Directly Correlated Photographic and Electrical Studies of Lightning from near Thunderstorms. *Proc. R. Soc. London Ser. A*, **191**:485–503 (1947).

Malan, D. J., and B. F. J. Schonland: The Distribution of Electricity in Thunderclouds. *Proc. R. Soc. London Ser. A*, **209**:158–177 (1951).

Nagai, Y., S. Kawamata, and Y. Edano: Observation of Preceding Leader and Its Downward Traveling Velocity in Utsunomiya District. *Res. Lett. Atmos. Electr.*, **2**:53–56 (1982).

Orville, R. E.: Spectrum of the Lightning Stepped Leader. *J. Geophys. Res.*, **73**:6999–7008 (1968).

Orville, R. E., and V. P. Idone: Lightning Leader Characteristics in the Thunderstorm Research International Program (TRIP). *J. Geophys. Res.*, **87**:11,177–11,192 (1982).

Phelps, C. T.: Comments on Brief Report by R. Klingbeil and D. A. Tidman: Theory and Computer Model of the Lightning Stepped Leader. *J. Geophys. Res.*, **79**:5669 (1974).

Pierce, E. T.: Electrostatic Field Changes Due to Lightning Discharges. *Q. J. R. Meteorol. Soc.*, **81**:211–228 (1955).

Schonland, B. F. J.: Progressive Lightning, Pt. 4. *Proc. R. Soc. London Ser. A*, **164**:132–150 (1938).

Schonland, B. F. J.: The Pilot Streamer in Lightning and the Long Spark. *Proc. R. Soc. London Ser. A*, **220**:25–38 (1953).

Schonland, B. F. J.: The Lightning Discharge. *Handb. Phys.*, **22**:576–628 (1956).

Schonland, B. F. J.: Lightning and the Long Electric Spark. *Adv. Sci.*, **19**:306–313 (1962).

Schonland, B. F. J., D. J. Malan, and H. Collens: Progressive Lightning, Pt. 2. *Proc. R. Soc. London Ser. A*, **152**:595–625 (1935).

Schonland, B. F. J., D. B. Hodges, and H. Collens: Progressive Lightning, Pt. 5, A Comparison of Photographic and Electrical Studies of the Discharge Process. *Proc. R. Soc. London Ser. A*, **166**:56–75 (1938a).

Schonland, B. F. J., D. J. Malan, and H. Collens: Progressive Lightning, Pt. 6. *Proc. R. Soc. London Ser. A*, **168**:455–469 (1938b).

Srivastava, C. M., and S. R. Khasgir: On the Maintenance of Current in the Stepped Leader Stroke of Lightning Discharge. *J. Sci. Ind. Res.*, **14B**:34–35 (1955).

Steptoe, B. J.: Some Observations on the Spectrum and Propagation of Atmospherics. Ph.D. thesis, University of London, 1958.

Szpor, S.: Review of the Relaxation Theory of the Lightning Stepped Leader. *Acta Geophys. Pol.*, **18**:73–77 (1970).

Szpor, S.: Critical Comparison of Theories of Stepped Leaders. *Arch. Elektrotech.*, **26**:291–299 (1977).

Thomson, E. M.: Characteristics of Port Moresby Ground Flashes. *J. Geophys. Res.*, **85**:1027–1036 (1980)

Thomson, E. M.: A Theoretical Study of Electrostatic Field Wave Shapes from Lightning Leaders. *J. Geophys. Res.*, **90**:8125–8135 (1985).

Thomson, E. M., M. A. Uman, and W. H. Beasley: Speed and Current for Lightning Stepped Leaders near Ground as Determined from Electric Field Records. *J. Geophys. Res.*, **90**:8136–8142 (1985).

Uman, M. A.: "Lightning," pp. 54, 211–214. McGraw-Hill, New York, 1969. See also, Dover, New York, 1984.

Uman, M. A., and D. K. McLain: Radiation Field and Current of the Lightning Stepped Leader. *J. Geophys. Res.*, **75**:1058–1066 (1970).

Wagner, C. F., and A. R. Hileman: The Lightning Stroke. *AIEE Trans.*, **77** (Pt. 3):229–242 (1958).

Wagner, C. F., and A. R. Hileman: The Lightning Stroke (2). *AIEE Trans.*, **80** (Pt. 3):622–642 (1961).

Weidman, C. D., and E.P. Krider: Submicrosecond Risetimes in Lightning Radiation Fields. "Lightning Technology," pp. 29–38. NASA Conference Publication 2128, FAA-RD-80-30, April 1980.

Weidman, C. D., E. P. Krider, and M. A. Uman: Lightning Amplitude Spectra in the Interval from 100 kHz to 20 MHz. *Geophys. Res. Lett.*, **8**:931–934 (1981).

Williams, D. P., and M. Brook: Magnetic Measurement of Thunderstorm Currents, I. Continuing Currents in Lightning. *J. Geophys. Res.*, **68**:3243–3247 (1963).

Workman, E. J., J. W. Beams, and L. B. Snoddy: Photographic Study of Lightning. *Physics*, **7**:345–379 (1936).

Chapter 6 | Attachment Process

6.1 INTRODUCTION

When the stepped leader approaches any conducting object such as a mound of earth, a tree, a transmission line tower, or an aircraft in flight, the electric field produced by the charge on the leader can be enhanced by the object to a level where electric discharges (called leaders, connecting leaders, connecting discharges, or, sometimes, streamers) are emitted from the object. The characteristics of these discharges are not well understood, but have been the subject of some theoretical analysis (e.g., Thum *et al.*, 1982) and of considerable discussion, particularly in the context of modeling lightning strikes to power lines where the attachment process plays a significant role in the design of overhead ground wire protection (e.g., Armstrong and Whitehead, 1968; Brown and Whitehead, 1969; Whitehead, 1977). One of the important parameters in the design of this type of lightning protection is the so-called "striking distance": the distance between the object to be struck and the tip of the downward-moving leader at the instant that the connecting leader is initiated from the object. It is usually assumed that at this instant the point of strike is determined. It follows that the actual junction point is somewhere between the object and the tip of the last leader step. Often, it is assumed to be midway between.

6.2 ANALYTICAL APPROACH AND MEASUREMENT

We now examine the attachment process as it relates to lightning strikes to ground or to objects attached to the ground. General reviews of this phenomena have been given by Golde (1967, 1977) who outlines the following analytical approach: a reasonable charge distribution is assumed for the leader channel, and the resultant fields on the ground or nearby objects are calculated. The leader is assumed to be at the striking distance when the field at some point exceeds a critical breakdown value that is determined from laboratory tests. Various authors have derived relations between the striking distance and the leader charge (e.g., Golde, 1945). The relationship of more practical value in the design of power line protection, however, is that of the striking distance to the peak current of the following return stroke since the probability of a stroke having a given current magnitude is reasonably well known (Section 7.2.2) whereas there is little statistical data on leader charge. To find the current–striking distance relationship, the peak current must be related to the stepped-leader charge. It is not clear from a physical point of view that these two quantities are related, since the stepped-leader charge may be spread over a rather large volume in various leader branches, whereas the peak stroke current is attained in a few microseconds in a short channel section that is attached to ground (Eriksson, 1978). Nevertheless, Berger (1972) has shown that there is a correlation between the measured return-stroke peak current at ground and the total charge transfer to ground in the first 1 msec or so. According to Berger (1972), the expression relating peak current I to charge transfer Q that best fits the available data for 89 negative first strokes is

$$I = 10.6Q^{0.7} \tag{6.1}$$

with I measured in kA and Q in coulombs. From this expression, a typical peak current of 25 kA corresponds to a total charge, assumed to originally reside on the leader, of 3.3 C. When Eq. (6.1) is combined with the relation between charge and breakdown field, a relation for striking distance d_s can be found in terms of peak current. For example, one of several theoretical analyses reviewed by Golde (1977) yields

$$d_s = 10I^{0.65} \tag{6.2}$$

where d_s is in meters and I in kA.

Several theoretical curves relating striking distance to peak current are discussed by Golde (1977), and recent simultaneous measurements of peak current and striking distance obtained on a tower in South Africa are given by Eriksson (1978). These theories and measurements are illustrated in Fig. 6.1. While the simultaneous measurements of peak current and striking distance have been relatively few, a number of streak and still photographs from which the striking distance can be estimated have been published. Golde (1947) has reproduced a streak-camera photograph obtained by Malan and Collins (1937) in which the last of three downward-moving leader steps is met 50 m above ground by the longest of three upward-going leaders apparently originating from the same point on the ground. Berger (1977) and Berger and Vogelsanger (1966) give a streak photograph, reproduced in Fig. 6.2, in which the stepped leader ends about 40 m above and 40 m horizontally away from a tower (point A). The connection between the leader and the tower top

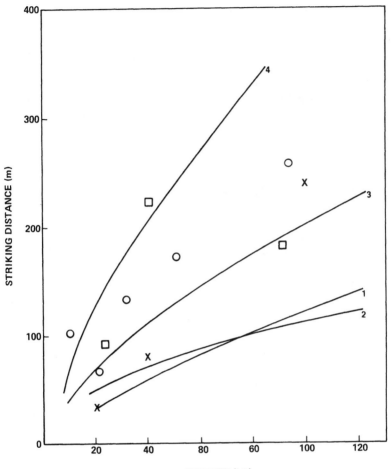

Fig. 6.1 Striking distance vs return-stroke peak current for objects attached to ground [curve 1, Golde (1945); curve 2, Wagner (1963); curve 3, Love (1973); curve 4, Ruhling (1972); X, theory of Davis (1962); ○, estimates from two-dimensional photographs by Eriksson (1978); □, estimates from three-dimensional photography by Eriksson (1978)]. Adapted from Golde (1977) and Eriksson (1978).

is made by an upward-moving positive discharge from the tower that branches at point B. An oscillogram of the tower current associated with this streak photograph is given by Berger (1977) and by Berger and Vogelsanger (1966). Typically, these measured tower currents rise to a few kiloamperes in a few microseconds and then to tens of kiloamperes in the next few

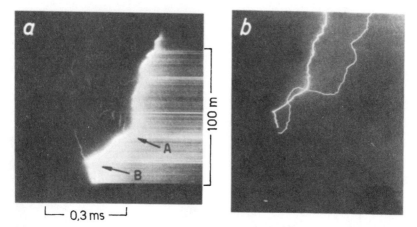

Fig. 6.2 (a) Streak photograph of the attachment process to tower number 2 on Mount San Salvatore in Lugano, Switzerland. (b) Still photograph of the flash from a and another flash that struck the tower below its top. Previously published by Berger (1977) and Berger and Vogelsanger (1966). Courtesy K. Berger and the Swiss High Voltage Research Committee (FKH), Zurich.

microseconds. It is not possible with the available data to identify the part of the current that is due to the upward connecting discharge. Orville and Idone (1982), from their streak-camera photographs infer upward leader lengths for three discharges of 20, 20, and 30 m.

On close still photographs there is often an identifiable split or loop in the observed channel that, as evidenced from laboratory spark studies, is the point where upward and downward leaders join. Occasionally, upward branching, associated with the upward leader propagation, is evident below the split, and downward branching above. Many photographs of lightning to ground or to structures show a pronounced kink and change in direction of the channel near the ground or structure. Below the kink, the channel is generally straight. The analysis of this type of photograph to determined the striking distance is subject to error from the subjective judgments that must be made as to the kink location on a two-dimensional photograph of a three-dimensional channel. The still photograph of lightning to a mountainside 810 m away shown in Fig. 6.3 has an inferred channel junction point (Krider and Ladd, 1975) about 16 m above ground with unconnected upward leaders of 8 and 10 m height about 10 and 15 m away, respectively, from the main channel. The flash on the right in Fig. 6.3 was 1560 m away and had an unconnected upward leader of 10 m height about 15 m away from the main channel, although it is difficult to see this unconnected leader in the reproduction. Orville (1968) has published the remarkable photograph

Fig. 6.3 A photograph originally published by Krider and Ladd (1975) showing unconnected upward-moving leaders in two lightning discharges striking mountainous terrain in southern Arizona.

shown in Fig. 6.4 of a strike to a 7-m-high European ash tree taken from a distance of 60 m. In the photograph the locations of the downward and the upward branching indicate a meeting point about 12 m above the treetop. The tree was unharmed by the lightning that followed the main trunk to ground. The photograph of lightning to a beach shown in Fig. 6.5 exhibits upward and downward branching on the main channel as well as unconnected upward-going leaders. Hagenguth (1947) describes a photograph of a strike to a patch of woods in a lake in which the channel split about 3 m above the water and remained split to a maximum separation of over 1 m to a height of about 9 m. Golde (1973, plate 3) shows a photograph of a strike to the chimney of a low building in which, at a height of 9 m above the chimney, the channel splits into at least three separate paths. A photograph of a positive lightning (Chapter 11) with a loop in the channel near ground is given in Fig. 11.1.

Fig. 6.4 A photograph taken by R. E. Orville of lightning to a European ash tree in Lugano, Switzerland showing upward branching associated with the upward-going leader near the tree-top and downward branching associated with the downward-going stepped leader farther above the tree (near the top of the photograph). From Orville (1968), *Science* **162** : 666. Copyright 1968 by the AAAS.

Fig. 6.5 A photograph of lightning to the sand at Manasquan Beach, New Jersey taken in July 1934 from a distance of about 30 m by Robert Edwards. No other information is available. Courtesy, Galloway, New York.

While photographs of unconnected upward-going leaders from ground are relatively rare, such leaders, as well as connecting leaders, have been more commonly photographed from tall towers or buildings (e.g., McEachron, 1939; Hagenguth and Anderson, 1952; Berger and Vogelsanger, 1966; Berger, 1967, 1977). These might be expected to be longer than the comparable leaders near ground. Unconnected discharges from tall structures occurring when there is a nearby ground or cloud discharge are in the range 20–100 m, although Berger (1967) has pointed out that positively charged

Fig. 6.6 A photograph of lightning to a TV tower guy wire originally published by Krider and Alejandro, *Weatherwise*, **36**:71–75 (1983). Reprinted with permission of the Helen Dwight Reid Educational Foundation. Copyright © 1983 by Heldref Publications, 4000 Albemarle Street, N.W., Washington, D.C. 20016.

upward-going leaders are typically faint and difficult to photograph so that there may be many shorter unphotographed connecting leaders than the photographed range of 20–100 m would indicate.

Krider and Alejandro (1983) have published a photograph, shown in Fig. 6.6, of a strike to a guy wire of a 80 m TV tower in which the channel, at about tower top height, takes an abrupt horizontal turn about 70 m from the guy wire. The horizontal channel contains three sections of which the last 35 m is interpreted as the striking distance. The strike point is 14 m below the top of the tower. An unconnected leader of 96 m height is present at the top of the tower.

Guo and Krider (1982) show from electric field and photoelectric measurements that, in rare instances, two separate upward-going leaders can attach to two different stepped-leader branches producing two return strokes separated in time by tens of microseconds. Both return strokes traverse the same leader channel above the appropriate leader branch point. Guo and Krider (1982) distinguish between this phenomena and the case in which multiple upward discharges connect with the same branch of a stepped leader (Schonland, 1956).

From all available information, it is reasonable to conclude that striking distances are generally between about 10 and a few hundred meters. Further, the available experimental data indicate that there is an increase in striking distance with increasing current, in support of the theory given above, but the scatter in the data is sufficiently large that one should consider a specific functional relationship such as Eq. (6.2) as only a rough guide.

Upward-connecting leaders that meet positive flashes to towers are discussed in Section 11.2. These negatively charged upward-going discharges are considerably longer than their positively charged counterparts.

Upward-initiated lightning, both natural and artificial, is discussed in Chapter 12. The leaders in this type of lightning propagate upward, generally from tall objects for kilometers and may or may not connect with downward- or horizontally moving discharges in the cloud. The upward-moving leaders associated with upward-initiated lightning may well just be a longer version of the connecting leader discussed in this chapter. On the other hand, upward-propagating leaders exhibit stepping similar to that of downward-propagating stepped leaders, whereas there does not appear to be evidence in the literature that the connecting discharges do.

REFERENCES

Armstrong, H. R., and E. R. Whitehead: Field and Analytical Studies of Transmission Line Shielding. *IEEE Trans., Part III*, **PAS-87**: 270–281 (1968).

Berger, K.: Novel Observations on Lightning Discharges. *J. Franklin Inst.*, **283**: 478–525 (1967).

Berger, K.: Methoden und Resultate der Blitzforschung auf dem Monte San Salvator bei Lugano in den Jahren 1963-1971. *Bull. Schweiz. Elektrotech. Ver.*, **63**:1403-1422 (1972).

Berger, K.: The Earth Flash. *In* "Lightning, Vol. 1, Physics of Lightning" (R. H. Golde, ed.). Academic Press, New York, 1977.

Berger, K., and E. Vogelsanger: Messungen und Resultate der Blitzforschung der Jahre 1955-1963 auf dem Monte San Salvatore. *Bull. Schweiz. Elektrotech. Ver.*, **56**:2-22 (1965).

Berger, K., and E. Vogelsanger: Photographische Blitzuntersuchungen der Jahre 1955-1963 auf dem Monte San Salvatore. *Bull. Schweiz. Elektrotech. Ver.*, **57**:599-620 (1966).

Brown, G. W., and E. R. Whitehead: Field and Analytical Studies of Transmission Line Shielding: Part II. *IEEE Trans. Power Appar. Syst.*, **PAS-88**:617-626 (1969).

Choy, L. A., and M. Darveniza: A Sensitivity Analysis of Lightning Performance Calculations for Transmission Lines. *IEEE Trans. Power Appar. Syst.*, **PAS-90**:1443-1451 (1971).

Currie, J. R., L. A. Choy, and M. Darveniza: Monte Carlo Determination of the Frequency of Lightning Strokes and Shielding Failures on Transmission Lines. *IEEE Trans. Power Appar. Syst.*, **PAS-90**:2305-2312 (1971).

Darveniza, M., F. Popolansky, and E. R. Whitehead: Lightning Protection of UHV Lines. *Electra*, **41**:39-69 (1975).

Davis, R.: Frequency of Lightning Flashover on Overhead Lines. *In* "Gas Discharges and the Electricity Supply Industry (J. S. Forest, P. R. Howard, and D. J. Littler, eds.), pp. 125-138. Butterworths, London, 1962.

Edano, Y., S. Kawamata, and Y. Nagai: Observation of Simultaneous-Bi-Stroke Flashes. *Res. Lett. Atmos. Electr.*, **2**:49-52 (1982).

Eriksson, A. J.: A Discussion on Lightning and Tall Structures. CSIR Special Report ELEK 152, National Electrical Engineering Research Institute, Pretoria, South Africa, July, 1978. See also, Lightning and Tall Structures. *Trans. South Afr. Inst. Electr. Eng.*, **69**:2-16 (1978).

Fieux, R. P., C. H. Gary, B. P. Hutzler, A. R. Eybert-Berard, P. L. Hubert, A. C. Meesters, P. H. Perroud, J. H. Hamelin, and J. M. Person: Research on Artificially Triggered Lightning in France. *IEEE Trans. Power Appar. Syst.*, **PAS-97**:725-733 (1978).

Fitzgerald, D. R.: Probable Aircraft "Triggering" of Lightning in Certain Thunderstorms. *Mon. Weather Rev.*, **95**:835-842 (1967).

Golde, R. H.: The Frequency of Occurrence and the Distribution of Lightning Flashes to Transmission Lines. *AIEE Trans.*, **64**:902-910 (1945).

Golde, R. H.: Occurrence of Upward Streamers in Lightning Discharges. *Nature (London)*, **160**:395-396 (1947).

Golde, R. H.: The Attractive Effects of a Lightning Conductor. *J. Inst. Electr. Eng. (London)*, **9**:212-213 (1963).

Golde, R. H.: The Lightning Conductor. *J. Franklin Inst.*, **283**:451-477 (1967).

Golde, R. H.: "Lightning Protection." Arnold, London, 1973.

Golde, R. H.: The Lightning Conductor. *In* "Lightning, Vol. II, Lightning Protection" (R. H. Golde, ed.), pp. 545-576. Academic Press, New York, 1977.

Golde, R. H.: Lightning and Tall Structures. *Proc. Inst. Electr. Eng. (London)*, **125**:347-351 (1978).

Guo, C., and E. P. Krider: The Optical and Radiation Field Signatures Produced by Lightning Return Strokes. *J. Geophys. Res.*, **87**:8913-8922 (1982).

Hagenguth, J. H.: Photographic Study of Lightning. *Trans. Am. Inst. Electr. Eng.*, **66**:577-585 (1947).

Hagenguth, J. H., and J. G. Anderson: Lightning to the Empire State Building, Pt. 3. *AIEE Trans.*, **71** (Pt. 3):641-649 (1952).

Horii, K., and H. Sakurano: Observation on Final Jump of the Discharge in the Experiment of Artificially Triggered Lightning. *IEEE Trans. Power Appar. Syst.*, **PAS-104**:2910-2917 (1985).

Idone, V. P., and R. W. Henderson: An Unusual Lightning Ground Strike. *Weatherwise*, **35**:223-224 (1982).

Idone, V. P., R. E. Orville, P. Hubert, L. Barret, and A. Eybert-Berard: Correlated Observations of Three Triggered Lightning Flashes. *J. Geophys. Res.*, **89** (D1):1385-1394 (1984).

Krider, E. P., and S. B. Alejandro: Lightning—An Unusual Case Study. *Weatherwise*, **36**:71-75 (1983).

Krider, E. P., and C. G. Ladd: Upward Streamers in Lightning Discharges to Mountainous Terrain. *Weather*, **30**:77-81 (1975).

Liew, A. C., and M. Darveniza: Calculation of the Lightning Performance of Unshielded Transmission Lines. *IEEE Trans. Power Appar. Syst.*, **PAS-101**:1471-1477 (1982a).

Liew, A. C., and M. Darveniza: Lightning Performance of Unshielded Transmission Lines, *IEEE Trans. Power Appar. Syst.*, **PAS-101**:1478-1482 (1982b).

Love, E. R.: Improvements on Lightning Stroke Modelling and Applications to the Design of EVH and UHV Transmission Lines. M.S. thesis, University of Colorado, Boulder, Colorado, 1973.

McEachron, K. B.: Lightning to the Empire State Building. *J. Franklin Inst.*, **227**:175-203 (1939).

McEachron, K. B., and W. A. Morris: The Lightning Stroke: Mechanism of Discharge. *Gen. Electr. Rev.*, **39**:487-496 (1936).

Malan, D. J., and H. Collens: Progressive Lightning III—The Fine Structure of Return Lightning Strokes. *Proc. R. Soc. London Ser. A*, **162**:175-203 (1937).

Orville, R. E.: Photograph of a Close Lightning Flash. *Science*, **162**:666-667 (1968).

Orville, R. E., and V. P. Idone: Lightning Leader Characteristics in the Thunderstorm Research International Program (TRIP). *J. Geophys. Res.*, **87**:11,177-11,192 (1982).

Pierce, E. T.: Triggered Lightning and Its Application to Rockets and Aircraft. 1972 Lightning and Static Electricity Conference, AFAL-TR-72-325, Air Force Avionics Laboratory, Wright-Patterson Air Force Base, Ohio, 1972.

Rühling, F.: Modelluntersuchungen über den Schutzraum und ihre Bedeutung für Gebäude-blitzableiter. *Bull. Schweiz. Elektrotech. Ver.*, **63**:522-528 (1972).

Schonland, B. F. J.: The Lightning Discharge. *Handb. Phys.*, **22**:576-628 (1956).

Thum, P. C., A. C. Liew, and C. M. Wong: Computer Simulation of the Initial Stages of the Lightning Protection Mechanism. *IEEE Trans. Power Appar. Syst.*, **PAS-101**:4370-4377 (1982).

Vonnegut, B.: Electrical Behavior of an Airplane in a Thunderstorm. A. L. Little, Inc., Cambridge, Massachusetts, February, 1965. Defense Documentation Center, AD-614-914, Clearinghouse for Federal Scientific and Technical Information.

Wagner, C. F.: Relation between Stroke Current and Velocity of the Return Stroke. *AIEE Trans.*, **83** (Pt. 3):606-617 (1963).

Wagner, C. F.: Lightning and Transmission Lines. *J. Franklin Inst.*, **283**:558-594 (1967).

Whitehead, E. R.: Protection of Transmission Lines. *In* "Lightning, Vol. II, Lightning Protection" (R. H. Golde, ed.), pp. 697-745. Academic Press, New York, 1977.

Chapter 7 | Return Stroke

7.1 INTRODUCTION

The return stroke that lowers negative charge to earth has been the most researched and, consequently, is the best understood of all the processes that make up a flash to earth. Return strokes that lower positive charge to earth, about which considerably less is known, are discussed separately in Chapter 11. In the present chapter we consider primarily the characteristics of negative return strokes. Studies of the return stroke have been motivated both by practical considerations (it is the return stroke current that produces most of the damage attributable to lightning) and by the fact that of all the processes comprising a lightning flash, the return stroke lends itself most easily to measurement (it is the optically brightest lightning process visible outside the cloud and it produces the largest and most easily identified electromagnetic signature).

7.2 MEASUREMENTS

The experimental data relating to return strokes can be arbitrarily divided into four general categories: (1) electric and magnetic fields, (2) current, (3) speed, and (4) luminosity and optical spectra. We now consider individually each of these four types of experimental data and the information they provide about the return stroke.

7.2.1 ELECTRIC AND MAGNETIC FIELDS

Return-stroke electric field variations on a millisecond time scale have been published, for example, by Brook *et al.* (1962), Kitagawa *et al.* (1962), Krehbiel *et al.* (1979), and Lin *et al.* (1979). Electric fields on this time scale are shown in Figs. 1.9, 1.12, and 4.3 and have been used to determine the magnitude and location of the charge lowered to earth by individual strokes (Sections 3.2 and 7.2.2) as well as to determine values for millisecond-scale channel currents (Section 9.3).

Return-stroke vertical electric and horizontal magnetic field variations on a microsecond and submicrosecond time scale have been reported, for example, by Fisher and Uman (1972), Uman *et al.* (1975), Tiller *et al.* (1976), Weidman and Krider (1978, 1980, 1982a–c, 1984), Lin *et al.* (1979), Weidman (1982), Cooray and Lundquist (1982, 1985), and Master *et al.* (1984). Frequency spectra of the measured fields have been published by Serhan *et al.* (1980), Weidman *et al.* (1981), Preta *et al.* (1985), and Weidman and Krider (1986). Drawings of typical microsecond-scale measured fields as a function of distance for both first and subsequent strokes are shown in Figs. 7.1a and b. Lin *et al.* (1979) give statistical data for the characteristics of the overall return-stroke field waveforms that are identified in Fig. 7.1, and some of these data are included in Table 7.1. Simultaneously measured close and distant waveforms are shown in Fig. 7.2. Measured distant electric field waveforms from first strokes are shown in Figs. 4.5 and 5.1.

The electric fields of strokes within a few kilometers, shown in Figs. 7.1a and 7.2, are, after the first few tens of microseconds, dominated by the electrostatic component of the total electric field, the only field component which is nonzero after the stroke current has ceased to flow. (The individual field components that comprise the electric and magnetic fields are discussed in more detail in Sections 7.3 and A.3.) The close magnetic fields at similar times are dominated by the magnetostatic component of the total magnetic field, the component that produces the magnetic field humps shown in Figs. 7.1a and 7.2. Distant electric and magnetic fields have essentially identical waveshapes and are bipolar, as illustrated in Fig. 7.1b. Both electric and magnetic waveshapes are composed essentially of the radiation component of the respective total fields.

The initial field peak evident in the waveforms of Figs. 7.1 and 7.2 is the dominant feature of the electric and magnetic field waveforms beyond about 10 km, is a significant feature of waveforms from strokes between a few and about 10 km, and can be identified, with some effort, in waveforms from strokes as close as a kilometer. The initial peak field is due to the radiation component of the total field and, hence, as discussed in Sections 7.3 and A.3, decreases inversely with distance in the absence of significant propagation effects (Lin *et al.*, 1979, 1980). Thus the initial peak fields produced by different return strokes at known distances can be range normalized for comparison, for example, to 100 km by multiplying the measured field peaks by $D/10^5$ where D is the stroke distance in meters. Statistics on the normalized initial peak electric field, derived from various studies, are presented in Table 7.1 and a histogram of normalized initial peak electric field for

Table 7.1

Statistics on Return-Stroke Vertical Electric Field from
Strokes Lowering Negative Charge to Earth

	First strokes			Subsequent strokes		
	Number of strokes	Mean	SD	Number of strokes	Mean	SD
Initial peak, normalized to 100 km (V/m)						
Master *et al.* (1984)	112	6.2	3.4	237	3.8	2.2
Krider and Guo (1983)	69	11.2	5.6	84	4.6	2.6
	31	8.8	4.0	31	6.0	1.9
Cooray and Lundquist (1982)	553	5.3	2.7			
McDonald *et al.* (1979)	54	5.4	2.1	119	3.6	1.3
McDonald *et al.* (1979)	52	10.2	3.5	153	5.4	2.6
Tiller *et al.* (1976)	75	9.9	6.8	163	5.7	4.5
Lin *et al.* (1979)						
[KSC]	51	6.7	3.8	83	5.0	2.2
[Ocala]	29	5.8	2.5	59	4.3	1.5
Taylor (1963)	47	4.8				
Zero-crossing time (μsec)						
Lin and Uman (1979)	46	54	18	77	36	17
Cooray and Lunquist (1985)						
[Sweden]	102	49	12	94	39	8
[Sri Lanka]	91	89	30	143	42	14
Zero-to-peak risetime (μsec)						
Master *et al.* (1984)	105	4.4	1.8	220	2.8	1.5
Cooray and Lundquist (1982)	140	7.0	2.0			
Lin *et al.* (1979)						
[KSC]	51	2.4	1.2	83	1.5	0.8
[Ocala]	29	2.7	1.3	59	1.9	0.7
Tiller *et al.* (1976)	120	3.3	1.0	163	2.3	0.9
Lin and Uman (1973)	12	4.0	2.2	83	1.2	1.1
Fisher and Uman (1972)	26	3.6	1.8	26	3.1	1.9
10-90% risetime (μsec)						
Master *et al.* (1984)	105	2.6	1.2	220	1.5	0.9
Slow front duration (μsec)						
Master *et al.* (1984)	105	2.9	1.3			
Cooray and Lundquist (1982)	82	5.0	2.0			
Cooray and Lundquist (1985)	104	4.6	1.5			
Weidman and Krider (1978)	62	4.0	1.7	44	0.6	0.2
	90	4.1	1.6	120	0.9	0.5
Slow front, amplitude as percentage of peak						
Master *et al.* (1984)	105	28	15			
Cooray and Lundquist (1982)	83	40	11			
Cooray and Lunquist (1985)	108	44	10			
Weidman and Krider (1978)	62	52	20	44	20	10
	90	40	20	120	20	10
Fast transition, 10-90% risetime (nsec)						
Master *et al.* (1984)	102	970	680	217	610	270
Weidman and Krider (1978)	38	200	100	80	200	40
	15	200	100	34	150	100
Weidman and Krider (1980, 1984); Weidman (1982)	125	90	40			

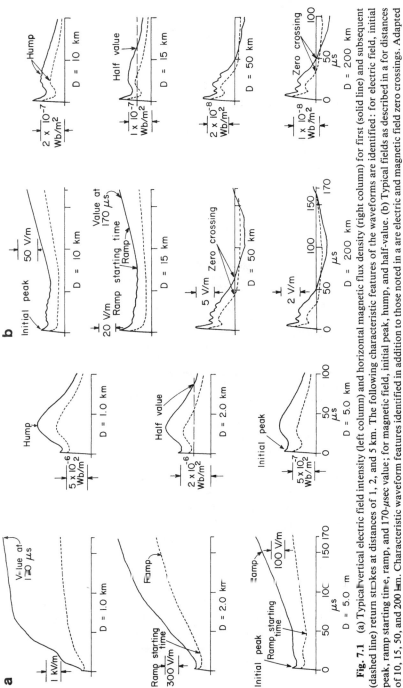

Fig. 7.1 (a) Typical vertical electric field intensity (left column) and horizontal magnetic flux density (right column) for first (solid line) and subsequent (dashed line) return strokes at distances of 1, 2, and 5 km. The following characteristic features of the waveforms are identified: for electric field, initial peak, ramp starting time, ramp, and 170-μsec value; for magnetic field, initial peak, hump, and half-value. (b) Typical fields as described in a for distances of 10, 15, 50, and 200 km. Characteristic waveform features identified in addition to those noted in a are electric and magnetic field zero crossings. Adapted from Lin *et al.* (1979)

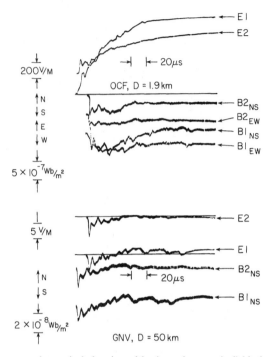

Fig. 7.2   Return-stroke vertical electric and horizontal magnetic fields from a two-stroke flash that occurred at 200108 UT on September 9, 1974, at a distance of 1.9 km from the Ocala (OCF) station and 50 km from the Gainesville (GNV) station. Magnetic fields are indented at top and electric fields are indented at bottom solely for purposes of clarity. Only one magnetic field component, the north–south, was available for the Gainesville station. All waveforms are on a 20 μsec per division scale. Adapted from Lin *et al.* (1979).

lightning within about 20 km over Florida soil is given in Fig. 7.3. The mean of the electric field initial peak value, normalized to 100 km, is generally found to be in the range 6–8 V/m for first strokes and 4–6 V/m for subsequent strokes. Higher observed mean values may be an indication of the fact that small strokes were not observed due to an equipment threshold trigger level that was set too high (e.g., Krider and Guo, 1983). Peckham *et al.* (1984) have shown experimentally that for the fixed threshold value characteristic of their equipment, the mean of the normalized initial peak field increased with distance from about 7 V/m in the 25- to 75-km range to about 9 V/m in the 100- to 150-km range because at the greater ranges more of the smaller fields fell below the threshold and were lost from the distribution. On the other hand, some compensation for this effect is caused by the fact that distant waveforms that propagate over poorly conducting

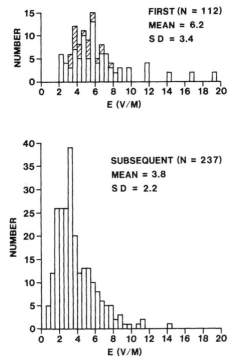

Fig. 7.3 Peak vertical electric field of the return stroke normalized to 100 km. The cross-hatched data represent separate channels to ground for strokes subsequent to the first in a flash. Adapted from Master *et al.* (1984).

earth will have their peak fields attenuated due to propagation over a non-perfectly conducting surface, as discussed later in this section. From their measured data on the initial field peaks, Krider and Guo (1983) have calculated that the electromagnetic power radiated at the time of those peaks has a mean of 2×10^{10} W for first strokes and 3×10^9 W for subsequent strokes.

The zero-crossing time for distant waveforms illustrated in Fig. 7.1b has been measured by Lin *et al.* (1979) in Florida and by Cooray and Lundquist (1985) in Sweden and in Sri Lanka. Mean zero-crossing times for first strokes in Florida and in Sweden were near 50 μsec whereas in Sri Lanka, in the Tropics, the mean was near 90 μsec, a difference that Cooray and Lundquist (1985) suggest may be due to differing meteorological conditions in the different locals. Subsequent stroke zero-crossing times were similar for the three locations. The data are summarized in Table 7.1.

Details of the shape of the return-stroke field rise to peak and the fine structure after the initial peak are shown in Fig. 7.4, and some measured characteristics are summarized in Table 7.1. Note that also shown in Fig. 7.4 are stepped-leader pulses occurring prior to the return stroke field. As illustrated in Fig. 7.4 and as can be seen in the experimental data of Figs. 5.1c, 4.5d, and 7.2, first return-strokes fields have a "slow front" (below the dotted line in Fig. 7.4 and labeled F) that rises in a few microseconds to an appreciable fraction of the peak field magnitude. Master *et al.* (1984) found a mean slow front duration of 2.9 μsec with about 30% of the initial peak field being due to the slow front. Cooray and Lundquist (1982, 1985) report corresponding values of about 5 μsec and 40%. Weidman and Krider (1978) find a mean slow front duration of about 4.0 μsec with 40–50% of the initial field peak attributable to the slow front. Additional data are given in Table 7.1.

The slow front is followed by a "fast transition" to peak field (labeled R in Fig. 7.4) with a 10–90% risetime of about 0.1 μsec when the field propagation path is over salt water. Weidman and Krider (1980, 1984) and Weidman (1982) report a mean risetime of 90 nsec with a standard deviation of 40 nsec for 125 first strokes. As illustrated in Fig. 7.4, fields from subsequent strokes have fast transitions similar to those of first strokes except that these transitions occupy most of the rise to peak, while the fronts are of shorter duration than for first strokes, typically 0.5–1 μsec and only comprise about 20% of the total rise to peak (Weidman and Krider, 1978). Additional data on the fast transition are found in Table 7.1.

The high frequency content of the return-stroke fields is preferentially degraded in propagating over a finitely conducting earth (Uman *et al.*, 1976; Lin *et al.*, 1979; Weidman and Krider, 1980, 1984; Cooray and Lundquist, 1983) so that the fast transition and other rapidly changing fields can be adequately observed only if the propagation path from the lightning to the antenna is over salt water, a relatively good conductor. It is for that reason that the fast transition time observed by Master *et al.* (1984) for lightning in the 1- to 20-km range over land is an order of magnitude greater than that observed over salt water by Weidman and Krider (1980, 1984) and Weidman (1982) (see Table 7.1). Lin *et al.* (1979) report that normalized peak fields are typically attenuated 10% in propagating over 50 km of Florida soil and 20% in propagating 200 km. Uman *et al.* (1976) reported on field risetimes observed both near a given stroke and 200 km from it. For typical strokes, zero-to-peak risetimes (see Table 7.1) are increased of the order of 1 μsec in propagating 200 km across Florida soil.

In measuring zero-to-peak risetime, as well as other waveform characteristics related to the beginning of the waveform, care must be taken to identify properly that beginning. For the articles listed in Table 7.1 that were published before 1982, risetime was measured with a system that displayed only a few microseconds of the field prior to oscilloscope triggering

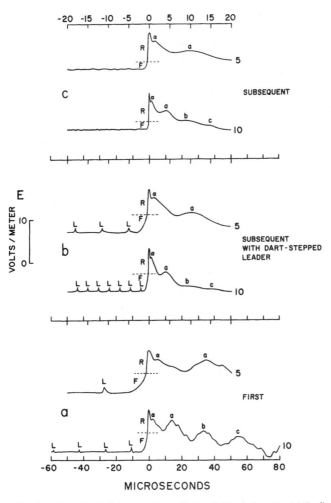

Fig. 7.4 Sketches of the detailed shapes of the radiation fields produced by (a) the first return stroke, (b) a subsequent return stroke preceded by a dart-stepped leader, and (c) a subsequent stroke preceded by a dart leader in lightning discharges to ground. The field amplitudes are normalized to a distance of 100 km. The small pulses characteristic of leader steps L are followed by a slow front F and an abrupt, fast transition to peak R. Following the fast transition, there is a small secondary peak or shoulder α and large subsidiary peaks, a, b, c. Characteristics of this fine structure following the peak are discussed by Weidman and Krider (1978) for Florida return strokes and by Cooray and Lundquist (1985) for strokes in Sri Lanka where the time between subsidiary peaks is reported to be generally less than in Florida. The lower trace in each case shows the field on a time scale of 20 μsec/division, and the upper trace is at 5 μsec/division. The origin of the time axis is chosen at the peak field in each trace. Adapted from Weidman and Krider (1978).

(an example is shown in Fig. 7.2; see also Section C.1, Fig. C.3b), making it impossible to be certain that the start of the waveform was observed. For example, the start of the waveform would not be observed for a relatively long slow front and a relatively high trigger level.

The electric field Fourier amplitude spectrum to 20 MHz for return strokes at a range of about 50 km is given in Fig. 7.5. Spectra to 700 kHz for return strokes within 10 km are shown in Fig. 7.6. For close lightning, the spectra below 10^4 Hz are dominated by induction and electrostatic fields, while the distant spectra, primarily radiation fields, exhibit a peak between 10^3 and 10^4 Hz, as shown in Figs. 7.5 and 7.6. Between 10^4 and 10^6 Hz all spectra vary inversely with frequency, while above 10^6 Hz the decrease with increasing frequency is faster. Some of the falloff near 10^7 Hz may be due to propagation effects associated with a salt water surface exhibiting waves. Except for the data given in Figs. 7.5 and 7.6, all of the return-stroke frequency spectra found in the literature (e.g., Taylor, 1963; 11 measurements discussed by Dennis and Pierce, 1964) have an upper frequency limit below 100 kHz.

LeVine and Krider (1977) made narrow-band measurements of RF emissions at 3, 30, 139, and 295 MHz correlated with wideband electric field measurements. They found that first strokes have strong radiation at all of

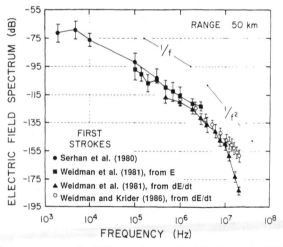

Fig. 7.5 Frequency spectra for 13 first strokes over land (Serhan *et al.*, 1980), 5 first return strokes over salt water (Weidman *et al.*, 1981), and 20 first strokes over salt water (Weidman and Krider, 1986) at a distance of about 50 km. The decibel values given are 20 times the logarithm (base 10) of the magnitude of the Fourier transform of the electric field intensity. Solid symbols indicate mean values; vertical bars, standard deviations. Spectra are essentially of radiation fields.

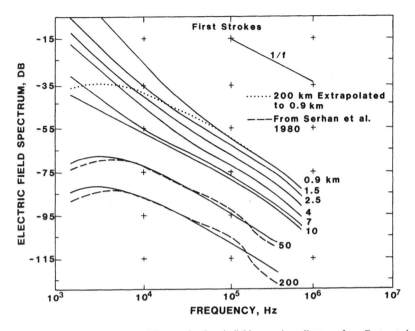

Fig. 7.6 Frequency spectra of first stroke electric fields at various distances from Preta *et al.* (1985). Solid lines represent spectra of model waveforms constructed from typical electric field data measured over both land and salt water so that the waveforms exhibit no propagation effects. Dashed lines, data at 50 and 200 km, are the measured spectra of Serhan *et al.* (1980) over land which show the effects of propagation at the higher frequencies. The dotted line represents the radiation field at 200 km extrapolated by the inverse distance relation to 0.9 km and illustrates the fact that, even at 0.9 km, the fields above about 10^5 are radiation. For return strokes within 10 km the frequency spectra below 10^4 Hz are dominated by induction and electrostatic fields. The decibel values given are 20 times the logarithm (base 10) of the magnitude of the Fourier transform of the electric field intensity.

these frequencies, but that the radiation does not peak until 10–30 μsec after the start of the wideband return-stroke electric field waveform, and suggested that, therefore, the RF may be due to the effects of branches in first strokes and cloud processes. Supporting this suggestion is their observation that subsequent strokes, which usually do not have branches, generate little HF or VHF radiation. The observations of Levine and Krider (1977) are in reasonable agreement with those of Takagi (1969a,b) who found a spectrum of delays between return stroke fields and RF at 60, 150, and 420 MHz with a peak in the first stroke delay distribution at about 10 μsec and peaks in the subsequent stroke delay distribution at about 10 and 50 μsec. On the other hand, Brook and Kitagawa (1964) reported that RF radiation at 420 and 850 MHz was always delayed 60–100 μsec in the 50% of the return strokes for which RF was detected and suggest, therefore, that the RF at those frequencies was due to breakdown at the top of the return stroke channel. Cooray (1986) and LeVine *et al.* (1986) argue that the observed time delay between RF emissions and the wideband electric field are due to propagation effects in that the RF radiation from the lower part of the channel is more strongly attenuated in propagating over a finitely conducting earth than is RF radiation from higher channel sections, which, due to the finite upward speed of the return stroke, are radiated later. Cooray (1986) supports his calculation of this suggested propagation effect with 3-MHz RF and wideband electric field measurements for lightning over both salt water and over earth. In the former case there is no time delay in the appearance of the RF emissions; in the latter case the RF increases slowly to peak following the start of the wideband return stroke field. The time-to-peak RF emission for lightning over land is reported to vary inversely as the distance to the source.

RF emission often exhibits a quiet period following the return-stroke propagation phase. Malan (1958) first reported these periods of decreased RF activity which lasted 5–20 msec, as illustrated in Fig. 1.12. Clegg and Thomson (1979) have reviewed seven previous studies of the quiet period, and, from their own work, report that the duration of the quiet period at 10 MHz follows closely the K-change interval distribution (Section 10.4), having a most probable interval in the range 6–9 msec.

7.2.2 CURRENT

The most complete description of lightning return stroke currents at the base of the stroke channel is due to Berger and co-workers in Switzerland (e.g., Berger, 1955a,b; 1962, 1967a, 1972, 1980; Berger and Vogelsanger, 1965, 1969). Their results are reviewed in English by Berger (1967b) and Berger *et al.* (1975). The currents were derived from measurements of the voltages induced in resistive shunts located at the tops of 2 towers each 55 m above the summit of Mt. San Salvatore in Lugano. The summit of the Mt. San Salvatore is 915 m above sea level and 640 m above the level of Lake Lugano, located at the base of the mountain. It is interesting to note that only about 10% of the total flashes observed at Mt. San Salvatore were due to negatively charged downward-moving stepped leaders, most discharges to the towers being initiated by upward-going leaders (Section 12.2).

The measurements of Garbagnati and co-workers in Italy (Garbagnati and Lo Pipero, 1970, 1973, 1982; Garbagnati *et al.*, 1975, 1978) were made using resistive shunts at the tops of 40-m television towers that were located on the tops of two mountains, each about 900 m above sea level. One station, in northern Italy, was within view of Mt. San Salvatore. The other station was in central Italy.

The studies of Eriksson (1978) were made on strikes to a 60-m tower resting on relatively flat ground in South Africa. The tower was insulated from ground and the lightning current measured at the bottom via a current transformer or a Rogowski coil. More than 50% of the observed flashes were initiated by the usual downward-moving negatively charged stepped leader. No positive flashes were observed. The most interesting feature of the South African work is that, in contrast to other tower measurements, very fast current risetimes were observed, as discussed later in this section.

The waveforms of currents measured at towers may well be different from those of strikes to the ground or small structures or from currents present in the channel above ground. Such differences are particularly to be expected during the rise to peak of the first stroke current, since (1) currents at this time must partially be due to an upward-going leader (Chapter 6 and Section 12.2) that presumably can be influenced by the height and perhaps other characteristics of the object being struck and (2) fast rates-of-change of current, characteristic of the final portion of the current rise to peak, may be altered by the electrical characteristics of the measuring circuit including the tower and by pickup on the measuring circuit of the electromagnetic fields of the close lightning (Berger and Vogelsanger, 1965). Berger and Vogelsanger (1965) state that because of these latter effects, risetimes shorter than 1 μsec cannot be measured accurately. Further, on theoretical grounds, first stroke currents in strikes to tall objects are expected to be larger, on the average, than those to normal ground (Sargent, 1972). This is so because a greater than average stepped-leader charge, associated with a greater than average return-stroke current, will produce a longer than average outward-going connecting leader from the tower, as discussed in Chapter 6. These long leaders will preferentially attach to the downward-moving leader keeping it from proceeding to ground and yielding strokes to the tower that have larger peak currents than the average stroke to ground. From available measurement, there does not appear to be a difference between the currents to towers and the currents to ground. (Eriksson, 1978).

The extensive measurements of Berger and co-workers for lightning initiated by downward-moving leaders are summarized in Figs. 7.7 and 7.8 and in Table 7.2. The Italian results are similar to those of Berger and co-workers and are summarized in Table 7.3. Note that flashes lowering positive charge are included in the statistics of Table 7.2. There is some question whether these positive flashes should be considered as initiated by downward-moving positive leaders or by upward-moving negative leaders, as discussed in Sections 11.2 and 12.2, since the upward-moving leaders associated with these discharges are relatively long.

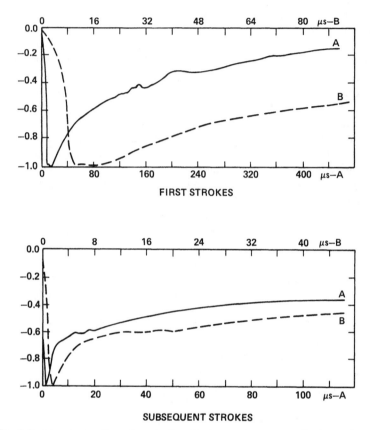

Fig. 7.7 Average negative return-stroke current waveshapes normalized to unity peak amplitude as reported by Berger *et al.* (1975).

In Fig. 7.7 are shown, on two time scales, the average current waveshapes for negative first and subsequent strokes observed in strikes to Mt. San Salvatore. In Fig. 7.8, statistical data for initial peak current, including positive strokes, are given for the Swiss data. The straight lines plotted on the cumulative probability paper in Fig. 7.8 represent a log normal distribution of peak current (Section B.3). This probability function approximates well the distribution of most current parameters listed in Table 7.2 and 7.3 (Berger *et al.*, 1975; Garbagnati and LoPipero, 1982).

Peak currents for first strokes are generally thought to have a median value in the range 20–40 kA with 200 kA occurring at about the 1% level, although there is some disagreement about the exact statistics (Szpor, 1969; Sargent, 1972; Popolansky, 1972; Berger *et al.*, 1975; Eriksson, 1978). Subsequent

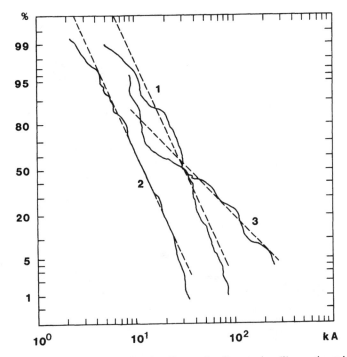

Fig. 7.8 Peak current distribution for (1) negative first stroke, (2) negative subsequent strokes, and (3) positive strokes from Berger *et al.* (1975). The straight dashed lines represent log normal distributions fit to the experimental data (see Section B.3).

stroke peak currents have a mean that is somewhat less than half of the first stroke values and are distributed similarly, as illustrated in Tables 7.2 and 7.3 and in Fig. 7.8.

As is evident from Tables 7.2 and 7.3, first stroke risetimes are considerably longer than those of subsequent strokes, some of this difference probably being due to the upward-going leaders that precede first strokes. Berger *et al.* (1975) find subsequent stroke risetimes for which the median value from 2 kA to peak is 1.1 μsec. The median value of the maximum rate of rise is 40 kA/μsec. Five percent of the 118 front times measured were less than 0.22 μsec, and 5% of the maximum rates of rise of current exceeded 120 kA/μsec. Eriksson (1978) found a maximum rate of rise of current of 180 kA/μsec for a second stroke in one of 11 flashes whose currents were measured. More important, Eriksson (1978) reports that when current is measured with an instrument that samples the waveform each 0.2 μsec, peak value for subsequent strokes is often reached within the first sample period,

Table 7.2

Lightning Current Parameters[a]

Number of events	Parameters	Unit	Percentage of cases exceeding tabulated value		
			95%	50%	5%
	Peak current (minimum 2 kA)				
101	Negative first strokes	kA	14	30	80
135	Negative subsequent strokes	kA	4.6	12	30
26	Positive first strokes (no positive subsequent strokes recorded)	kA	4.6	35	250
	Charge				
93	Negative first strokes	C	1.1	5.2	24
122	Negative subsequent strokes	C	0.2	1.4	11
94	Negative flashes	C	1.3	7.5	40
26	Positive flashes	C	20	80	350
	Impulse charge				
90	Negative first	C	1.1	4.5	20
117	Negative subsequent strokes	C	0.22	0.95	4.0
25	Positive first strokes	C	2.0	16	150
	Front duration (2 kA to peak)				
89	Negative first strokes	μsec	1.8	5.5	18
118	Negative subsequent strokes	μsec	0.22	1.1	4.5
19	Positive first stroke	μsec	3.5	22	200
	Maximum di/dt				
92	Negative first strokes	kA/μsec	5.5	12	32
122	Negative subsequent strokes	kA/μsec	12	40	120
21	Positive first strokes	kA/μsec	0.20	2.4	32
	Stroke duration (2 kA to half-value)				
90	Negative first strokes	μsec	30	75	200
115	Negative subsequent strokes	μsec	6.5	32	140
16	Positive first strokes	μsec	25	230	2000
	Integral ($i^2 dt$)				
91	Negative first strokes	A^2 sec	6.0×10^3	5.5×10^4	5.5×10^5
88	Negative subsequent strokes	A^2 sec	5.5×10^2	6.0×10^3	5.2×10^4
26	Positive first strokes	A^2 sec	2.5×10^4	6.5×10^5	1.5×10^7
	Time interval				
133	Between negative strokes	msec	7	33	130
	Flash duration				
94	Negative (including single-stroke flashes)	msec	0.15	13	1100
39	Negative (excluding single-stroke flashes)	msec	31	180	900
24	Positive (only single flashes)	msec	14	85	500

[a] Adapted from Berger *et al.* (1975).

Table 7.3

Current Parameters Reported by Garbagnati and Co-Workers in Italy for
Discharges Lowering Negative Charge[a]

	Downward		Upward	
	First strokes	Subsequent strokes	First strokes	Subsequent strokes
Number of strokes	42	33	61	142
Peak value (kA)	33 (0.25)	18 (0.22)	7 (0.23)	8 (0.22)
Maximum rate of rise (kA/μsec)	14 (0.36)	33 (0.39)	5 (0.76)	13 (0.52)
Time to crest (μsec) (3 kA to peak)	9 (0.40)	1.1 (0.33)	4 (0.55)	1.3 (0.35)
Time to half-value (μsec)	56 (0.32)	28 (0.50)	35 (0.37)	31 (0.35)
Impulse charge (C) (to end of impulse or 500 μsec)	2.8 (0.35)	1.4 (0.31)	0.5 (0.30)	0.6 (0.32)

[a] The values given are geometric means; the standard deviations in parentheses are expressed in logarithmic units consistent with the existence of a log normal parent distribution (see Section B.3). The discharges are arbitrarily divided into downward and upward solely on the basis of the early part of the first-stroke current waveshapes. In the downward category are the currents that are initially slowly rising but then take a relatively fast jump to peak. In the upward category are those with a smoother rise to peak and a more rounded peak. The latter category is assumed to have longer upward-going leaders (see Chapter 6) than the former, perhaps extending to the cloud.

and he therefore questions whether previous oscilloscopic measurement systems in which waveforms were recorded on film were adequate to resolve such fast wavefronts. For subsequent strokes in artifically initiated lightning the 10–90% risetimes had a median value of 0.3 μsec in the studies in France (Section 12.3.3.1) and 0.5 μsec in the experiments in New Mexico (Section 13.3.3.2).

Risetimes and rates of rise of current have been derived from measured return-stroke electric fields by Weidman and Krider (1980, 1982a–c). They find that both first and subsequent strokes have a mean maximum rate of change of current of about 180 kA/μsec, the upper end of the distribution measured on towers, with a maximum value about twice the mean. A direct comparison of tower and field measurements of current is given by Weidman and Krider (1984). More discussion of the relation between fields and currents is found in Section 7.3.

The charge lowered to earth by a stroke can be determined by integrating the current waveform over time. As indicated in Tables 7.2 and 7.3, values of a few coulombs for first strokes and about a coulomb for subsequent strokes are typical. Charge transfer has also been determined from remote electric field measurements (e.g., Brook *et al.*, 1962; Krehbiel *et al.*, 1979).

Charge values determined in these two ways are similar, although one would expect those determined using the latter technique to be slightly larger due to the effect of small currents that flow for milliseconds at the end of the stroke. These currents are generally not included in tower measurements of "impulse" charge. Additional discussion of the magnitude of stroke charges and their initial location in the cloud is found in Section 3.2.

Lightning currents that passed through aircraft in flight have been measured by Petterson and Wood (1968), Pitts (1982), Pitts and Thomas (1981, 1982), Thomas and Pitts (1983), Lee et al. (1984), Thomas (1985, 1986), Zaepfel et al. (1985), and Rustan (1986), although the aircraft were apparently involved with few if any return strokes. Currents in the tethers of barrage balloons flown at 600 m have been reported by Davis and Standring (1947).

7.2.3 SPEED

The definition of a return stroke speed must be, to some extent, arbitrary since the luminosity of the return stroke wavefront has a shape that varies with height, so that it is not obvious how to identify the same luminous feature at different heights (Hubert and Mouget, 1981; Jordan and Uman, 1983). Nevertheless, the error involved in identifying the time of initial exposure on streak photographs, as a basis for the speed measurements, is apparently not large, especially near ground. The pioneering work in the measurement of return stroke speed was done by Schonland and Collens (1934) and Schonland et al. (1935) in South Africa. They used a Boys camera, a two-lens streak camera (Section C.4, Fig. C.8), in their studies, as have most subsequent investigators. Schonland et al. (1935) found that first return stroke speeds at the channel base were typically near 1×10^8 m/sec, decreasing abruptly as each branch point was reached. The associated channel luminosity also decreased at each branch point. The speed at the top of the main channel was typically near 5×10^7 m/sec. Timetables for the return-stroke wavefront's arrival at various points on the main channel and branches are given for a number of first strokes by Schonland et al. (1935).

More recently, speed data for natural lightning have been published by Boyle and Orville (1976), Orville et al. (1978), and Idone and Orville (1982), while speeds of strokes in artificially triggered discharges (Chapter 12) have been given by Hubert and Mouget (1981) and Idone et al. (1984). The work of Idone and Orville (1982) is the most comprehensive to date on natural lightning. Seventeen first strokes and 46 subsequent stroke speeds are analyzed. The speed distribution derived by Idone and Orville (1982) is given in Fig. 7.9. The first stroke mean speed within about 1.3 km of ground was

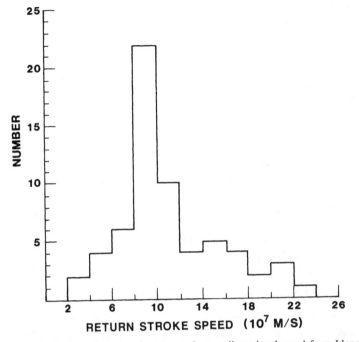

Fig. 7.9 Distribution of measured return-stroke two-dimensional speed from Idone and Orville (1982). Data include 17 first and 43 subsequent strokes within 1.3 km of ground. Mean speed is 1.1×10^8 m/sec.

9.6×10^7 m/sec; the subsequent stroke mean speed was 1.2×10^8 m/sec. Both first and subsequent stroke speeds decreased with height. Idone *et al.* (1984) report three-dimensional return-stroke speeds (data from all others except Hubert and Mouget, 1981, are two-dimensional) near ground of 56 subsequent strokes in artificially triggered flashes. The mean three-dimensional speed is 1.2×10^8 m/sec, with minimum and maximum values of 6.7×10^7 and 1.7×10^8 m/sec, respectively. The more recently measured return stroke speeds are generally higher than the earlier results (see Fig. 10 of Idone and Orville, 1982), a result that may be in part attributable to the fact that the recent measurements were made on channels close to the ground where the return stroke speed was near its maximum value.

Return stroke speed and return-stroke peak current might be expected to be related since each must depend to some extent on the charge per unit length on the leader channel and the electric potential at each point on the leader channel resulting from that charge. The existence of a relation between speed and peak current has important implications for the modeling of return stroke fields and currents since speed and current have been treated as independent to date in these modeling efforts, as indicated in Section 7.3. Lundholm (1957) and Wagner (1963) have developed theoretical expressions relating speed and peak current. Idone *et al.* (1984) and Hubert and Mouget (1981), for subsequent strokes in artificially initiated lightning (Section 12.3.3.1), find that their data are reasonably well fit by the expression attributed to Lundholm (1957).

$$v = c[1 + (W/i_p)]^{-1/2}$$

where v is the return stroke speed, c the speed of light, and i_p the peak current in kA, with the constant $W = 40$ kA. Idone *et al.* (1984) have also investigated the correlation of return stroke speed with both the preceding dart leader speed and the preceding interstroke interval. The leader and return stroke speeds are correlated for subsequent strokes in artificially initiated lightning, but, according to Idone *et al.* (1984), are not correlated for natural lightning. For the artificially initiated lightning, speed and interstroke interval are weakly correlated.

7.2.4 OPTICAL SPECTRA AND LUMINOSITY

Reviews of lightning spectroscopy have been published by Orville and Salanave (1970), Krider (1973), Orville (1977), Uman (1969, 1984), and Salanave (1980). Orville (1968a–c) has presented return-stroke optical spectra with a time resolution of a few microseconds (Section C.5, Fig. C.9). A time-resolved spectrum is shown in Fig. 7.10. In addition, Orville has published similar data for stepped leaders (Orville, 1968d; Section 5.4) and for dart leaders (Orville, 1975; Section 8.2). Physical properties of the lightning channel such as temperature, particle densities, and pressure have been determined from these spectra, as discussed next.

Orville (1968a–c) found that the peak temperature in 10 return stroke channels occurred within the first 10 μsec of the discharges and were of the order of 30,000 K. These peak temperatures were derived from an analysis of the time-resolved radiation from singly ionized nitrogen atoms by assuming that the radiating gas was optically thin and in local thermodynamic equilibrium. Since the time resolution was a few microseconds, if the initial temperatures varied more rapidly than the time resolution, the peak temperature values represent some sort of average over that time. The return stroke temperatures for all strokes studied were below 30,000 K after 10 μsec and were near or below 20,000 K after 20 μsec.

Electron densities in lightning return strokes were determined by Orville (1968a–c) by the technique of comparing the measured Stark width of the H_α line radiated by hydrogen atoms with theory. The Stark width of H_α is

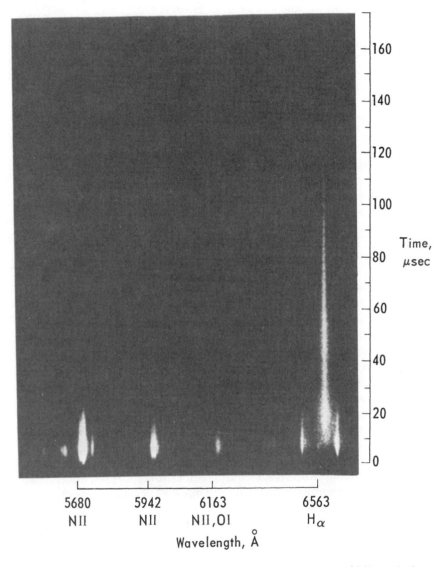

Fig. 7.10 A time-resolved spectrum of a short section of a first return stroke channel in the H_α region. The H_α line is due to the Balmer series of neutral hydrogen and does not occur early in the discharge because most of the hydrogen, present originally in water vapor, is ionized. H_α is broadened by the Stark effect. The NII lines are from singly ionized nitrogen, the OI line from neutral oxygen. The ion lines appear early in the discharge when the temperature is relatively high. A drawing of the spectrometer with which the data were obtained is shown in Fig. C.9. Spectrum adapted from Orville (1968a).

independent of the population of the bound atomic energy levels and is only a weak function of the electron temperature. According to Orville (1968a–c), the electron density averaged over the first 5 μsec of the return stroke has a value of about 10^{18} cm^{-3}. This electron density and the measured channel temperature averaged over a similar initial time interval yield a channel whose pressure is about 8 atm, an average value that is probably weighted toward lower pressures (Orville, 1968a–c). In any event, the initial channel pressure exceeds ambient, and the lightning channel must, therefore, expand, apparently reaching pressure equilibrium with the surrounding atmosphere in a time of the order of 10 μsec (Orville, 1968a–c). When pressure equilibrium is reached, the measured electron density is about 10^{17} cm^{-3}, and thereafter it stays constant for at least as long as the singly ionized nitrogen lines are recorded. According to the calculations of Drellishak (1964), a nitrogen plasma in local thermodynamic equilibrium at 1 atm pressure will have an electron density of about 10^{17} cm^{-3} for all temperatures between 13,000 and 35,000 K. Since nitrogen and air plasma can be expected to be very similar, the theoretical results provide good confirmation of the measured values.

Theory supporting the analysis techniques used in deriving physical properties from spectra is given by Uman (1969). Basically, he shows that for the line radiation studied, channels are optically thin and a reasonable approximation to local thermodynamic equilibrium is achieved, except perhaps during the initial microseconds, allowing meaningful temperature and related thermodynamic quantities to be determined. On the other hand, Hill (1972) has discussed the validity of the assumption that the return stroke channel is optically thin and has concluded that, for some spectral lines, this may not be the case.

An extensive spectroscopic study of lightning has been carried out at the Los Alamos Scientific Laboratories (Barasch, 1968, 1969, 1970; Connor, 1967, 1968; Connor and Barasch, 1968). A comparison between the spectral intensities of first return strokes, subsequent return strokes, stepped leaders, dart leaders, M-components, intracloud discharge, and cloud-to-air discharges is given in five narrow regions of the visible and near-infrared (Barasch, 1968, 1970). Stroke-resolved spectra recorded on film are used by Connor (1967, 1968) to examine the efficiency with which the discharge converts input energy to light (see below) and to measure the radial extent of the channel radiation. One unusual stroke appeared to radiate H_α to a radius of 120 m (Connor, 1968). By comparison, channel radii have been found photographically by Schonland (1937) to be in the 5.5- to 11.5-cm range, by Evans and Walker (1963) to be in the 1.5- to 6-cm range, and by Orville et al. (1974) to be in the 2- to 4-cm range. Orville et al. (1974) review some of the theory (Braginskii, 1958; Oetzel, 1968; Plooster, 1971; Hill, 1971, 1972; see

Section 15.3.1 and Fig. 15.11) and records of physical damage caused by lightning (Schonland, 1950; Hill, 1963; Uman, 1964; Taylor, 1965; Jones, 1968) that generally support the values they obtained for the return stroke channel radius after the initial $10 \mu sec$ or so of the channel expansion.

Return stroke luminosity has been measured with widefield photoelectric systems (Section C.3), which viewed essentially the whole visible channel and the cloud illuminated from within, by Krider (1966), Mackerras (1973), Turman (1977, 1978), Guo and Krider (1982, 1983), and Orville and Henderson (1984). Mackerras (1973) characterized the magnitude and shape of the light pulses arising from both ground and cloud flashes and estimated the efficiency of the conversion of input energy to luminosity as typically 1–3%. This compares reasonably well with the average value from Connor (1967) of 0.7% and from Krider et al. (1968) for a single stroke of 0.4%. Input energy per unit length is found to be of the order of 10^5 J/m from comparison with laboratory spark studies (Krider et al., 1978) and from consideration of the cloud charge energy dissipated (Conner, 1967; Mackerras, 1973; Section A. 1.6). Plooster (1971) and Hill (1977), however, have argued, on the basis of computer simulation studies of channel behavior when current is injected, that this input energy per unit length is one to two orders of magnitude too large (Section 15.3.1). Mackerras (1973) also determined the time duration of the largest light pulse, presumably from a return stroke, in each of 46 ground discharges. The total duration was about 1 msec. The median effective width, defined as the duration of a *rectangular* pulse having the same peak value and time integral as the measured pulse, was $200 \mu sec$ with a range from $40 \mu sec$ to 1.0 msec. Interestingly, the stroke light waveforms often exhibited a second peak that followed the initial primary peak by a few hundred microseconds.

Guo and Krider (1982, 1983), using the experimental setup shown in Fig. C.7, give absolute values for the light intensity of first and subsequent return strokes in the 0.4- to 1.1-μm wavelength range and present correlated electric field data. Peak optical power has a mean of 2.3×10^9 W for first strokes, 4.8×10^8 W for subsequent strokes preceded by dart leaders, and 5.4×10^8 W for subsequents preceded by dart-stepped leaders (Guo and Krider, 1982). The average time at which the peak optical power occurs is about $60 \mu sec$ after the initial electric field peak because the risetime of the optical signal is determined by the geometrical growth of the return stroke channel in the field of view (Guo and Krider, 1982). Guo and Krider (1982) also compare their data with similar measurements of Krider (1966), Mackerras (1973), and Turman (1978). The time and space averaged mean radiance near the channel bottom was found by Guo and Krider (1983) to be 1.3×10^6 W/m for first strokes and 3.9×10^5 W/m for subsequent strokes preceded by dart leaders, and, by Guo and Krider (1982), 1×10^6 W/m for first strokes and 2.5×10^5

and 4.5×10^5 W/m for subsequents preceded by normal and dart-stepped leaders, respectively. By making use of the streak-photography data of Jordan and Uman (1983), Guo and Krider (1983) estimate that the average peak radiance for subsequent strokes is 2 to 4 times the time and space averaged radiance given above. Paxton *et al.* (1986) have calculated the optical output expected from a typical first return stroke current and find good agreement with the measurements of Guo and Krider (1982, 1983). Krider and Guo (1983) show that the radio frequency power radiated by a return stroke at the peak field, a mean of 2×10^{10} W for first strokes and 3×10^9 W for subsequents, is two orders of magnitude greater than the original power radiated in the 0.4- to 1.1-μm band at the time of peak field. The optical power, however, dominates at later times.

Turman (1977, 1978) measured the peak optical return stroke power observed from a satellite in earth orbit and found a median of 10^9 W with 2% of the strokes above 10^{11} W. Turman (1977) coined the term "superbolt" to describe those powerful light emitters with peak powers in the 10^{11} to 10^{13} W range. Five flashes in 10^7 were found to exceed an optical power of 3×10^{12} W. Some or many of the superbolts may be positive strokes (Turman, 1977; Hill, 1978a,b; Uman, 1978; Chapter 11).

Orville and Henderson (1984) have published time-integrated absolute spectral data with 50-nm wavelength resolution. The absolute spectral irradiance from 375 to 880 nm for an average return stroke at a distance of about 15 km was 3.5×10^{-5} J/m^2 for the visible band (375–650 nm) and about one-third that value for the infrared (650–880 nm). Orville and Henderson (1984) compare their work with the time-resolved measurements in the region of the H$_\alpha$ line by Barasch (1970) and with the time-resolved measurements in the 0.4- to 1.1-μm band by Guo and Krider (1982) and show that there is reasonable agreement between the various data.

Jordan and Uman (1983) have investigated the light waveshapes of subsequent return strokes as a function of time and height using the technique of streak photography (Section C.4). The measured intensity profiles consisted of a fast rise to peak followed by a slower decrease to a nearly constant value, as shown in the example in Fig. 7.11. For nine strokes in three flashes, the initial peak luminosity decreased exponentially with height with a decay height of 0.6–0.8 km, and the risetime of the light increased with height from a value near ground of a few microseconds to about a factor of two greater at the cloud-base height of between 1 and 2 km. The relative light intensity 30 μsec after the initial peak was relatively constant with height and had an amplitude of 15–30% of the initial peak value near the ground and 50–100% of the initial peak value near the cloud base. The logarithm of the initial peak light intensity near ground was found to be linearly related to the initial peak electric field intensity.

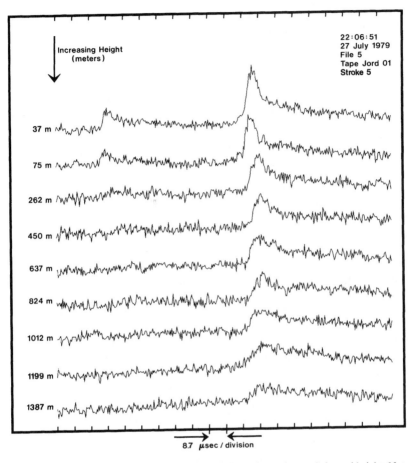

Fig. 7.11 Subsequent return-stroke relative light intensity vs time and channel height. Note that height above ground given on the left side of the figure is shown increasing in the downward direction. The noise on the light waveforms is due to the densitometer used in measuring intensity from the film image. The plots have been corrected for film nonlinearity. Adapted from Jordan and Uman (1983).

Ganesh et al. (1984) used a photoelectric system to isolate the light from a section of return stroke channel in an attempt to verify this latter result of Jordan and Uman (1983). Ganesh et al. (1984) were able to verify that the light and electric field are correlated but were not able to determine the appropriate analytical relationship. They present correlated light and electric field waveforms for both close and distant lightning as well as light risetime statistics. Risetimes were in the range from 1 to 60 μsec with the larger values probably being due to a combination of multiple scattering of the optical signal off aerosols and the viewing of higher portions of channel on the more distant lightning.

Idone and Orville (1985) have shown, for 39 subsequent strokes in two rocket-initiated lightning flashes (Section 12.3.3.2), that return-stroke peak light intensity is strongly correlated with return-stroke peak current. For a linear relationship, the individual correlation coefficients for the two flashes were $r = 0.97$ and $r = 0.92$, respectively (see Section B.6). Stroke peak currents ranged from 1.6 to 21 kA.

The effects of multiple scattering on optical signals generated by lightning sources within clouds has been considered by Thomason and Krider (1982). They show that the delay and the time broadening of an instantaneous light impulse generated within the cloud and observed outside can be several tens of microseconds or more and that essentially all of the visible light eventually escapes the clouds.

7.3 MODELING

We now consider models that can be used for prediction of the physical characteristics of the return stroke, with primary emphasis on models that relate the remote electric and magnetic fields to the channel currents. There are basically three levels of sophistication in the mathematical modeling of return strokes. (1) The most sophisticated model describes the detailed physics of the lightning channel in terms of equations of conservation of mass, momentum, and energy, equations of state, and Maxwell's equations. This type of model requires a detailed knowledge of physical parameters such as the ionization and recombination coefficients and of thermodynamic properties such as the thermal and electrical conductivities. Using this basic approach, one can attempt to predict the channel current as a function of height and time. Once the current has been derived, the remote electric and magnetic fields can be calculated from it (Uman *et al.*, 1975; Master *et al.*, 1983; Section A.3). Modeling of this type has been attempted for lightning return strokes by Strawe (1979) and holds considerable promise of providing a better understanding of the return stroke. Using models of this type but assuming an input current, Hill (1971, 1975) and Plooster (1971) have determined the thermodynamic properties of the channel and the channel radius as a function of time as well as properties of the shock wave generated by expansion of the hot channel (Sections 7.2.4 and 15.3.1). (2) A less sophisticated level of modeling involves mathematically describing the return stroke channel as a R-L-C transmission line with circuit elements that may vary with height and time. The intent again is to predict a channel current as a function of height and time, and then to use this current to calculate the fields. Price and Pierce (1977) and Little (1978, 1979) have used this approach for return strokes. (3) In the least sophisticated approach to modeling, a temporal and spatial form from the channel current is assumed and then used to calculate the remote fields. The assumed current is constrained in its characteristics by the properties of lightning currents measured at ground level and by the available data on the measured electric and magnetic fields and return stroke speed. Lin *et al.* (1980) have reviewed the literature on this last type of modeling for return strokes and have presented a return stroke model that better relates to the experimental data than previous models (e.g., Bruce and Golde, 1941; Wagner, 1960; Dennis and Pierce, 1964; Uman and McLain, 1969; Leise and Taylor, 1977; LeVine and Meneghini, 1978a,b). The model of Lin *et al.* (1980), which we will discuss later in this section, has been further refined by Master *et al.* (1981).

Uman *et al.* (1975) and Master and Uman (1983) have shown how to compute the remote electric and magnetic fields from Maxwell's equations given the current in a vertical channel above a perfectly conducting ground. The electric and magnetic fields at a location (r, ϕ, z) in a cylindrical coordinate system (see Fig. 7.12) from a short vertical section of channel dz' at height z' carrying a time-varying current $i(z', t)$ are (Master *et al.*, 1981; Master and Uman, 1983; Section A.3)

$$d\mathbf{E}(r, \phi, z, t) = \frac{dz'}{4\pi\varepsilon_0} \left\{ \left[\frac{3r(z - z')}{R^5} \int_0^t i(z', \tau - R/c) \, d\tau \right. \right.$$
$$+ \frac{3r(z - z')}{cR^4} i(z', t - R/c) + \frac{r(z - z')}{c^2R^3} \frac{\partial i(z', t - R/c)}{\partial t} \bigg] \mathbf{a}_r$$
$$+ \left[\frac{2(z - z')^2 - r^2}{R^5} \int_0^t i(z', \tau - R/c) \, d\tau \right. \tag{7.1}$$
$$+ \frac{2(z - z')^2 - r^2}{cR^4} i(z', t - R/c)$$
$$\left. \left. - \frac{r^2}{c^2R^3} \frac{\partial i(z', t - R/c)}{\partial t} \right] \mathbf{a}_z \right\}$$

$$d\mathbf{B}(r, \phi, z, t) = \frac{\mu_0 \, dz'}{4\pi} \left[\frac{1}{R^3} i(z', t - R/c) + \frac{r}{cR^2} \frac{\partial i(z', t - R/c)}{\partial t} \right] \mathbf{a}_\phi \tag{7.2}$$

where ε_0 and μ_0 are the free-space permittivity and permeability, respectively, and all geometrical factors are defined in Fig. 7.12. In Eq. (7.1) the terms containing the integral of the current, the charge transferred through dz', are called electrostatic fields and, because of their strong distance dependence,

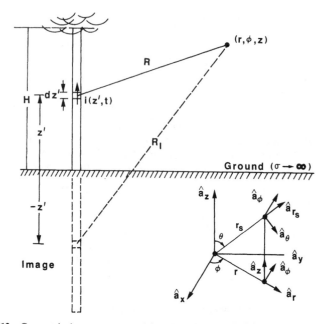

Fig. 7.12 Geometrical parameters used in calculating return stroke fields. At the bottom right, coordinates and unit vectors for spherical and cylindrical coordinate systems are defined.

are the dominant field component close to the source, the terms containing the derivative of the current are called radiation fields and are the dominant field component far from the source, and the terms containing current are called intermediate or induction fields. In Eq. (7.2) the first term is called the induction or magnetostatic term and is the dominant field component near the source; the second term is the radiation field and is the dominant field component far from the source. The effects of the perfectly conducting ground plane are included by postulating an image current beneath the plane as shown in Fig. 7.12. The electric and magnetic fields of the image are obtained by substituting R_I for R and $-z'$ for z' in Eqs. (7.1) and (7.2). Once the expression for the fields of a short channel section and its image is formulated, the fields for the total channel are found by integration over the channel.

An important special case is that of the fields on the ground because it is at ground level that most measurements have been made. For a vertical channel between the heights of H_B (usually zero, or ground level) and H_T above a perfectly conducting earth, these are (Uman *et al.*, 1975; Master and Uman, 1983; Section A.3)

$$\mathbf{E}(r, \phi, 0, t) = \frac{1}{2\pi\varepsilon_0}\left[\int_{H_B}^{H_T} \frac{2z'^2 - r^2}{R^5}\int_0^t i(z', \tau - R/c)\, d\tau\, dz'\right.$$

$$+ \int_{H_B}^{H_T} \frac{2z'^2 - r^2}{cR^4} i(z', t - R/c)\, dz' \qquad (7.3)$$

$$\left. - \int_{H_B}^{H_T} \frac{r^2}{c^2 R^3} \frac{\partial i(z', t - R/c)}{\partial t}\, dz'\right]\mathbf{a}_z$$

$$\mathbf{B}(r, \phi, 0, t) = \frac{\mu_0}{2\pi}\left[\int_{H_B}^{H_T} \frac{r}{R^3} i(z', t - R/c)\, dz'\right.$$

$$\left. + \int_{H_B}^{H_T} \frac{r}{cR^2} \frac{\partial i(z', t - R/c)}{\partial t}\, dz'\right]\mathbf{a}_\phi \qquad (7.4)$$

In Eq. (7.3) the first term is the electrostatic field, the second the intermediate or induction field, and the third the radiation field. In Eq. (7.4) the first term is the magnetostatic or induction field and the second is the radiation field. Far from the source where $r \cong R$, the radiation fields have an r^{-1} distance dependence, that is, the electric and the magnetic radiation fields decrease inversely with range if they are propagating over a perfectly conducting earth. Note that in much of the lightning literature and elsewhere in this book the horizontal distance between the field observation point and the lightning strike point is called D, and thus that we use the cylindrical coordinate r and the distance D interchangably. In addition to the r^{-1} distance dependence far from the source, the electric and magnetic radiation fields have identical waveshapes with $E_z/B_\phi = c$, the speed of light. While Eqs. (7.1)–(7.4) have been given in cylindrical coordinates, they can easily be transformed to other coordinate systems, spherical coordinates probably being the most used to date (e.g., Uman *et al.*, 1975; Section A.3).

Lin *et al.* (1980) have proposed a return-stroke current model that yields electric and magnetic fields in good agreement with the measured data of Lin *et al.* (1979) discussed in Section 7.2.1. In that model the return stroke current is decomposed into the three components shown in Fig. 7.13: (1) a breakdown pulse that propagates up the previous leader channel and is responsible for the initial peak electric and magnetic fields shown in Figs. 7.1, 7.2, 7.4, 4.5, and 5.1. The breakdown pulse can be constant with height as shown in Fig. 7.13 or can decay in an arbitrary manner with height (Master *et al.*, 1981; Uman *et al.*, 1982) to account, for example, for the decrease of light output with height observed by Jordan and Uman (1983); (2) a current due to the discharge of the charge stored in the leader corona envelope, that charge being released at each height after the breakdown pulse passes; and

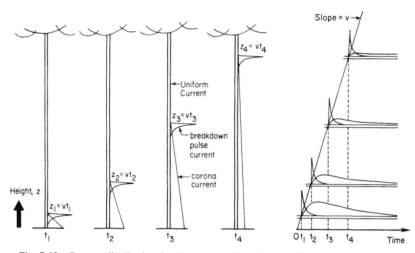

Fig. 7.13 Current distribution for the model of Lin et al. (1980). The breakdown pulse current is assumed constant with height, and the velocity of the breakdown pulse current is assumed constant at v. Current profiles are shown at four different times t_1 through t_4, when the return stroke wavefront and the breakdown pulse current are at four different heights z_1 through z_4, respectively.

(3) a uniform current that can be viewed as a continuation of the preceding steady leader current and that is responsible for the ramps in the electric fields identified in Fig. 7.1.

Lin et al. (1980) indicate how each of these three current components can be extracted from measurements taken simultaneously close to and far away from the return stroke. Model currents obtained in this way are listed by Uman et al. (1982). Of particular interest is the relation between the breakdown current pulse $i(t)$, assumed responsible for the peak current i_p at the ground, and the initial few microseconds of either the vertical electric field intensity $E_z(t)$ or the horizontal magnetic flux density $B_\phi(t)$ observed at ground level more than a few kilometers away from the discharge

$$i(t) = \frac{2\pi cr}{\mu_0 v} B_\phi\left(t + \frac{r}{c}\right) \tag{7.5}$$

where at those ranges and for the rapid field variation characteristic of those early times in the waveform, $E_z(t)$ and $B_\phi(t)$ are essentially radiation fields (thus related by $E_z/B_\phi = c$) and v is the return stroke speed. Equation (7.5) can be derived from the last term, the radiation field, of either Eq. (7.3) or Eq. (7.4) [e.g., Uman et al., 1975; Section A.4.3; Eq. (A.50)]. Thus the initial current can be determined from the field if the return stroke speed and distance are known. Some support for the validity of Eq. (7.5) is provided by Fieux *et al.* (1978), Djebari *et al.* (1981), and Weidman *et al.* (1986). Fieux *et al.* (1978) and Djebari *et al.* (1981) made simultaneous measurements of peak current, i_p, and peak magnetic and electric fields, B_p and E_p, respectively, 3 km from 53 subsequent strokes in rocket-initiated lightning flashes in France (Section 12.3.3.1). From the B_p, i_p measurements and Eq. (7.5) they find a mean speed of 1.3×10^8 m/sec with a standard deviation of 0.34×10^8 m/sec; from the E_p, i_p measurements, a mean speed of 1.7×10^8 m/s with a standard deviation of 0.43×10^8 m/sec. These speed determinations are consistent with the photographic measurements of Idone and Orville (1982) who, as noted in Section 7.2.3, found a mean speed of 1.2×10^8 m/sec for natural subsequent return strokes. Weidman *et al.* (1986) performed a similar experiment in Florida in 1985 (Section 12.3.3.4) except that the electric field measurement was made 50 m from the base of a 20-m-high metal structure to which the rocket-initiated subsequent strokes attached and they recorded the time derivatives of the electric field and of the current rather than the field and the current. They found good overall correspondence between the waveshapes of the field derivative and the current derivative. The speeds derived for 31 strokes using the peak electric field derivative and peak current derivative [employing the time derivative of Eq. (7.5) at peak derivative with an assumed constant speed] had both a mean and a median equal to about the speed of light. Twenty-five of the strokes had calculated speeds in the relatively narrow range between 2.6 and 3.6×10^8 m/sec while the peak derivatives varied over a factor of 5. The return stroke speeds obtained by Weidman *et al.* (1986) were about a factor of two greater than those typically measured using photographic techniques (see above). Weidman *et al.* (1986) suggest two possible explanations. (1) The speeds calculated are to be associated with the very bottom of the channel including the metal structure since the peak derivatives occur in 0.1 μsec or less when the return stroke can have propagated only tens of meters. The inference of this view is that both speed determinations from the peak field and peak current which occur later in time than the peak derivatives and from optical measurements effectively consider larger overall channel sections than the derivative measurements and, since the speed decreases with height (e.g., Idone and Orville, 1982), yield lower speeds. Note that Eq. (7.5) is, strictly speaking, not valid unless the speed is constant, but it may nevertheless be a reasonable approximation for a slowly varying speed. (2) The current measured at ground is only half of the total current that produces the measured field because upward- and downward-propagating current waves originate at or near the top of the metal structure. Since both current waves produce similar fields, which are additive, but only one current wave is measured, the speed derived from Eq. (7.5) is about a factor of 2 too high.

As follows from the above and Eq. (7.5), maximum rate of change of initial current can be determined from field measurements if a speed is assumed. Weidman and Krider (1980, 1982) find, for about 100 measurements, a mean maximum rate of change of current of about 180 kA/μsec for an assumed speed of 1×10^8 m/sec, as noted in Section 7.2.2. Current rate-of-change values of the order of 10^{11} A/sec, while typical for currents derived from measured fields, are at the upper limit of measurements made on towers, although as noted in Section 7.2.2 there is some question about the validity of most tower measurements if the risetimes are in the submicrosecond range. A direct comparison of tower and field measurements of current is given by Weidman and Krider (1984).

The model of Lin *et al.* (1980), both with and without the breakdown current pulse decreasing with height, has been used by Master *et al.* (1981) to compute the electric and magnetic fields to be expected in the air. The logarithmic relationship between return-stroke peak light near ground and initial peak electric field observed by Jordan and Uman (1983) coupled with the observed decrease in light with height (Section 7.2.4) implies that the breakdown current pulse decreases with height but not as fast as does the light. On the other hand, Idone and Orville (1985) found peak current and peak light near the ground to be linearly related (Section 7.2.4). It is interesting to note that the fields at the ground are relatively unaffected by whether or not the breakdown current pulse decreases with height since they are radiated during the first few microseconds of the return stroke when the pulse is close to ground and therefore cannot have changed much in amplitude. Jordan and Uman (1983) also found that the return-stroke light output after its initial peak became uniform as a function of height, implying that the treatment of the corona current by Lin *et al.* (1980) (the model corona current decreases with height as illustrated in Fig. 7.13) is not completely adequate.

The model currents used by Master *et al.* (1981) and listed in detail by Uman *et al.* (1982) have been used by Uman *et al.* (1982) to compute the spectrum of the electric field of a severe (5 times typical current) first return stroke at both 50 km and at 50 m for comparison with the spectrum of the electric field from a nuclear weapon exploded above the atmosphere (exoatmospheric NEMP).

Cooray and Lundquist (1983) have calculated the effects of the finitely conducting earth in modifying the initial portion of the vertical electric field waveforms from the values expected over an infinitely conducting earth and find good agreement with the risetime measurements of Uman *et al.* (1976) discussed in Section 7.2.1.

The horizontal electric field E_H near ground level associated with the finite ground conductivity (there is no horizontal field if the conductivity is infinite) can be calculated from the vertical electric field E_z using the "wavetilt" formula in the frequency domain (Master and Uman, 1984)

$$\frac{E_H(j\omega)}{E_z(j\omega)} = \frac{1}{\sqrt{\varepsilon_r + \sigma/j\omega\varepsilon_0}} \tag{7.6}$$

where ε_r is the relative permittivity of the earth, σ the ground conductivity, and ω the angular frequency. Equation (7.6) is applicable to essentially plane-wave radiation generated by a source near ground for which $\varepsilon_r \gg 1$, that is, it is a reasonable approximation for distant lightning or for the early microseconds of close lightning when the return stroke wavefront is near the ground. From Eq. (7.6) it follows that, for the most common types of earth, the horizontal electric fields crudely resemble the time derivatives of the vertical fields with peak values between 0.1 and 0.01 of the vertical field peaks.

In the modeling discussed above, the lightning channel is treated as if it were a straight vertical antenna. Actual lightning is characterized by tortuosity on a scale from less than 1 m to over 1 km (e.g., Evans and Walker, 1963; Hill, 1968). Hill (1969) and LeVine and Meneghini (1978a), using simple models, have investigated the effects of channel tortuosity on distant radiation fields. LeVine and Meneghini (1978a) find that the waveforms computed for the case of the tortuous channel have more fine structure than those for a straight channel, resulting in a frequency spectrum for the tortuous channel that, for the same channel current, has a larger amplitude at frequencies above about 100 kHz.

REFERENCES

MEASURED FIELDS

Baum, C. E., E. L. Breen, J. P. O'Neill, C. B. Moore, and D. L. Hall: Measurement of Electromagnetic Properties of Lightning with 10 Nanosecond Resolution. "Lightning Technology," pp. 39–84. NASA Conference Publication 2128, FAA-RD-80-30, April 1980.

Baum, R. K.: Airborne Lightning Characterization. "Lightning Technology," pp. 153–172. NASA Conference Publication 2128, FAA-RD-80-30, April 1980.

Brook, M., and N. Kitagawa: Radiation from Lightning Discharges in the Frequency Range 400 to 1000 Mc/s. J. Geophys. Res., 69:2431–2434 (1964).

Brook, M., N. Kitagawa, and E. J. Workman: Quantitative Study of Strokes and Continuing Currents in Lightning Discharges to Ground. J. Geophys. Res., 67:649–657 (1962).

Clegg, R. J., and E. M. Thomson: Some Properties of EM Radiation from Lightning. J. Geophys. Res., 84:719–724 (1979).

Cooray, V.: Further Characteristics of Positive Radiation Fields from Lightning in Sweden. J. Geophys. Res., 89:11,807–11,815 (1984).

Cooray, V.: Temporal Behavior of Lightning HF Radiation at 3 MHz near the Time of First Return Strokes. J. Atmos. Terr. Phys., 48:73–78 (1986).

Cooray, V., and S. Lundquist: On the Characteristics of Some Radiation Fields from Lightning and Their Possible Origin in Positive Ground Flashes. *J. Geophys. Res.*, **87**:11,203-11,214 (1982).

Cooray, V., and S. Lundquist: Effects of Propagation on the Rise Times and the Initial Peaks of Radiation Fields from Return Strokes. *Radio Sci.*, **18**:409-415 (1983).

Cooray, V., and S. Lundquist: Characteristics of the Radiation Fields from Lightning in Sri Lanka in the Tropics. *J. Geophys. Res.*, **90**:6099-6109 (1985).

Djebari, B., J. Hamelin, C. Leteinturier, and J. Fontaine: Comparison between Experimental Measurements of the Electromagnetic Field Emitted by Lightning and Different Theoretical Models—Influence of the Upward Velocity of the Return Stroke. Electromagnetic Compatability Symposium, 4th, Zurich, March, 1981. Available from T. Dvorak, ETH Zentrum-IKT, CH-8092 Zurich, Switzerland.

Fisher, R. J., and M. A. Uman: Measured Electric Fields Risetimes for First and Subsequent Lightning Return Strokes. *J. Geophys. Res.*, **77**:399-406 (1972).

Gupta, S. N.: Distribution of Peaks in Atmospheric Radio Noise. *IEEE Trans. Electromagn. Compat.*, **EMC-15**:100-103 (1973).

Hart, J. E.: VLF Radiation from Multiple Stroke Lightning. *J. Atmos. Terr. Phys.*, **29**:1011-1014 (1967).

Himley, R. O.: VLF Radiation from Subsequent Return Strokes in Multiple Stroke Lightning. *J. Atmos. Terr. Phys.*, **31**:749-753 (1969).

Kitagawa, N., M. Brook, and E. J. Workman: Continuing Currents in Cloud-to-Ground Lightning Discharges. *J. Geophys. Res.*, **67**:637-647 (1962).

Krehbiel, P. R., M. Brook, and R. A. McCrory: An Analysis of the Charge Structure of Lightning Discharges to Ground. *J. Geophys. Res.*, **84**:2432-2455 (1979).

Krider, E. P., and C. Guo: The Peak Electromagnetic Power Radiated by Lightning Return Strokes. *J. Geophys. Res.*, **88**:8471-8474 (1983).

LeVine, D. M., and E. P. Krider: The Temporal Structure of HF and VHF Radiation during Florida Lightning Return Strokes. *Geophys. Res. Lett.*, **4**:13-16 (1977).

LeVine, D. M., L. Gesell, and M. Kao: Radiation from Lightning Return Strokes over a Finitely Conducting Earth. *J. Geophys. Res.*, **91**:11,897-11,908 (1986).

Lin, Y. T., and M. A. Uman: Electric Radiation Fields of Lightning Return Strokes in Three Isolated Florida Thunderstorms. *J. Geophys. Res.*, **78**:7911-7915 (1973).

Lin, Y. T., M. A. Uman, J. A. Tiller, R. D. Brantley, W. H. Beasley, E. P. Krider, and C. D. Weidman: Characterization of Lightning Return Stroke Electric and Magnetic Fields from Simultaneous Two-Station Measurements. *J. Geophys. Res.*, **84**:6307-6314 (1979).

Lin, Y. T., M. A. Uman, and R. B. Standler: Lightning Return Stroke Models. *J. Geophys. Res.*, **85**:1571-1583 (1980).

McDonald, T. B., M. A. Uman, J. A. Tiller, and W. H. Beasley: Lightning Location and Lower-Ionosphere Height Determination from Two-Station Magnetic Field Measurements. *J. Geophys. Res.*, **84**:1727-1734 (1979).

Malan, D. J.: Radiation from Lightning Discharges and Its Relation to the Discharge Process. *In* "Recent Advances in Atmospheric Electricity" (L. G. Smith, ed.), pp. 557-563. Pergamon, New York, 1958.

Master, M. J., M. A. Uman, W. H. Beasley, and M. Darveniza: Lightning Induced Voltages on Power Lines: Experiment. *IEEE Trans. PAS*, **PAS-103**:2519-2529 (1984).

Nakai, T.: On the Time and Amplitude Properties of Electric Field near Sources of Lightning in the VLF, LF, and HF Bands. *Radio Sci.*, **12**:389-396 (1977).

Nanevicz, J. E., and E. F. Vance: Analysis of Electrical Transients Created by Lightning. SRI4026, NASA Contractor Report 159308, July 1980.

Peckham, D. W., M. A. Uman, and C. E. Wilcox: Lightning Phenomenology in the Tampa Bay Area. *J. Geophys. Res.*, **89**:11,789-11,805 (1984).

Pitts, F. L.: Electromagnetic Measurement of Lightning Strikes to Aircraft. AIAA 19th Aerospace Sciences Meeting, St. Louis, Missouri, January 1981.

Pitts, F. L., and M. E. Thomas: 1980 Direct Strike Lightning Data. NASA Technical Memorandum 81946, Langley Research Center, Hampton, Virginia, February 1981.

Takagi, M.: VHF Radiation from Ground Discharges. *In* "Planetary Electrodynamics" (S. C. Coroniti and J. Hughes, eds.), pp. 535-538. Gordon & Breach, New York, 1969a.

Takagi, M.: VHF Radiation from Ground Discharges. *Proc Res. Inst. Atmos., Nagoya Univ. (Japan)*, **16**:163-168 (1969b).

Takagi, N., and T. Takeuti: Oscillating Bipolar Electric Field Changes Due to Close Lightning Return Strokes. *Radio Sci.*, **18**:391-398 (1983).

Tiller, J. A., M. A. Uman, Y. T. Lin, R. D. Brantley, and E. P. Krider: Electric Field Statistics for Close Lightning Return Strokes near Gainesville, Florida. *J. Geophys. Res.*, **81**:4430-4434 (1976).

Uman, M. A.: Lightning Return Stroke Electric and Magnetic Fields. *J. Geophys. Res.*, **90**:6121-6130 (1985).

Uman, M. A., D. K. McLain, R. J. Fisher, and E. P. Krider: Electric Field Intensity of the Lightning Return Stroke. *J. Geophys. Res.*, **78**:3523-3529 (1973).

Uman, M. A., R. D. Brantley, J. A. Tiller, Y. T. Lin, E. P. Krider, and D. K. McLain: Correlated Electric and Magnetic Fields from Lightning Return Strokes. *J. Geophys. Res.*, **80**:373-376 (1975).

Uman, M. A., C. E. Swanberg, J. A. Tiller, Y. T. Lin, and E. P. Krider: Effects of 200 km Propagation on Florida Lightning Return Stroke Electric Fields. *Radio Sci.*, **11**:985-990 (1976).

Weidman, C. D., and E. P. Krider: Submicrosecond Risetimes Lightning Radiation Fields. "Lightning Technology," pp. 29-38. NASA Conference Publication 2128, FAA-RD-80-30, April 1980.

Weidman, C. D., and E. P. Krider: The Fine Structure of Lightning Return Stroke Wave Forms. *J. Geophys. Res.*, **83**:6239-6247 (1978). Correction, *J. Geophys. Res.*, **87**:7351 (1982).

Weidman, C. D., and E. P. Krider: Submicrosecond Risetimes in Lightning Return Stroke Fields. *Geophys. Res. Lett.*, **7**:955-958 (1980). Correction, *J. Geophys. Res.*, **87**:7351 (1982).

Weidman, C. D., and E. P. Krider: Correction, *J. Geophys. Res.*, **87**:7351 (1982).

Weidman, C. D.: The Submicrosecond Structure of Lightning Radiation Fields. Ph.D. thesis, University of Arizona, Tucson, Arizona, 1982.

Weidman, C. D., and E. P. Krider: Variations à l'Échelle Submicroseconde des Champs Électromagnétiques Rayonnés par la Foudre. *Ann. Telecommun.*, **39** (5-6):165-174 (1984).

FREQUENCY SPECTRA

Barlow, J. S., G. W. Frey, Jr., and J. B. Newman: Very Low Frequency Noise Power from the Lightning Discharge. *J. Franklin Inst.*, **17**:115-162 (1954)

Bradley, P. A.: The Spectra of Lightning Discharges at Very Low Frequencies. *J. Atmos. Terr. Phys.*, **26**:1069-1073 (1964).

Bradley, P. A.: The VLF Energy Spectra of First and Subsequent Return Strokes of Multiple Lightning Discharges to Ground. *J. Atmos. Terr. Phys.*, **27**:1045-1053 (1965).

Croom, D. L.: The Spectra of Atmospherics and Propagation of Very Low Frequency Radio Waves. Ph.D. thesis, University of Cambridge, Cambridge, England, 1961.

Croom, D. L.: The Frequency Spectra and Attenuation of Atmospherics in the Range 1-15 kc/s. *J. Atmos. Terr. Phys.*, **26**:1015-1046 (1964).

Dennis, A. S., and E. T. Pierce: The Return Stroke of the Lightning Flash to Earth as a Source of VLF Atmospherics. *Radio Sci.*, **68D**:779-794 (1964).

Galejs, J.: Amplitude Statistics of Lightning Discharge Currents and ELF and VLF Radio Noise. *J. Geophys. Res.*, **72**:2943-2953 (1967).

Garg, M. B., K. C. Mathpal, J. Rai, and N. C. Varshneya: Frequency Spectra of Electric and Magnetic Fields of Different Forms of Lightning. *Ann. Geophys.*, **38**:177-188 (1982).

Harwood, J., and B. N. Harden: The Measurement of Atmospheric Radio Noise by an Aural Comparison Method in the Range 15-500 kc/s. *Proc. Inst. Electr. Eng. (London)*, **B107**:53-59 (1960).

Horner, F.: Narrow Band Atmospherics from Two Local Thunderstorms. *J. Atmos. Terr. Phys.*, **21**:13-25 (1961).

Horner, F.: Atmospherics of near Lightning Discharges. *In* "Radio Noise of Terrestrial Origin" (F. Horner, ed.), pp. 16-17. American Elsevier, New York, 1962.

Horner, F.: Radio Noise from Thunderstorms. *In* "Advances in Radio Research" (J. A. Saxton, ed.), pp. 122-215. Academic Press, New York, 1964.

Horner, F., and P. A. Bradley: The Spectra of Atmospherics from near Lightning. *J. Atmos. Terr. Phys.*, **26**:1155-1166 (1964).

Le Boulch, M., and J. Hamelin: Rayonnement en Ondes Métriques et Décimétriques des Orages. *Ann. Telecommun.*, **40**:277-313 (1985).

Levine, D., and E. P. Krider: The Temporal Structure of HF and VHF Radiation during Florida Lightning Return Strokes. *Geophys. Res. Lett.*, **4**:13-16 (1977).

Marney, G. O., and K. Shanmugam: Effect on Channel Orientation on the Frequency Spectrum of Lightning Discharges. *J. Geophys. Res.*, **76**:4198-4202 (1971).

Maxwell, E. L.: Atmospheric Noise from 20 Hz to 30 KHz. *Radio Sci.*, **2**:637-644 (1967).

Maxwell, E. L., and D. L. Stone: Natural Noise Fields from 1 cps to 100 kc. *IEEE Trans. Antenna Propag.*, **AP-11**:339-343(1963).

Obayashi, T.: Measured Frequency Sprectra of VLF Atmospherics. *J. Res. Nat. Bur. Stand.*, **64D**:41-48 (1960).

Preta, J., M. A. Uman, and D. G. Childers: Comment on "The Electric Field Spectra of First and Subsequent Lightning Return Strokes in the 1- to 200-km Range" by Serhan *et al. Radio Sci.*, **20**:143-145 (1985).

Serhan, G. I., M. A. Uman, D. G. Childers, and Y. T. Lin: The RF Spectra of First and Subsequent Lightning Return Strokes in the 1-200 km Range. *Radio Sci.*, **15**:1089-1094 (1980).

Shumpert, T. H., M. A. Honnell, and G. K. Lott, Jr.: Measured Spectral Amplitude of Lightning Sferics in the HF, VHF, and UHF Bands. *IEEE Trans. Electromagn. Compat.*, **EMC-24**:368-372 (1982).

Steptoe, B. J.: Some Observations on the Spectrum and Propagation of Atmospherics. Ph.D. thesis, Univeristy of London, London, 1958.

Taylor, W. L.: Radiation Field Characteristics of Lightning Discharges in the Band 1 kc/s to 100 kc/s. *J. Res. Nat. Bur. Stand.*, **67D**:539-550 (1963).

Taylor, W. L., and A. G. Jean: Very Low Frequency Radiation Spectra of Lightning Discharges. *J. Res. Bur. Stand.*, **63D**:199-204 (1959).

Watt, A. D., and E. L. Maxwell: Characteristics of Atmospherics Noise from 1 to 100 kc. *Proc. Inst. Radio Eng.*, **45**:55-62 (1957).

Weidman, C. D.: The Submicrosecond Structure of Lightning Radiation Fields. Ph.D. thesis, University of Arizona, Tucson, Arizona, 1982.

Weidman, C. D., and E. P. Krider: The Amplitude Spectra of Lightning Radiation Fields in the Interval from 1 to 20 MHz. *Radio Sci.*, **21**:964-970 (1986).

Weidman, C. D., E. P. Krider, and M. A. Uman: Lightning Amplitude Spectra in the Interval from 100 kHz to 20 MHz. *Geophys. Res. Lett.*, **8**:931-934 (1981).

Williams, J. C.: Thunderstorms and VLF Radio Noise. Ph.D. thesis, Harvard University, Cambridge, Massachusetts, 1959.

CURRENT

Berger, K.: Lightning Research in Switzerland. *Weather*, **2**:231-238 (1947).

Berger, K.: Die Messeinrichtungen für die Blitzforschung auf dem Monte San Salvatore. *Bull. Schweiz. Elektrotech. Ver.*, **46**:193-204 (1955a).

Berger, K.: Resultate de Blitzmessungen der Jahre 1947-1954 auf dem Monte San Salvatore. *Bull. Schweiz. Elektrotech. Ver.*, **46**:405-424 (1955b).

Berger, K.: Front Duration and Current Steepness of Lightning Strokes to Earth. *In* "Gas Discharges and the Electricity Supply Industry" (J. S. Forrest, P. R. Howard, and D. J. Littler, eds.), pp. 63-73. Butterworths, London, 1962.

Berger, K.: Gewitterforschung auf dem Monte San Salvatore. *Elektrotechnik*, **Z-A82**:249-260 (1967a).

Berger, K.: Novel Observations on Lightning Discharges: Results of Research on Mount San Salvatore. *J. Franklin Inst.*, **283**:478-525 (1967b).

Berger, K.: Methoden und Resultate der Blitzforschung auf dem Monte San Salvatore bei Lugano in den Jahren 1963-1971. *Bull. Schweiz. Elektrotech. Ver.*, **63**:1403-1422 (1972).

Berger, K.: Extreme Blitzströme und Blitzschutz. *Bull. ASE/USC*, **71**:460-464 (1980).

Berger, K., and E. Vogelsanger: Messungen und Resultate der Blitzforschung der Jahre 1955-1963, auf dem Monte San Salvatore. *Bull. Schweiz. Electrotech. Ver.*, **56**:2-22 (1965).

Berger, K., and E. Vogelsanger: New Results of Lightning Observations. *In* "Planetary Electrodynamics" (S. C. Coroniti and J. Hughes, eds.), pp. 489-510. Gordon & Breach, New York, 1969.

Berger, K., R. B. Anderson, and H. Kroninger: Parameters of Lightning Flashes. *Electra*, **80**:23-37 (1975).

Davis, R., and W. G. Standring: Discharge Currents Associated with Kite Balloons. *Proc. R. Soc. London Ser. A*, **191**:304-322 (1947).

Eriksson, A. J.: Lightning and Tall Structures. *Trans. South Afr. IEE*, **69** (Pt. 8):238-252 (1978).

Eriksson, A. J.: The Lightning Ground Flash—An Engineering Study. Ph.D. thesis, Univeristy of Natal, South Africa, December 1979. Available on CSIR Special Report ELEK 189 dated 1980 from National Electrical Engineering Research Institute, P.O. Box 398, Pretoria, 0001 South Africa.

Eriksson, A. J.: The CSIR Lightning Research Mast-Data for 1972-1982. NEERI Internal Report No. Ek/9/82, National Electrical Engineering Research Institute, P.O. Box 395, Pretoria, 0001 South Africa, August 1982.

Fieux, R. P., C. H. Gary, B. P. Hutzler, A. R. Eybert-Berard, P. L. Hubert, A. C. Meesters, P. H. Perroud, J. H. Hamelin, and J. M. Person: Research on Artificially Triggered Lightning in France. *IEEE Trans. Power Appar. Syst.*, **PAS 97**:725-733 (1978).

Garbagnati, E., and G. B. Lo Pipero: Stazione Sperimentale per il Rilievo delle Caratteristiche dei Fulmini. *Electrotecnica*, **LVII**:288-297 (1970).

Garbagnati, E., and G. B. Lo Piparo: Nuova Stazione Automatica per il Rilievo delle Caratteristiche dei Fulmini. *Energ. Elettr.*, **6**:375-383 (1973).

Garbagnati, E., and G. B. Lo Piparo: Parameter von Blitzströmen. *Elektrotech. Z. etz-a*, **103**:61-65 (1982).

Garbagnati, E., E. Giudice, G. B. Lo Piparo, and U. Magagnoli: Relieve delle Caratteristiche dei Fulmini in Italia. Risultati Ottenuti Negli Anni 1970–1973. *Electtrotecnica*, **LXII**: 237–249 (1975).

Garbagnati, E., E. Giudice, and G. B. Lo Piparo: Messung von Blitzströmen in Italien— Ergebnisse einer Statistischen Auswertung. *Elektrotech. Z. etz-a*, **99**: 664–668 (1978).

Gilchrist, J. H., and J. B. Thomas: A Model for the Current Pulses of Cloud-to-Ground Lightning Discharges. *J. Franklin Inst.*, **299**: 199–210 (1975).

Hagenguth, J. H., and J. G. Anderson: Lightning to the Empire State Building. *AIEE*, **71** (Pt. 3): 641–649 (1952).

Hylten-Cavallius, N., and A. Stromberg: The Amplitude, Time to Half-Value, and the Steepness of the Lightning Currents. *ASEA J.*, **29**: 129–134 (1956).

Hylten-Cavallius, N., and A. Stromberg: Field Measurement of Lightning Currents. *Elteknik*, **2**: 109–113 (1959).

Lee, L. D., C. B. Finelli, M. E. Thomas, and F. L. Pitts: 1980 to 1982 Statistical Analysis of Direct-Strike Lightning Data. NASA Technical Memorandum 2252, Langley Research Center, Hampton, Virginia, January 1984.

Lewis, W. W., and C. M. Foust: Lightning Investigation on Transmission Lines, Pt. 7. *Trans. AIEE*, **64**: 107–115 (1945).

McCann, D. G.: The Measurement of Lightning Currents in Direct Strokes. *Trans. AIEE*, **63**: 1157–1164 (1944).

McEachron, K. B.: Lightning to the Empire State Building. *J. Franklin Inst.*, **227**: 149–217 (1939).

McEachron, K. B.: Wave Shapes of Successive Lightning Current Peaks. *Electr. World*, **56**: 428–431 (1940).

McEachron, K. B.: Lightning to the Empire State Building. *Trans. AIEE*, **60**: 885–889 (1941).

Nakahori, K., T. Egawa, and H. Mitani: Characteristics of Winter Lightning Currents in Hokuriku District. *IEEE Trans Power Appar. Syst.*, **PAS-101**: 4407–4412 (1982).

Petterson, B. J., and W. R. Wood: Measurements of Lightning Strikes to Aircraft. Report No. SC-M-67-549, Sandia Laboratories, Albuquerque, New Mexico, 1968.

Pitts, F. L.: Electromagnetic Measurement of Lightning Strikes to Aircraft. *AIAA J. Aircraft*, **19**: 246–250 (1982).

Pitts, F. L., and M. E. Thomas: 1980 Direct Strike Lightning Data. NASA Technical Memorandum 81946, Langley Research Center, Hampton, Virginia, February 1981.

Pitts, F. L., and M. E. Thomas: 1981 Direct Strike Lightning Data. NASA Technical Memorandum 83273, Langley Research Center, Hampton, Virginia, March 1982.

Planning Research Corp., JFK Space Center: Historical Log of Lightning Strikes at NASA/KSC Launch Areas May to October 1981. Prepared by Planning Research Corp., Systems Services Co. for Design Engineering, JFK Space Center, NASA, PRC Control No. 344-1217, May 1982.

Popolansky, F.: Frequency Distribution of Amplitudes of Lightning Currents. *Electra*, **22**: 139–147 (1972).

Rustan, P. L.: The Lightning Threat to Aerospace Vehicles. *AIAA J. Aircraft*, **23**: 62–67 (1986).

Sargent, M. A.: The Frequency Distribution of Current Magnitudes of Lightning Strokes to Tall Structures. *IEEE Trans. Power Appar. Syst.*, **PAS-91**: 2224–2229 (1972).

Szpor, S.: Comparison of Polish versus American Lightning Records. *IEEE Trans. Power Appar. Syst.*, **PAS-88**: 646–652 (1969).

Thomas, M. E., and F. L. Pitts: 1982 Direct Strike Lightning Data. NASA Technical Memorandum 84626, Langley Research Center, Hampton, Virginia, March 1983.

Thomas, M. E.: 1983 Direct Strike Lightning Data. NASA Technical Memorandum 86426, Langley Research Center, Hampton, Virginia, August 1985.

Thomas, M. E., and H. E. Carney: 1984 Direct Strike Lightning Data. NASA Technical Memorandum 87690, Langley Research Center, Hampton, Virginia, March 1986.

Uman, M. A., D. K. McLain, R. F. Fisher, and E. P. Krider: Currents in Florida Lightning Strokes. *J. Geophys. Res.*, **78**:3530-3537 (1973).

Weidman, C. D., and E. P. Krider: The Fine Structure of Lightning Return Stroke Wave Forms. *J. Geophys. Res.*, **83**:6239-6247 (1978). Correction, *J. Geophys. Res.*, **87**:7351 (1982).

Weidman, C. D., and E. P. Krider: Submicrosecond Rise Time in Lightning Return Stroke Fields. *Geophys. Res. Lett.*, **7**:955-958 (1980). Correction, *J. Geophys. Res.*, **87**:7351 (1982).

Weidman, C. D., and E. P. Krider: Correction, *J. Geophys. Res.*, **87**:7351 (1982).

Weidman, C. D., and E. P. Krider: Variations à l'Échelle Submicroseconde des Champs Électromagnétiques Rayonnés par la Foudre. *Ann. Telecommun.*, **39**:165-174 (1984).

Zaepfel, K. P., B. D. Fisher, and M. S. Ott: Direct Strike Lightning Photographs, Swept-Flash Attachment Patterns, and Flight Conditions for Storm Hazards '82. NASA Technical Memorandum 86347, Langley Research Center, Hampton, Virginia, February 1985.

SPEED

Boyle, J. S., and R. E. Orville: Return Stroke Velocity Measurement in Multistroke Lightning Flashes. *J. Geophys. Res.*, **81**:4461-4466 (1976).

Hubert, P., and G. Mouget: Return Stroke Velocity Measurements in Two Triggered Lightning Flashes. *J. Geophys. Res.*, **86**:5253-5261 (1981).

Idone, V. P., and R. E. Orville: Lightning Return Stroke Velocities in the Thunderstorm Research International Program (TRIP). *J. Geophys. Res.*, **87**:4903-4915 (1982).

Idone, V. P., R. E. Orville, P. Hubert, L. Barret, and A. Eybert-Berard: Correlated Observations of Three Triggered Lightning Flashes. *J. Geophys. Res.*, **89**:1385-1394 (1984).

Jordan, D., and M. A. Uman: Variations of Light Intensity with Height and Time from Subsequent Return Strokes. *J. Geophys. Res.*, **88**:6555-6562 (1983).

Lundholm, R.: Induced Overvoltage Surges on Transmission Lines. *Chalmers Tek. Hoegsk. Handl.*, **188**:1-117 (1957).

Orville, R. E., G. G. Lala, and V. P. Idone: Daylight Time Resolved Photographs of Lightning. *Science*, **201**:59-61 (1978).

McEachron, K. B.: Photographic Study of Lightning. *AIEE Trans.*, **66**:577-585 (1947).

Radda, G. J., and E. P. Krider: Photoelectric Measurements of Lightning Return Stroke Propagation Speeds. *Trans. Am. Geophys. Union*, **56**:1131 (1974).

Schonland, B. F. J., and H. Collens: Progressive Lightning. *Proc. R. Soc. London Ser. A*, **143**:654-674 (1934).

Schonland, B. F. J., D. J. Malan, and H. Collens: Progressive Lightning II. *Proc. R. Soc. London Ser. A*, **152**:595-625 (1935).

Wagner, C. F.: The Relation between Stroke Current and the Velocity of the Return Stroke. *Trans. AIEE Power Appar. Syst.*, **82**:609-617 (1963).

OPTICAL SPECTRA AND LUMINOSITY

Barasch, G. E.: The 1965 ARPA-AEC Joint Lightning Study at Los Alamos, Vol. 2, The Lightning Spectrum as Measured by Collimated Detectors; Atmospheric Transmission; Spectral Intensity Radiated. Los Alamos Science Laboratory Report LA-3755, February 9, 1968.

Barasch, G. E.: The 1965 ARPA-AEC Joint Lightning Study at Los Alamos, Vol. 4, Discrimination against False Triggering of Air Fluorescence Detection Systems by Lightning. Los Alamos Science Laboratory Report LA-3757, April 3, 1969.

Barasch, G. E.: Spectral Intensities Emitted by Lightning Discharges. *J. Geophys. Res.*, **75**:1049–1057 (1970).

Braginskii, S. I.: Theory of the Development of a Spark Channel. *Sov. Phys. JETP (Engl. Transl.)*, **34**:1068–1074 (1958).

Brook, M., and N. Kitagawa: Electric Field Changes and the Design of Lightning-Flash Counters. *J. Geophys. Res.*, **65**:1927–1931 (1960).

Connor, T. R.: The 1965 ARPA–AEC Joint Lightning Study at Los Alamos, Vol. 1, The Lightning Spectrum; Charge Transfer in Lightning; Efficiency of Conversion of Electrical Energy into Visible Radiation. Los Alamos Science Laboratory Report LA-3754, December 5, 1967.

Connor, T. R.: Stroke-and-Space-Resolved Slit Spectra of Lightning. Los Alamos Science Laboratory Report LA-3754 Addendum, August 2, 1968.

Conner, T. R., and G. E. Barasch: The 1965 ARPA–AEC Joint Lightning Study at Los Alamos, Vol. 3, Comparison of the Lightning Spectrum as Measured by All-Sky and Narrow-Field Detectors; Propagation of Light from Lightning to All-Sky Detectors. Los Alamos Science Laboratory Report A-3756, July 25, 1968.

Drellishak, K. S.: Partition Functions and Thermodynamic Properties of High Temperature Gases. Defense Documentation Center AD 428 210, January 1964.

Edano, Y., S. Kawamata, and Y. Nagai: Observation of Simultaneous-Bi-Stroke Flashes. *Res. Lett. Atmos. Electr.*, **2**:49–52 (1982).

Ganesh, C., M. A. Uman, W. H. Beasley, and D. M. Jordan: Correlated Optical and Electric Field Signals Produced by Lightning Return Strokes. *J. Geophys. Res.*, **89**:4905–4909 (1984).

Guo, C., and E. P. Krider: The Optical and Radiation Field Signatures Produced by Lightning Return Strokes. *J. Geophys. Res.*, **87**:8913–8922 (1982).

Guo, C., and E. P. Krider: The Optical Power Radiated by Lightning Return Strokes. *J. Geophys. Res.*, **88**:8621–8622 (1983).

Hill, E. L., and J. D. Robb: Spectroscopic and Thermal Temperatures in the Return Stroke of Lightning. *J. Geophys. Res.*, **74**:3426–3430 (1969).

Hill, R. D.: Determination of Charges Conducted in Lightning Strokes. *J. Geophys. Res.*, **68**:1365–1375 (1963).

Hill, R. D.: Channel Heating in Return Stroke Lightning. *J. Geophys. Res.*, **76**:637–645 (1971).

Hill, R.D.: Optical Absorption in the Lightning Channel. *J. Geophys. Res.*, **77**:2642–2647 (1972).

Hill, R. D.: Comments on "Quantitative Analysis of a Lightning Return Stroke for Diameter and Luminosity Changes as a Function of Space and Time" by Richard E. Orville, John H. Helsdon, Jr., and Walter H. Evans. *J. Geophys. Res.*, **80**:1188 (1975).

Hill, R. D.: Energy Dissipation in Lightning. *J. Geophys. Res.*, **82**:4967–4968 (1977).

Hill, R. D.: Comment on "Detection of Lightning Superbolts" by B. N. Turman. *J. Geophys. Res.*, **83**:1381–1382 (1978a).

Hill, R. D.: Reply, *J. Geophys. Res.*, **83**:5524 (1978b).

Idone, V. P., and R. E. Orville: Lightning Return Stroke Velocities in the Thunderstorm Research International Program (TRIP). *J. Geophys. Res.*, **87**:4903–4916 (1982).

Idone, V. P., and R. E. Orville: Correlated Peak Relative Light Intensity and Peak Current in Triggered Lightning Subsequent Return Strokes. *J. Geophys. Res.*, **90**:6159–6164 (1985).

Jones, R. C.: Return Stroke Core Diameter. *J. Geophys. Res.*, **73**:809–814 (1968).

Jordan, D. M., and M. A. Uman: Variation in Light Intensity with Height and Time from Subsequent Lightning Return Strokes. *J. Geophys. Res.*, **88**:6555–6562 (1983).

Kitagawa, N., and M. Kobayashi: Field Changes and Variations of Luminosity Due to Lightning Flashes. *In* "Recent Advances in Atmospheric Electricity" (L. G. Smith, ed.), pp. 485–501. Pergamon, Oxford, 1959.

Krider, E. P.: Some Photoelectric Observations of Lightning. *J. Geophys. Res.*, **71**:3095-3098 (1966).

Krider, E. P.: Lightning Spectroscopy. *Nucl. Instrum. and Methods*, **110**:411-419 (1973).

Krider, E. P., and C. Guo: The Peak Electromagnetic Power Radiated by Lightning Return Strokes. *J. Geophys. Res.*, **88**:8471-8474 (1983).

Krider, E. P., and G. Marcek: A Simplified Technique for the Photography of Lightning in Daylight. *J. Geophys. Res.*, **77**:6017-6020 (1972).

Lundquist, S., and V. Scuka: Some Time Correlated Measurements of Optical and Electromagnetic Radiation from Lightning Flashes. *Ark. Geophys.*, **5**:585-593 (1970).

Mackerras, D.: Photoelectric Observations of the Light Emitted by Lightning Flashes. *J. Atmos. Terr. Phys.*, **35**:521-535 (1973).

Oetzel, G. N.: Computation of the Diameter of a Lightning Return Stroke. *J. Geophys. Res.*, **73**:1889-1896 (1968).

Orville, R. E.: A High-Speed Time-Resolved Spectroscopic Study of the Lightning Return Stroke: Part I. A Quantitative Analysis. *J. Atmos. Sci.*, **25**:827-838 (1968a).

Orville, R. E.: A High-Speed Time-Resolved Spectroscopic Study of the Lightning Return Stroke: Part II. A Quantitative Analysis. *J. Atmos. Sci.*, **25**:839-851 (1968b).

Orville, R. E.: A High-Speed Time-Resolved Spectroscopic Study of the Lightning Return Stroke: Part III. A Time-Dependent Model. *J. Atmos. Sci.*, **25**:852-856 (1968c).

Orville, R. E.: Spectrum of the Lightning Stepped Leader. *J. Geophys. Res.*, **73**:6999-7008 (1968d).

Orville, R. E.: Spectrum of the Lightning Dart Leader. *J. Atmos. Sci.*, **32**:1829-1837 (1975).

Orville, R. E.: Lightning Spectroscopy. *In* "Lightning, Vol. 1, Physics of Lightning" (R. H. Golde, ed.), pp. 281-306. Academic Press, New York, 1977.

Orville, R. E.: Daylight Spectra of Individual Lightning Flashes in the 370-690 nm Region. *J. Appl. Meteorol.*, **19**:470-473 (1980).

Orville, R. E., and R. W. Henderson: Absolute Spectral Irradiance Measurements of Lightning from 375 to 880 nm. *J. Atmos. Sci.*, **41**:3280-3187 (1984).

Orville, R. E., and L. E. Salanave: Lightning Spectroscopy-Photographic Techniques. *Appl. Opt.*, **9**:1775-1781 (1970).

Orville, R. E., M. A. Uman, and A. M. Sletten: Temperature and Electron Density in Long Air Sparks. *J. Appl. Phys.*, **38**:895 (1967).

Paxton, A. H., R. L. Gardner, and L. Baker: Lightning Return Stroke. A Numerical Calculation of the Optical Radiation. *Phys. Fluids*, **29**:2736-2741 (1986).

Plooster, M. N.: Numerical Model of the Return Stroke of the Lightning Discharge. *Phys. Fluids*, **14**:2124-2133 (1971).

Prueitt, M. L.: The Excitation Temperature of Lightning. *J. Geophys. Res.*, **68**:803 (1963).

Salanave, L. E.: "Lightning and Its Spectrum." Univ. of Arizona Press, Tucson, Arizona, 1980.

Schonland, B. F. J.: "The Flight of Thunderbolts," p. 63. Oxford Univ. Press, London and New York, 1950.

Scuka, V.: Electronic Optical System for Lightning Research. *Ark. Geophys.*, **38**:569-584 (1969).

Taylor, A. R.: Diameter of Lightning as Indicated by Tree Scars. *J. Geophys. Res.*, **70**:5693-5695 (1965).

Thomason, L. W., and E. P. Krider: The Effects of Clouds on the Light Produced by Lightning. *J. Atmos. Sci.*, **39**:2051-2065 (1982).

Turman, B. N.: Detection of Lightning Superbolts. *J. Geophys. Res.*, **83**:2566-2568 (1977).

Turman, B. N.: Analysis of Lightning Data from DMSP Satellite. *J. Geophys. Res.*, **83**:5019-5024 (1978).

Uman, M. A.: The Peak Temperature of Lightning. *J. Atmos. Terr. Phys.*, **26**:123 (1964a).

Uman, M. A.: The Diameter of Lightning. *J. Geophys. Res.*, **69**:583–585 (1964b).

Uman, M. A.: Quantitative Lightning Spectroscopy. *IEEE Spectrum*, **3** (August):102–110 (1966). See also, **3** (October):154 (1966).

Uman, M. A.: On the Determination of Lightning Temperature. *J. Geophys. Res.*, **74**:949–957 (1969).

Uman, M. A.: Criticism of "Comment on 'Detection of Lightning Superbolts'" by B. N. Turman. *J. Geophys. Res.*, **83**:5523 (1978).

Uman, M. A.: "Lightning," pp. 138–180. McGraw-Hill, New York, 1969. See also, Dover, New York, 1984.

Uman, M. A., and R. E. Orville: Electron Density Measurement in Lightning from Stark-Broadening of H_α. *J. Geophys. Res.*, **69**:5151–5154 (1964).

Uman, M. A., and R. E. Orville: The Opacity of Lightning. *J. Geophys. Res.*, **70**:5491 (1965).

Vonnegut, B., O. H. Vaughan, Jr., and M. Brook: Photographs of Lightning from the Space Shuttle. *Bull. Am. Meteorol. Soc.*, **64**:150–151 (1983).

Wolfe, W. L.: Aircraft-Borne Lightning Sensor. *Opt. Eng.*, **22**:456–459 (1983).

MODELS

Albright, N. W., and D. A. Tidman: Ionizing Potential Waves and High Voltage Breakdown Streamers. *Phys. Fluids*, **15**:86–90 (1972).

Barreto, E., H. Jurenka, and S. I. Reynolds: The Formation of Small Sparks. *J. Appl. Phys.*, **48**:4510–4520 (1977).

Braginskii, S. I.: Theory of the Development of a Spark Channel. *Sov. Phys. JEPT (Engl. Transl.)*, **34**:1068–1074 (1958).

Bruce, C. E. R., and R. H. Golde: The Lightning Discharge. *J. Inst. Electr. Eng. (London)*, **88**:487–520 (1941).

Cooray, V., and S. Lundquist: Effects of Propagation on the Rise Times and the Initial Peaks of Radiation Fields from Return Strokes. *Radio Sci.*, **18**:409–415 (1983).

Cravath, A. M., and L. B. Loeb: The Mechanism of the High Velocity of Propagation of Lightning Discharges. *Physics* (now, *J. Appl. Phys.*), **6**:125–127 (1935).

Dennis, A. S., and E. T. Pierce: The Return Stroke of the Lightning Flash to Earth as a Source of VLF Atmospherics. *Radio Sci.*, **68D**:779–794 (1964).

Djebari, B., J. Hamelin, C. Leteinturier, and J. Fontaine: Comparison between Experimental Measurements of the Electromagnetic Field Emitted by Lightning and Different Theoretical Models—Influence of the Upward Velocity of the Return Stroke. Electromagnetic Compatability Symposium, 4th, Zurich, March, 1981. Available from T. Dvorak, ETH Zentrum-IKT, CH-8092 Zurich, Switzerland.

Evans, W. H., and R. L. Walker: High Speed Photographs of Lightning at Close Range. *J. Geophys. Res.*, **68**:4455–4461 (1963).

Fowler, R. G.: "Nonlinear Electron Acoustic Waves, Part I., Adv. Electronics Electron Physics," Vol. 35. Academic Press, New York, 1974. See also, "Nonlinear Electron Acoustic Waves, Part II, Adv. Electronics Electron Physics," Vol. 41. Academic Press, New York, 1976.

Fowler, R. G.: Lightning. *Appl. At. Collision Phys.*, **5**:31–67 (1982).

Gardner, R. L.: Effect of the Propagation Path on Lightning-Induced Transient Fields. *Radio Sci.*, **16**:377–384 (1981).

Gallimberti, I.: The Mechanism of the Long Spark Formation. *J. Phys.*, **40**:C7/193–250 (1979).

Gorbachev, L. P., and V. F. Federov: Electromagnetic Radiation from a Return Streamer of Lightning. *Geomagn. Aeron.*, **17**:641-642 (1977).

Hill, E. L.: Electromagnetic Radiation from Lightning Strokes. *J. Franklin Inst.*, **263**:107-109 (1957).

Hill, R. D.: Electromagnetic Radiation from the Return Stroke of a Lightning Discharge. *J. Geophys. Res.*, **71**:1963-1967 (1966).

Hill, R. D.: Analysis of Irregular Paths of Lightning Channels. *J. Geophys. Res.*, **73**:1897-1905 (1968).

Hill, R. D.: Electromagnetic Radiation from Erratic Paths of Lightning Strokes. *J. Geophys. Res.*, **74**:1922-1929 (1969).

Hill, R. D.: Channel Heating in Return Stroke Lightning. *J. Geophys. Res.*, **76**:637-645 (1971).

Hill, R. D.: Optical Absorption in the Lightning Channel. *J. Geophys. Res.*, **77**:2642-2647 (1972).

Hill, R. D.: Comments on "Quantitative Analysis of a Lightning Return Stroke for Diameter and Luminosity Changes as a Function of Space and Time" by Richard E. Orville, John H. Helsdon, Jr., and Walter H. Evans. *J. Geophys. Res.*, **80**:1188 (1975).

Hill, R. D.: Comments on "Numerical Simulation of Spark Discharges in Air." *Phys. Fluids*, **20**:1584-1586 (1977a).

Hill, R. D.: Energy Dissipation in Lightning. *J. Geophys. Res.*, **82**:4967-4968 (1977b).

Himley, R. O.: VLF Radiation from Subsequent Return Strokes in Multiple Stroke Lightning. *J. Atmos. Terr. Phys.*, **31**:749-753 (1969).

Iwata, A.: Calculations of Waveforms Radiating from Return Strokes. *Proc. Res. Inst. Atmos., Nagoya Univ.*, **17**:115-123 (1970).

Jones, D. L.: Electromagnetic Radiation from Multiple Return Strokes of Lightning. *J. Atmos. Terr. Phys.*, **32**:1077-1093 (1970).

Jones, R. D., and H. A. Watts: Close-In Magnetic Fields of a Lightning Return Stroke. Sandia Laboratory, SAND75-0114, 1975.

Kline, L. E., and J. G. Siambis: Computer Simulation of Electrical Breakdown in Gases. *Phys. Rev. A*, **5**:794-805 (1972).

Klingbeil, R., D. A. Tidman, and R. F. Fernsler: Ionizing Gas Breakdown Waves in Strong Electric Fields. *Phys. Fluids*, **15**:1969-1973 (1972).

Lefferts, R. E.: A Statistical Simulation of Ground-Wave Atmospherics Generated by Lightning Return Strokes. *Radio Sci.*, **13**:121-130 (1978).

Lefferts, R. E.: Probabilistic Model for the Initial Peaks of Ground Wave Atmospherics Generated by Lightning Return Strokes. *Radio Sci.*, **14**:1017-1026 (1979).

Leise, J. A., and W. L. Taylor: A Transmission Line Model with General Velocities for Lightning. *J. Geophys. Res.*, **82**:391-396 (1977).

LeVine, D. M., and R. Meneghini: Simulation of Radiation from Lightning Return Strokes: The Effects of Tortuosity. *Radio Sci.*, **13**:801-809 (1978a).

LeVine, D. M., and R. Meneghini: Electromagnetic Fields Radiated from a Lightning Return Stroke: Application of an Exact Solution to Maxwell's Equations. *J. Geophys. Res.*, **83**:2377-2384 (1978b).

LeVine, D. M., L. Gesell, and M. Kao: Radiation from Lightning Return Strokes over a Finitely Conducting Earth. *J. Geophys. Res.*, **91**:11,897-11,908 (1986).

Lin, Y. T., M. A. Uman, and R. B. Standler: Lightning Return Stroke Models. *J. Geophys. Res.*, **85**:1571-1583 (1980).

Little, P. F.: Transmission Line Representation of a Lightning Return Stroke. *J. Phys. D: Appl. Phys.*, **11**:1893-1910 (1978).

Little, P. F.: The Effect of Altitude on Lightning Hazards to Aircraft. "Proceedings of the 15th European Conference on Lightning Protection," Vol. 2. Institute of High Voltage Research, Uppsala University, Sweden, 1979.

Loeb, L. B.: Ionizing Waves of Potential Gradient. *Science*, **148**:1417-1426 (1965).

Master, M. J., and M. A. Uman: Electric and Magnetic Fields Associated with Establishing a Finite Electrostatic Dipole: An Exercise in the Solution of Maxwell's Equations. *Am. J. Phys.*, **51**:118-126 (1983).

Master, M. J., M. A. Uman, Y. T. Lin, and R. B. Standler: Calculations of Lightning Return Stroke Electric and Magnetic Fields above Ground. *J. Geophys. Res.*, **86**:12,127-12,132(1981).

Meneghini, R.: Application of the Lienard-Wiechert Solution to a Lightning Return Stroke Model. *Radio Sci.*, **19**:1485-1498 (1984).

Norinder, H.: Lightning Currents and Their Variations. *J. Franklin Inst.*, **220**:69-92 (1935).

Oetzel, G. N.: Computation of the Diameter of a Lightning Return Stroke. *J. Geophys. Res.*, **73**:1889-1896 (1968).

Papet-LeVine, J.: Electromagnetic Radiation and Physical Structure of Lightning Discharges. *Ark. Geophys.*, **3**:391-400 (1961).

Pierce, E. T.: Atmospherics from Lightning Flashes with Multiple Strokes. *J. Geophys. Res.*, **65**:1867-1871 (1960).

Plooster, M. N.: Numerical Simulation of Spark Discharges in Air. *Phys. Fluids*, **14**:2111-2123 (1971a).

Plooster, M. N.: Numerical Model of the Return Stroke of the Lightning Discharge. *Phys. Fluids*, **14**:2124-2133 (1971b).

Price, G. H., and E. T. Pierce: The Modeling of Channel Current in the Lightning Return Stroke. *Radio Sci.*, **12**:381-388 (1977).

Rai, J.: Current and Velocity of the Return Lightning Stroke. *J. Atmos. Terr. Phys.*, **40**:1275-1285 (1978).

Rai, J., and P. K. Bhattacharya: Impulse Magnetic Flux Density Close to the Multiple Return Strokes of a Lightning Discharge. *J. Phys. D: Appl. Phys.*, **4**:1252-1256 (1971).

Rao, M.: Notes on the Corona Currents in a Lightning Discharge and the Emission of ELF Waves. *Radio Sci.*, **2**:1394 (1967).

Rao, M., and H. Bhattacharya: Lateral Corona Currents from the Return Stroke Channel and Slow Field Change after the Return Stroke in a Lightning Discharge. *J. Geophys. Res.*, **71**:2811-2814 (1966).

Rao, M., and S. R. Khastgir: The Physics of the Return Stroke and the Time-Variation of Its Current in a Lightning Discharge. *Trans. Bose Res. Inst.*, **29**:19-24 (1966).

Srivastava, K. M. L.: Return Stroke Velocity of a Lightning Discharge. *J. Geophys. Res.*, **71**:1283-1286 (1966).

Srivastava, K. M. L., and B. A. P. Tantry: VLF Characteristic of Electromagnetic Radiation from the Return Stroke of Lightning Discharge. *Indian J. Pure Appl. Phys.*, **4**:272-275 (1966).

Strawe, D. F.: Non-Linear Modeling of Lightning Return Strokes. "Proceedings of the Federal Aviation Administration/Florida Institute of Technology Workshop on Grounding and Lightning Technology," pp. 9-15. March 6-8, 1979, Melbourne, Florida, Report FAA-RD-79-6.

Suzuki, T.: Propagation of Ionizing Waves in Glow Discharge. *J. Appl. Phys.*, **48**:5001-5007 (1977).

Takagi, N., and T. Takeuti: Oscillating Bipolar Electric Field Changes Due to Close Lightning Return Strokes. *Radio Sci.*, **18**:391-398 (1983).

Turcotte, D. L., and R. S. B. Ong: The Structure and Propagation of Ionizing Wave Fronts. *J. Plasma Phys.*, **2**:145-155 (1968).

Uman, M. A.: Lightning Return Stroke Electric and Magnetic Fields. *J. Geophys. Res.*, **90**:6121-6130 (1985).

Uman, M. A., and D. K. McLain: Magnetic Field of Lightning Return Stroke. *J. Geophys. Res.*, 74:6899-6910 (1969).

Uman, M. A., and D. K. McLain: Lightning Return Stroke Current from Magnetic and Radiation Field Measurement. *J. Geophys. Res.*, 75:5143-5147 (1970).

Uman, M. A.., D. K. McLain, and E. P. Krider: The Electromagnetic Radiation from a Finite Antenna. *Am. J. Phys.*, 43:33-38 (1975).

Uman, M. A., M. J. Master, and E. P. Krider: A Comparison of Lightning Electromagnetic Fields with the Nuclear Electromagnetic Pulse in the Frequency Range 10^4 to 10^7 Hz. *IEEE Trans. Electromagn. Compat.*, **EMC-24**:410-416 (1982).

Volland, H.: A Wave Guide Model of Lightning Currents. *J. Atmos. Terr. Phys.*, 43:191-204 (1981a).

Volland, H.: Wave Form and Spectral Distribution of the Electromagnetic Field of Lightning Currents. *J. Atmos. Terr. Phys.*, 43:1027-1041 (1981b).

Volland, H.: Simulation of a Lightning Channel by a Prolate Spheroid. *Radio Sci.*, 17:445-452 (1982).

Wagner, C. F.: Determination of the Wave Front of Lightning Stroke Currents from Field Measurements. *AIEE Trans.*, **79** (Pt. 3):581-589 (1960).

Wagner, C. F.: Relation between Stroke Current and Velocity of the Return Stroke. *AIEE Trans.*, **82** (Pt. 3):606-617 (1963).

Wagner, C. F., and A. R. Hileman: The Lightning Stroke. *AIEE Trans.*, **77** (Pt. 3):229-242 (1958).

Wagner, C. F., and A. R. Hileman: The Lightning Stroke. *AIEE Trans.*, **80** (Pt. 3):623-642 (1961).

Wagner, C. F., and A. R. Hileman: Surge Impedance and Its Application to the Lightning Stroke. *AIEE Trans.*, **80** (Pt. 3):1011-1022 (1962).

Weidman, C., J. Hamelin, C. Leteinturier, and L. Nicot: Correlated Current Derivative (dI/dt) and Electric Field Derivative (dE/dt) Emitted by Triggered Lightning. Paper presented orally at the 11th International Aerospace and Ground Conference on Lightning and Static Electricity, June 24-26, 1986, Dayton, Ohio and also a personal communication, 1986.

Weidman, C. D., and E. P. Krider: The Fine Structure of Lightning Return Stroke Wave Forms. *J. Geophys. Res.*, 83:6239-6247 (1978). Correction, *J. Geophys. Res.*, 87:7351 (1982).

Weidman, C. D., and E. P. Krider: Submicrosecond Rise Times in Lightning Return Stroke Fields. *Geophys. Res. Lett.*, 7:955-958 (1980). Correction, *J. Geophys. Res.*, 87:7351 (1982).

Weidman, C. D., and E. P. Krider: Correction, *J. Geophys. Res.*, 87:7351 (1982).

Weidman, C. D., and E. P. Krider: Variations à l'Échelle Submicroseconde des Champs Électromagnétiques Rayonnés par la Foudre. *Ann. Telecommun.*, 39:165-174 (1984).

Winn, W. P.: A Laboratory Analog to the Dart Leader and Return Stroke of Lightning. *J. Geophys. Res.*, 70:3265-3270 (1965).

Winn, W. P.: Ionizing Space-Charge Waves in Gases. *J. Appl. Phys.*, 38:783-790 (1967).

Chapter 8 | Dart Leader

8.1 INTRODUCTION

Return strokes after the first in a lightning discharge to ground are usually initiated by dart leaders. The dart leader deposits charge on the remains of the primary current-carrying channel of the previous stroke, which may include one or more major branches if the previous stroke was a first stroke, and in that way renders that channel at a high potential relative to the earth, setting the stage for a subsequent return stroke. Dart leaders are so named because on streak-camera photographs they appear to be luminous sections of channel, or darts, some tens of meters in length propagating toward earth, generally without branching.

8.2 OPTICALLY DETERMINED PROPERTIES

The length of the luminous dart has been determined from the width of the dart leader image on streak photographs. Orville and Idone (1982) show an example of how that measurement is made in their Fig. 3b, and Idone *et al.* (1984) in their Fig. 9 give an additional example for the case of a dart leader in a rocket-initiated flash (Section 12.3.3.2). Schonland and Collens (1934) were the first to measure the dart length and found values from 25 to 112 m with a mean of 54 m for 9 leaders. Orville and Idone (1982) have provided a similar analysis for 11 dart leaders. They made two measurements, one at the upper end and one at the lower end of the channel, for each leader. For the 22 values they found a mean of 34 m with a range of 7–75 m. Dart lengths were a mean of 15 m shorter for six leaders photographed through a red filter than for five that were not, implying that the measurement is dependent on the spectral band of the light recorded by the film. The measured dart length must also, to some extent, be a function of the film properties and its processing and of the background light level. In most cases, the dart lengths found by Orville and Idone (1982) were smaller in the lower channel section. Idone and Orville (1984), for 19 dart leaders in artificially initiated flashes (Section 12.3.3.2), found a mean dart length of 50 m with a range from 15 to 90 m.

The length of the dart is determined both by the time that a given small section of the dart leader channel emits light after being turned on by the leader wavefront and by the velocity of the wavefront. If, for example, the decay time of the light is constant, then the faster the dart leader moves earthward, the longer the dart length is. Orville and Idone (1982) have shown that dart length and dart leader speed are indeed correlated. An additional explanation for this observed correlation is that the more energetic leaders not only are faster but also that they provide more channel heating, resulting in a slower decaying channel. In this latter view, the luminosity decrease is related to the temperature decrease in the channel. The temperature decreases because of thermal conduction, thermal convection, and radiation losses (see also discussion in Section 8.4). Originally, Schonland (1938) had suggested that the dart channel luminosity existed for a time of the order of a lifetime of the atomic excited energy states contributing to the luminosity. For a dart speed of 10^7 m/sec, the lifetime would have to be 5 μsec for a 50-m dart. Since the spectral lines emitted by dart leaders studied by Orville (1975) are from energy states having lifetimes near 10^{-2} μsec (Wiese et al., 1966), Schonland's view is not correct.

A survey of measurements of dart-leaders propagation speed is given in Table 8.1. A histogram showing the data on dart leader speed measured by Orville and Idone (1982) is given in Fig. 8.1. For all the sources listed in Table 8.1, the ranges of speed are in good agreement: the lower limit is 1–3 $\times$ 10^6 m/sec, while the upper limit is 21–23 $\times$ 10^6 m/sec. All mean values are also similar, near 1 $\times$ 10^7 m/sec, with the exception of that of Schonland et al. (1935), whose mean of 5.5 $\times$ 10^6 m/sec, about half that of the other investigators, is lower because about one-third of their 55 measurements fall in the 1–3 $\times$ 10^6 m/sec range.

Table 8.1

Summary of Dart-Leader Mean Speeds from Various Investigators[a]

Investigators	Sample size	Mean (10^6 m/sec)
Schonland et al. (1935)	55	5.5
McEachron (1939)	17	11.0
Brook and Kitagawa (Winn, 1965)	103	9.7
Berger (1967)	80	9.0
Hubert and Mouget (1981)	10^b	11.0^b
Orville and Idone (1982)	21	11.0

[a] Adapted from Orville and Idone (1982).
[b] These values were obtained from two artificially triggered lightning flashes (Section 12.3.3.1).

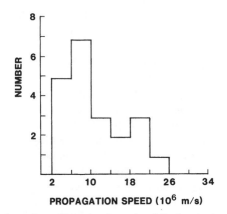

Fig. 8.1 Distribution of two-dimensional speeds of 21 dart leaders that occurred in 10 Florida flashes (14 leaders) and 2 New Mexico flashes (7 leaders). Speeds were measured in the lowest 0.8 km. Mean speed is 11×10^6 m/sec, with a range from 2.9 to 23×10^6 m/sec. Adapted from Orville and Idone (1982) and Idone *et al.* (1984).

Schonland *et al.* (1935) and Schonland (1956) report that high dart leader speeds are associated with short intervals of time from the previous stroke and that low speeds are associated with long intervals, these data being summarized in Table 8.2. Measurements made by Brook and Kitagawa (Winn, 1965) on 103 dart leaders confirm this observation. These data are shown in Fig. 8.2. Schonland *et al.* (1935) and Schonland (1956) also found that, in general, the longer the time interval between strokes, the more intense the luminosity from the subsequent stroke, as also documented in Table 8.2. Contrary to these observations, Idone *et al.* (1984), for 32 dart leaders in rocket-initiated discharges, find a positive correlation between dart-leader propagation speed and the ensuing return-stroke peak current, and for 56 dart leaders in the same study they can find little apparent correlation between

Table 8.2

Properties of Dart Leaders and Subsequent Return Strokes[a]

Number of dart leaders	Time interval from previous strokes (sec)	Mean interval (sec)	Speed (m/sec)	Mean speed (m/sec)	Intensity of return stroke	Mean intensity
5	0.005–0.12	0.044	$15–22 \times 10^6$	19×10^6	0.3–0.8	0.46
5	0.07–0.48	0.17	$1.7–2.8 \times 10^6$	2×10^6	1.2–5.0	2.2

[a] First stroke intensity is unity. Adapted from Schonland (1956) and Schonland *et al.* (1935).

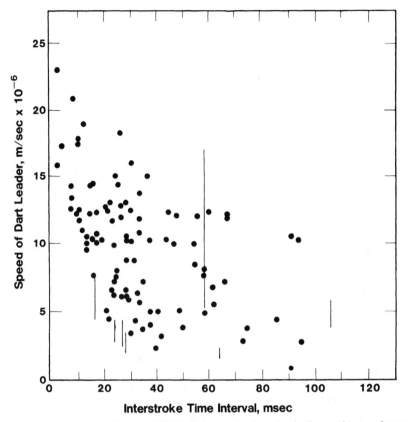

Fig. 8.2 Data of M. Brook and N. Kitagawa showing how dart leaders tend to travel more slowly down older channels. Vertical lines indicate velocity range of those dart leaders whose velocities were not constant. Shortest interstroke time is 3 msec. Adapted from Winn (1965).

speed and interstroke interval (Section 12.3.3.2). In laboratory experiments Winn (1965) has shown that voltage pulses applied to defunct 0.1-m spark channels produced luminous waves that traveled more slowly along the older channels. Orville and Idone (1982) report that for strokes within a given flash there is a positive correlation between dart leader luminosity and the following return stroke luminosity.

Schonland *et al.* (1935) found that dart-leader speeds often decrease as the leader approaches ground and did not observe any cases where the speeds increased. Orville and Idone (1982) found 4 cases out of a sample of 16 whose speeds increased toward ground. Since the photographic measurements were two-dimensional, it is possible that some of the apparent speed changes were due to channel geometry. For example, for a constant leader speed and a channel that is roughly vertical near ground, and nonvertical above, as is usually the case, the two-dimensional speed will be lower for the nonvertical portion than for the vertical, making the leader appear to increase in speed as it travels downward.

Idone and Orville (1985) have estimated dart-leader peak currents for 22 leaders in 2 rocket-initiated flashes (Section 12.3.3.2) using two different optical techniques. (1) The ratio of the dart leader to return stroke current is taken equal to the ratio of dart leader to return stroke speed, which assumes a simple model in which an equal charge per unit length is involved in each process. The speed ratio and the return stroke current are measured, allowing a calculation of the dart leader current. (2) The relation between return-stroke peak current I_R and return-stroke peak relative light intensity L_R in each of two flashes ($L_R = 1.5I_R^{1.6}$ and $L_R = 6.4I_R^{1.1}$, respectively) is applied to the dart-leader relative light intensities in that flash to determine the dart leader current. The two techniques produce very similar results, a mean current of 1.8 kA for (2) and 1.6 kA for (1). Individual values ranged from 100 A to 6 kA. The ratio of dart leader to return stroke current ranged from 0.03 to 0.3 with a mean of 0.17 from method (2) and 0.16 from method (1). The largest dart-leader to return-stroke current ratios were associated with the largest return stroke currents and relative intensities. Idone and Orville (1985) discuss the validity of the techniques used to find dart leader currents, which, as they state, are certainly open to question.

Guo and Krider (1985) found that 39 of 726 multiple-stroke flashes observed at the NASA Kennedy Space Center, Florida had one or more dart leaders whose light output per unit length was comparable to that of the following return stroke. Dart leaders with such anomalous light outputs were not observed by Idone and Orville (1985) who reported, in the study discussed above, that the ratio of maximum light output from a dart leader to the following return stroke varied from 0.02 to 0.23, with a mean of 0.1.

In Chapter 6 we discussed the upward-going leaders that rise from ground or from objects attached to ground to meet downward-moving stepped leaders. No evidence has been published to indicate that similar connecting discharges, or ones that propagate up the previous return stroke channel, occur for natural dart leaders. Orville and Idone (1982) conclude from their streak-camera observations that dart leaders propagate completely to the ground, but that it is possible that upward-connecting discharges exist if the discharges have a length less than the 10-m resolution of the experiment. On the other hand, for dart leaders in artificially initiated lightning (Section 12.3), Idone *et al.* (1984) report that they observed upward-connecting discharges of 20- to 30-m meeting dart leaders in two separate flashes.

When the time interval between strokes is long, the dart leader may change from a continuously moving leader to a stepped leader of relatively high average speed (for a stepped leader, although less than the dart leader speed existing in the channel above), relatively short step length, and relatively short time interval between steps. The rapid stepping follows the channel of the previous stroke. Sometimes the dart leader or dart-stepped leader will revert to a normal stepped leader, in which case the leader does not follow the channel of the previous stroke. Properties of dart-stepped leaders observed in South Africa are given in Table 8.3. Average speeds vary between about 0.5 and 1.7×10^6 m/sec, step lengths are typically 10 m and the time interval between steps is near 10 μsec. Orville and Idone (1982) have analyzed four dart-stepped leaders in detail, including the variation of the step length and interstep time with height and the luminous structure of the steps. One of the dart-stepped leaders shows a decrease in propagation speed and an increase in interstep time as ground is approached. Another exhibits the opposite effect. Several of the individual steps showed a distinct bright tip at the bottom that fanned out into a symmetrically diffuse structure in the upper portion of the steps. Overall, the step lengths and interstep times are in good agreement with those given in Table 8.3. The interstep times of both Schonland (1956) and Orville and Idone (1982) are consistent with the time between electric field pulse identified by Krider *et al.* (1977) as being associated with dart-stepped-leader steps. For a period of 200 μsec prior to the return stroke, Krider *et al.* (1977) found a mean time interval of 6.5 and 7.8 μsec for dart-stepped-leader electric field pulses in Florida and Arizona, respectively. Examples of these electric field pulses are shown in Fig. 7.4b. Comparable mean values for normal stepped leaders were 15.9 and 25.3 μsec in Florida and Arizona, respectively (see also Section 5.3). Fields from normal stepped leaders are shown in Figs. 5.1 and 7.4c.

Table 8.3

Average Speed, Average Step Length, and Average Time Interval between
Steps for Dart-Stepped Leaders Preceding the Second
Stroke of a Multiple-Stroke Flash[a]

Flash designation	Speed (m/sec)	Step length (m)	Time interval (μsec)
67	1.2×10^6	9.0	7.4
64	1.1	10.0	9.0
75	1.0	7.4	7.4
130	1.7	25.0	15.0
657	1.7	13.0	7.8
X7	0.48	12.0	25.0

[a] Adapted from Schonland (1956).

The optical spectra of five dart leaders have been measured by Orville (1975) with 9-μsec resolution using the spectrometer shown in Fig. C.9. The spectra were characterized by line radiation from singly ionized nitrogen and oxygen, similar to the spectra of the return stroke. Peak temperatures in excess of 20,000 K were calculated from the relative intensities of the spectral lines. Orville (1975) found the dart leaders to be about a factor of 10 less intense in their peak radiation than the return stroke peak, both being viewed in a 13-m-high channel section. Barasch (1970) measured the absolute intensities of dart leaders in five wavelength regions using widefield photoelectric detectors and narrow passband interference filters. He also found dart leaders to have similar spectral characteristics to return strokes, but reported that dart leader emissions that were 20 to 100 times less intense than those from the return stroke. Orville (1975) argues that the Barasch value is not meaningful because Barasch's wide field-of-view and millisecond time resolution resulted in the comparison of the radiation from a short dart with the radiation from the total length of the return stroke, thus making the ratio of the emitted radiations smaller than was actually the case.

8.3   ELECTRICALLY DETERMINED PROPERTIES

The electric field changes associated with dart leaders have been described by Malan and Schonland (1951), some of whose data are shown in Fig. 8.3. Malan and Schonland (1951) reported that at a range of 5–8 km, the first dart leaders in multistroke flashes produced positive field changes and later ones hook-shaped field changes starting with negative polarity. The dart leaders for which negative field changes are clearly observed are marked by the arrows in Fig. 8.3. Malan and Schonland (1951) argued that the data can be interpreted as due to succeeding leaders originating from higher charge volumes in the cloud and found that, with the assumption of a vertical discharge, the typical increase in height between successive leaders was about 0.7 km. These results were obtained by comparing the measurement with the theoretical curves shown in Fig. 5.2 for a uniformly charged vertical channel descending from a point charge source.

From measurements in which the photographically obtained dart leader speed below the cloud and the measured total time duration of the dart-leader field change were used to compute the leader length, Brook et al. (1962) report "apparent" leader heights (assuming the channel to be vertical) between 2 and 13 km with an apparent increase in height between leaders of about 0.3 km. These data are shown in Fig. 8.4.

Krehbiel et al. (1979) argue that the observed dart-leader field waveshapes are in part due to leader geometry and that horizontal leader development

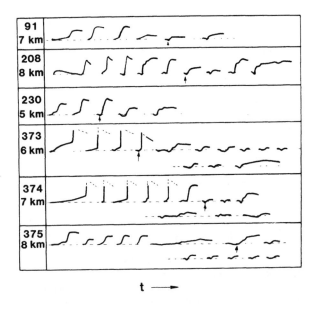

t ⟶

Fig. 8.3   Electric field changes due to successive leader–return stroke sequences in multiple-stroke flashes. The first field changes on the left are from stepped leaders and first return stroke, the remainder from dart leaders and subsequent return strokes. Flash 208 exhibits a BIL field change (see discussion in Section 4.2 and Fig. 4.1) except that the time constant of the system was such as to allow the signal to decay during the I period. Adapted from Malan and Schonland (1951).

is significant in determing those waveshapes. Multiple-station measurements are necessary to determine the geometry of the leader. Malan and Schonland's (1951) measurements were made from a single station. Krehbiel et al. (1979), using multiple-station measurements, found successive return-stroke charge locations displaced horizontally in the cloud (see Fig. 3.3) and hence successive dart leaders extending more horizontally than vertically. Thus it is reasonable to interpret the "apparent" heights given by Brook et al. (1962) as total dart leader lengths. On the other hand, if the dart leaders studied by Malan and Schonland (1951) were not vertical, their calculated heights are not meaningful since the model used in the calculations is invalid.

Schonland et al. (1938) have inferred from the ratios of 46 dart-leader and 26 stepped-leader field changes to their associated return-stroke field changes that dart leader channels tend to be uniformly charged. The comments given in Section 5.4 on the implication of a unity ratio for stepped leaders and return strokes are applicable.

The charge lowered by the dart leader can be determined from either measurement of the dart-leader field change or of the combined dart-leader

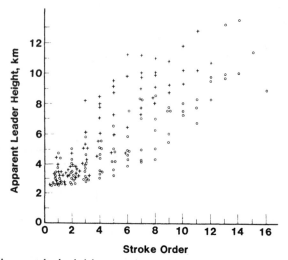

Fig. 8.4 Apparent leader height vs stroke order for flashes without continuing current intervals (circles) and flashes with continuing current intervals (plus signs). Continuing current is discussed in Chapter 9. The first leader height is arbitrarily obtained by either subtracting from the second leader apparent height the average apparent height difference between successive leaders in that flash or, if that value is below the cloud base, using the second leader apparent height. As noted in the text, apparent height is best interpreted, in light of present knowledge, as total leader length since there are probably significant channel deviation from vertical in the cloud. Adapted from Brook *et al.* (1962).

and return-stroke field change if a model for the leader charge distribution is assumed (Section A.1.3). The former does not appear to have been done. Brook *et al.* (1962) report that the minimum charge brought down by strokes subsequent to the first is 0.21 C, the most frequent value lying between 0.5 and 1 C. Thus the dart leader would appear to carry less charge than does the stepped leader (Section 5.4). If 1 C were lowered onto the dart leader channel in a millisecond or so, consistent with the speeds in Fig. 8.1 and the lengths in Fig. 8.4, the resulting dart leader current would be of the order of 1 kA. Currents between 100 A and 6 kA, with a mean of about 1.7 kA, were found for 22 dart leaders by Idone and Orville (1985) using an optical technique, as discussed in Section 8.2.

We have noted in the previous section that measurements of the time between dart-stepped-leader electric field pulses, a mean of 6.5 μsec in Florida and 7.8 μsec in Arizona, are consistent with optical measurements. Further, Krider *et al.* (1977) report that the electric fields of individual steps in dart-stepped leaders are essentially the same as individual stepped-leader pulses (Section 5.3) and are very similar to the small regular pulses produced by intracloud discharges (Krider *et al.*, 1975; Section 13.6). The mean ratio of the largest dart-stepped pulse to the following return stroke peak is about 0.1, similar to that ratio for a normal stepped leader (Section 5.3).

The dart leader produces considerable radiation at VHF (Takagi, 1969a, b; LeVine and Krider, 1977; Rustan, 1979; Rust *et al.*, 1979; Hayenga, 1979; Rustan *et al.*, 1980) and in the microwave region from 400 to over 2000 MHz (Brook and Kitagawa, 1964; Rust *et al.*, 1979). This radiation emanates primarily from the cloud rather than from the channel to ground (Proctor, 1971, 1976; Rustan, 1979; Hayenga, 1979, 1984; Rustan *et al.*, 1980). The dart leader RF observed by LeVine and Krider (1977) started an average of 265 μsec before the return stroke. Takagi (1969a,b) reports that the dart leader radiation begins 100–1000 μsec prior to the return stroke. At frequencies above about 100 MHz, the dart leader radiation ceased about 100 μsec prior to the return stroke according to LeVine and Krider (1977) and Brook and Kitagawa (1964), while at 3 MHz LeVine and Krider (1977) found it often continued up to and during the return stroke. Takagi (1969a,b) finds that about one-third of the dart leaders cease radiating in the frequency range 60–420 MHz by 50 μsec prior to the return stroke, one-third cease at about the return stroke, and the remaining one-third radiate during the return stroke. Perhaps some aspect of the discussion in the next to last paragraph of Section 7.2.1 is applicable to these observations.

8.4 SOME THEORETICAL CONSIDERATIONS

Uman and Voshall (1968) have presented calculations to show that no special mechanism is needed to keep the lightning channel in a state such that a dart leader can be initiated after a typical interstroke time of tens of milliseconds. Previously, Loeb (1966), in a general discussion of dart leader properties, had suggested that K-changes (Chapter 10) were necessary to keep the channel conducting between strokes and other investigators (e.g., McEachron, 1940; Brook *et al.*, 1962) had suggested that it might be necessary for a small, undetected current to flow in the channel between strokes, although McCann (1944) determined that channel currents fell below 0.1 A during that time. Uman and Voshall (1968) showed that in the absence of input energy to the channel, the channel temperature decays sufficiently slowly that the channel is still an order of magnitude above ambient temperature and in a transition stage between conductor and insulator after a typical interstroke period of 50 msec. The channel temperature determined the degree of ionization and hence the conductivity. The channel temperature decay is viewed as a heat transfer problem in that energy must leave the channel before the channel temperature can decrease. The cooling rate is found to be appropriately slow for channels of centimeter radius, those thought to be characteristic of a return stroke (Section 7.2.4).

The behavior of the dart leader is influenced by the conductivity, radius, and heavy-particle density of the defunct previous stroke channel. The channel conductivity and heavy-particle density are probably the controlling factors in dart leader behavior since the channel radius is not likely to change much with time. The channel heavy-particle density for a temperature near 3000 K, the temperature near which air makes the transition between conductor insulator and the temperature which exists in the channel near the end of the interstroke interval according to Uman and Voshall (1968), is about an order of magnitude less than that existing outside the channel. Ionization rates and charged particle velocities are inversely proportional to the heavy-particle density. For a given electric field associated with the dart leader wavefront, the low value of heavy-particle density in the channel relative to regions outside the channel serves to make the channel a preferred path for the dart leader whether or not there is appreciable conductivity remaining in the channel. Residual conductivity will make the channel additionally preferred.

The observations of the previous paragraph are consistent with the laboratory experiments of Winn (1965) discussed in Section 8.2.

Jurenka and Barreto (1985) review the literature on various theoretical attempts to understand the propagation of ionizing luminuous waves along weakly ionized spark channels with application to the dart leader, and they present a fluid dynamic analysis of the propagation characteristics of electron density gradients (electron shock waves) along defunct lightning channels such as those theorized to exist at the end of the interstroke period by Uman and Voshall (1968). Using a system of fluid equations (continuity, momentum, and energy conservation), Jurenka and Barreto (1985) show that an electron shock wave, basically a sharp increase in the electron density at the wavefront, can propagate into a weakly ionized gas without appreciable attenuation and with a speed exceeding the electron acoustic velocity of about 10^6 m/sec in the gas and hence similar to observed dart leader speeds.

Some speculation on the electric field, electron density, electron drift velocity, and electron ionization coefficient present in the dart leader wavefront is given by Uman (1969). He concludes that the dart-leader wavefront electric field is less than a meter in extent with a wavefront electric field near 10^7 V/m.

REFERENCES

Abbas, I., and P. Bayle: Non-Equilibrium between Electrons and Field in a Gas Breakdown Ionizing Wave, I, Macroscopic Model. *J. Phys. D*, **14**:549–560 (1981).

Albright, N. W., and D. A. Tidman: Ionizing Potential Waves of High-Voltage Breakdown Streamers. *Phys. Fluids*, **15**:86–90 (1972).

Barasch, G. E.: Spectral Intensities Emitted by Lightning Discharges. *J. Geophys. Res.*, **75**:1049-1057 (1970).

Berger, K., and E. Vogelsanger: Photographische Blitzuntersuchungen der Jahre 1955-1965 auf dem Monte San Salvatore. *Bull. Schweiz. Elektrotech. Ver.*, **57**:599-620 (1966).

Brook, M., and N. Kitagawa: Radiation from Lightning Discharges in the Frequency Range 400 to 1000 Mc/s. *J. Geophys Res.*, **69**:2431-2434 (1964).

Brook, M., N. Kitagawa, and E. J. Workman: Quantitative Study of Strokes and Continuing Currents in Lightning Discharges to Ground. *J. Geophys. Res.*, **67**:649-659 (1962).

Fowler, R. G.: Non-Linear Electron Acoustic Waves, Part I. *Adv. Electron. Electron Phys.*, **35**:1-86 (1974).

Fowler, R. G.: Non-Linear Electron Acoustic Waves, Part II. *Adv. Electron. Electron Phys.*, **41**:1-72 (1976).

Guo, C., and E. P. Krider: Anomalous Light Output from Lightning Dart Leaders. *J. Geophys. Res.*, **90**:13,073-13,075 (1985).

Hayenga, C. O.: Positions and Movement of VHF Lightning Sources Determined with Microsecond Resolution by Interferometry. Ph.D. thesis, University of Colorado, Boulder, Colorado, 1979.

Hayenga, C. O.: Characteristics of Lightning VHF Radiation near the Time of the Return Stroke. *J. Geophys. Res.*, **89**:1403-1410 (1984).

Hubert, P., and G. Mouget: Return Stroke Velocity Measurements in Two Triggered Lightning Flashes. *J. Geophys. Res.*, **86**:5253-5261 (1981).

Idone, V. P., and R. E. Orville: Correlated Peak Relative Light Intensity and Peak Current in Triggered Lightning Subsequent Return Strokes. *J. Geophys. Res.*, **90**:6159-6164 (1985).

Idone, V. P., R. E. Orville, P. Hubert, L. Barret, and A. Eybert-Berard: Correlated Observations of Three Triggered Lightning Flashes. *J. Geophys. Res.*, **89**:1385-1394 (1984).

Jurenka, H., and E. Barreto: Study of Electron Waves in Electrical Discharge Channels. *J. Appl. Phys.*, **53**:3581-3590 (1982).

Jurenka, H., and E. Barreto: Electron Waves in the Electrical Breakdown of Gases, with Application to the Dart Leader in Lightning. *J. Geophys. Res.*, **90**:6219-6224 (1985).

Kekez, M. M., and P. Savic: Contributions to Continuous Leader Channel Development. *In* "Electrical Breakdown and Discharges in Gases, Part A" (E. E. Kunhardt and L. H. Luessen, eds.), pp. 419-455. Plenum, New York, 1983.

Klingbeil, R., D. A. Tidman, and R. F. Fernsler: Ionizing Gas Breakdown Waves in Strong Electric Fields. *Phys. Fluids*, **15**:1969-1973 (1972).

Krehbiel, P. R., M. Brook, and R. McCrory: Analysis of the Charge Structure of Lightning Discharges to Ground. *J. Geophys. Res.*, **84**:2432-2456 (1979).

Krider, E. P., G. J. Radda, and R. C. Noggle: Regular Radiation Field Pulses Produced by Intracloud Discharges. *J. Geophys. Res.*, **80**:3801-3804 (1975).

Krider, E. P., C. D. Weidman, and R. C. Noggle: The Electric Fields Produced by Lightning Stepped Leaders. *J. Geophys. Res.*, **82**:951-960 (1977).

Loeb, L. B.: Ionizing Waves of Potential Gradient. *Science*, **148**:1417-1426 (1965).

Loeb, L. B.: The Mechanism of Stepped and Dart Leaders in Cloud-to-Ground Lightning Strokes. *J. Geophys. Res.*, **71**:4711-4721 (1966).

LeVine, D. M., and E. P. Krider: The Temporal Structure of HF and VHF Radiations during Florida Lightning Return Strokes. *Geophys. Res. Lett.*, **4**:13-16 (1977).

McCann, G. D.: The Measurement of Lightning Currents in Direct Strokes. *Trans. AIEE*, **63**:1157-1164 (1944).

McEachron, K. B.: Lightning to the Empire State Building. *J. Franklin Inst.*, **227**:149-217 (1939).

McEachron, K. B.: Wave Shapes of Successive Lightning Current Peaks. *Electr. World, Jan.* **10**:56-59 (1940).

Malan, D. J., and B. F. J. Schonland: The Distribution of Electricity in Thunderclouds. *Proc. R. Soc. London Ser. A*, **209**:158-177 (1951).

Orville, R. E.: Spectrum of the Lightning Dart Leader. *J. Atmos. Sci.*, **32**:1829-1837 (1975).

Orville, R. E., and V. P. Idone: Lightning Leader Characteristics in the Thunderstorm Research International Program (TRIP). *J. Geophys. Res.*, **87**:11,177-11,192 (1982).

Orville, R. E., G. G. Lala, and V. P. Idone: Daylight Time-Resolved Photographs of Lightning. *Science*, **201**:59-61 (1978).

Proctor, D. E.: A Hyperbolic System for Obtaining VHF Radio Pictures of Lightning. *J. Geophys. Res.*, **76**:1478-1489 (1971).

Proctor, D. E.: A Radio Study of Lightning. Ph.D. thesis, University of Witwatersrand, Johannesburg, South Africa, 1976.

Rao, M.: The Dependence of the Dart Leader Velocity on the Interstroke Time Interval in a Lightning Flash. *J. Geophys. Res.*, **75**:5868-5872 (1970).

Rust, W. D., P. R. Krehbiel, and A. Shlanta: Measurements of Radiation from Lightning at 2200 MHz. *Geophys. Res. Lett.*, **6**:85-88 (1979).

Rustan, P. L.: Properties of Lightning Derived from Time Series Analysis of VHF Radiation Data. Ph.D. thesis, University of Florida, Gainesville, Florida, 1979.

Rustan, P. L., M. A. Uman, D. G. Childers, W. H. Beasley, and C. L. Lennon: Lightning Source Locations from VHF Radiation Data for a Flash at Kennedy Space Center. *J. Geophys. Res.*, **85**:4893-4903 (1980).

Schonland, B. F. K.: Progressive Lightning, Pt. 4, The Discharge Mechanism. *Proc. R. Soc. London Ser. A*, **164**:132-150 (1938).

Schonland, B. F. J.: The Lightning Discharge. *Handb. Phys.*, **22**:576-628 (1956).

Schonland, B. F. J., and H. Collens: Progressive Lightning, Pt. 1. *Proc. R. Soc. London Ser. A*, **152**:654-674 (1934).

Schonland, B. F. J., D. J. Malan, and H. Collens: Progressive Lightning, Pt. 2. *Proc. R. Soc. London Ser. A*, **152**:595-625 (1935).

Schonland, B. F. J., D. B. Hodges, and H. Collens: Progressive Lightning, Pt. 5, A Comparison of Photographic and Electrical Studies of the Discharge Process. *Proc. R. Soc. London Ser. A*, **166**:56-75 (1938).

Takagi, M.: VHF Radiation from Ground Discharges. *In* "Planetary Electrodynamics" (S. C. Coroniti and J. Hughes, eds.), pp. 543-570. Gordon & Breach, New York, 1969a.

Takagi, M.: VHF Radiation from Ground Discharges. *Proc. Res. Inst. Atmos., Nagoya Univ., Japan*, **16**:163-168 (1969b).

Thomson, E. M.: A Theoretical Study of Electrostatic Field Wave Shapes from Lightning Leaders. *J. Geophys. Res.*, **90**:8125-8135 (1985).

Uman, M. A.: "Lightning," pp. 223-224. McGraw-Hill, New York, 1969. See also, Dover, New York, 1984.

Uman, M. A., and R. E. Voshall: Time Interval between Lightning Strokes and the Initiation of Dart Leaders. *J. Geophys. Res.*, **73**:497-506 (1968).

Wiese, W. L., M. W. Smith, and B. M. Glennon: "Atomic Transition Probabilities—Hydrogen through Neon," Vol. 1. U.S. Dept. of Commerce, National Bureau of Standards, NSRDS-NBS4, 59-61, 1966. (Available from Government Printing Office, Washington, D.C., 20402.)

Winn, W. P.: A Laboratory Analogy to the Dart Leader and Return Stroke of Lightning. *J. Geophys. Res.*, **70**:3256-3270 (1965).

Chapter 9 | Continuing Current

9.1 INTRODUCTION

The return stroke in a negative ground flash traverses the preceding leader channel from the ground to the cloud in a time of the order of 100 μsec. During that time the stroke current at the channel base increases to its peak value and then decays to about one-tenth of that value. Following the return-stroke propagation phase, a channel base current of the order of 1 kA with a duration of the order of 1 msec may flow. Hagenguth and Anderson (1952), who first observed this current component in measurements made on flashes to the Empire State Building, referred to it as an "intermediate" current. Channel current in the 100-A range, which sometimes follows the intermediate current, is called "continuing current," although the distinction between intermediate current and relatively short duration continuing current probably has no physical basis. Kitagawa et al. (1962) and Brook et al. (1962) have further subdivided the continuing current in negative ground flashes into two classes: (1) "long," lasting in excess of 40 msec (the typical interstroke interval found in New Mexico), and (2) "short," having a duration less than 40 msec. This subdivision, again, is not likely to have a physical basis in the sense that a continuing current of 20 msec duration is probably caused by the same physical processes as one of 60 msec duration. Kitagawa et al. (1962) and Brook et al. (1962) have also introduced the following two definitions, which are used commonly in the literature: (1) "hybrid" flash, a flash containing at least one long-continuing current component, and (2) "discrete" flash, a flash without long-continuing current but which may contain short-continuing current components.

Examples of electric field changes associated with continuing current in negative ground flashes are given in Figs. 1.9, 4.3, and 9.1. In much of the literature the existence of a continuing current is inferred only from the single-station measurement of a slow steady electric field increase following the more rapid field change characteristic of a return stroke. That is, there is neither a corresponding measurement of channel luminosity, such as is

167

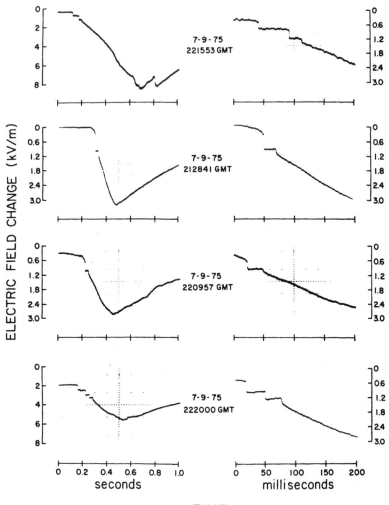

Fig. 9.1 The electrostatic fields on two time scales produced by four ground flashes, each containing a continuing current component. The field changes due to the continuing current are shown with higher time resolution on the right. Adapted from Livingston and Krider (1978).

illustrated in Fig. 1.9 (see also Fig. 1.7) nor multiple-station field measurement, such as are plotted in Fig. 4.3, to rule out the possibility that the field change was due to an in-cloud process. Most of the available statistics on negative continuing current concerns either (1) only long-continuing currents (e.g., Kitagawa *et al.*, 1962; Brook *et al.*, 1962; Fuquay *et al.*, 1967) or

(2) all continuing currents but without a definition of the shortest allowed interval (e.g., Livingston and Krider, 1978). When statistics are given for the duration of all interstroke field changes that could be associated with continuing current (e.g., Thomson, 1980), there is always a question as to which fraction of the shorter field changes was actually from continuing current.

Continuing currents in negative ground flashes are of significance in that (1) most flashes contain at least one short- or long-continuing current interval and (2) roughly 50% of all flashes contain a long-continuing current component and those flashes transfer to earth about twice the charge that flashes without a long-continuing current do. From a practical point of view, the effect of continuing current on objects struck by lightning is to cause potentially serious heating damage. For example, long-continuing currents burn holes in the metal skins of aircraft (e.g., Fisher and Plumer, 1977) and preferentially cause forest fires (Fuquay et al., 1967, 1972).

In this chapter we discuss only continuing current that lowers negative charge to ground. Continuing current in positive ground flashes is discussed separately in Chapter 11. Although positive ground flashes are much less common than negative ground flashes, continuing current in those positive flashes, generally composed of a single stroke followed by continuing current, apparently is more common and lowers a larger fraction of the total flash charge than for the case of negative flashes. Statistics on positive continuing currents are summarized in Table 11.1 and an electric field change apparently due to a positive continuing current is found in Fig. 11.2.

9.2 OCCURRENCE STATISTICS

Forty-six percent of the 193 flashes for which correlated streak photographs and electric field records were obtained by Kitagawa et al. (1962) and Brook et al. (1962) showed simultaneous continuing luminosity and slow electric field changes for times longer than 40 msec in at least one interval between strokes. About one-quarter of all interstroke intervals contained a long-continuing current. Examples of these data are shown in Fig. 1.9. Livingston and Krider (1978) found that 39% of 239 flashes to ground in Florida had continuing current. Examples of electric field changes indicative of continuing current from this study are shown in Fig. 9.1. Fuquay et al. (1967), in Montana, reported that half of 856 ground flashes had long-continuing current. Thomson (1980), in Papua New Guinea, found that 47% of 34 flashes exhibited a long-continuing current.

There is considerable discrepancy between different investigators regarding whether continuing current follows a single-stroke flash. Kitagawa et al. (1962)

reported that only 2 of 27 single-stroke flashes were followed by long-continuing current, while Malan (1954) found that about half of 71 single-stroke flashes were followed by a slow or F (final) field change he associated with continuing current and Pierce (1955) found 59% of 140 single-stroke flashes were followed by similar slow field changes which he termed S-changes.

There is somewhat better agreement between different investigators regarding the probability of the occurrence of continuing current after the final stroke of a multiple-stroke flash. Kitagawa *et al.* (1962), in New Mexico, found that about half of the long-continuing currents in multiple-stroke flashes occurred after the final stroke, whereas Malan (1954), in South Africa, reported that fields indicative of continuing currents occurred primarily after the final stroke in a flash, with only 6% of slow field changes exceeding 100 msec occurring between strokes. Probably a greater percentage of Malan's interstroke intervals contained continuing current, however, since Malan (1955) reported that 23% of interstroke field changes from lightning between 20 and 150 km were positive, and calculations by Brook *et al.* (1962) indicate that distant field changes of the magnitude observed by Malan should be interpreted as due to continuing current (see Section 10.3). Thomson (1980) found that 9 of 34 multiple-stroke flashes having a long-continuing current interval ended with that current. Malan (1954) reported that 43% of 181 two- and three-stroke flashes, 21% of 107 four-stroke flashes, 15% of 61 five-stroke flashes, and 6% of 110 flashes with greater than 6 stroke were followed by a F-field change. Pierce (1955) found S-changes following 44% of flashes with two to four strokes and 30% of flashes with five or more strokes. He also found that 26% of the intervals between strokes had field changes sufficiently large to be attributed to continuing current (Brook *et al.*, 1962). For flashes that did contain a continuing current interval, Brook *et al.* (1962) found that 4 of 5 two- and three-stroke flashes and 2 of 7 flashes with over seven strokes had a long-continuing current following the final stroke. Krehbiel *et al.* (1979) studied four ground flashes in New Mexico with a multiple-station electric field network (Fig. 4.2) and found long-continuing current following both a four- and a five-stroke flash (Fig. 4.3) and between the third and fourth stroke of another flash. The fourth flash may have been followed by a continuing current of 8 msec duration.

Thus various investigators' data on the probability of occurrence of continuing currents after the final stroke of a multiple-stroke flash are more or less in agreement with an apparent decrease in the probability of such an occurrence with an increasing number of strokes. On the other hand, the available data on the probability of continuing current following a single-stroke flash are not in good agreement.

9.3 CURRENTS, CHARGES, AND CHARGE LOCATIONS

Continuing currents have been measured directly in strikes to towers and have been inferred from remote measurements of interstroke electric and magnetic fields. Prior to the work of Krehbiel *et al.* (1979), continuing current determinations from field measurements were made using the charge source height and distance determined from a model in which the lightning channel was assumed to be vertical. As discussed later in this section, Krehbiel *et al.* (1979) found that the charge sources for continuing currents were distributed horizontally in the cloud. Thus, errors in the earlier calculations of currents and charges are present if, as is likely, charge distributions were in general similar to those observed by Krehbiel *et al.* (1979). Data on the effective height, which is best interpreted as the total length (Section 8.3), of dart leaders in flashes containing a long-continuing current component are shown by crosses in Fig. 8.4. For flashes with more than two or three strokes, it appears that horizontal channel lengths of many kilometers may be present. With this warning as to possible errors in charge and current values, we review the earlier data.

Brook *et al.* (1962) and Kitagawa *et al.* (1962) found long-continuing currents with durations up to 500 msec, the average duration being 150 msec. The charges lowered by long-continuing currents were found to be between 3.4 and 31.2 C, the average being about 12 C, whereas current values were in the relatively narrow range between 38 and 130 A. In this regard, Brook *et al.* (1962), in their Fig. 6, plot the charge lowered by long-continuing current vs the time duration of that current and show that there is a strong correlation between the two. The plotted data also constitute a frequency distribution of time durations with most between 40 and 100 msec. Flashes with long-continuing current intervals lowered twice the charge as those without such currents, according to Brook *et al.* (1962).

Williams and Brook (1963), also in New Mexico, used a magnetometer to measure the lightning magnetic field from which the continuing current was derived. For 14 continuing current intervals they found an average current of 184 A, an average charge transfer of 31 C, and an average duration of 184 msec.

Krehbiel *et al.* (1979) using the eight electric field measuring stations shown in Fig. 4.2 found continuing currents between 50 and 580 A for three discharges. The initial continuing current values were 580, 185, and 130 A, and all three continuing currents decreased with time. The charge sources for these currents were distributed horizontally in the cloud at about the same height as the stroke charges. Stroke and continuing current charge locations are illustrated in Fig. 3.4 and the electric fields at the eight stations for one flash are given in Fig. 4.3.

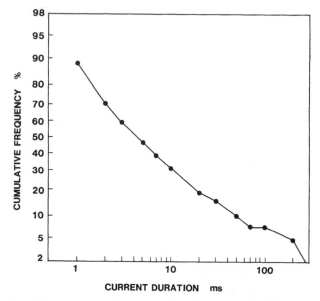

Fig. 9.2 Cumulative frequency distribution of durations of electric field changes following 223 strokes in Papua New Guinea (Thomson, 1980). All strokes, including those with field durations less than the 0.2-msec time resolution of the experiment, are included in the distribution. The spots indicate the bin size for the distribution. A straight line on the log normal probability paper on which the distribution is plotted indicates a log normal distribution (Section B.3).

Continuing currents measured directly by Berger and Vogelsanger (1965) in Switzerland following normal negative strikes to a tower were found to be of the order of 100–300 A. Half of the continuing current intervals lowered over 25 C of charge, with a maximum charge of about 80 C.

Thomson (1980) gives a cumulative frequency distribution for 223 field changes following strokes to ground which he interprets as due to continuing current. That distribution is reproduced in Fig. 9.2. Note that about 90% of the strokes are followed by a field duration exceeding 1 msec.

9.4   M-COMPONENTS

M-component is the name given to an increase in channel luminosity accompanied by a rapid electric field variation, itself called a M-electric field change. M-components occur when the channel is already faintly luminous. Downward-moving leaders have not been observed to precede

M-components. Early M-components may be confused with *branch components*, the increases in channel luminosity that occur between each branch and ground when the upward-propagating return stroke reaches that branch (e.g., Malan and Schonland, 1947), since the higher branches are obscured by the cloud. Illustrations of M-components are found in Figs. 1.9 and 9.3.

Malan and Schonland (1947) reported correlated photographic and electric field measurements from 37 cloud-to-ground flashes (199 strokes), most of which were at distances less than 6 km. Short duration, hook-shaped field changes were found to occur during the slow final portion of the stroke field change (termed R_c) which followed the fast portion associated with the return-stroke wavefront propagation (termed R_b). The discharge channel was observed photographically to be fairly luminous during the R_c field change with increases in that luminosity being correlated with field changes as illustrated in Fig. 9.3. The R_c field change is probably associated with intermediate or short-continuing current although an alternate possibility is discussed in Section 9.5. Malan and Schonland (1947) found that 60% of the return strokes examined did not have an R_c portion. These showed no M-components. The hooked-shaped, M-component field changes observed at close distances began with a negative field change and were followed without any pause by a larger positive field change. The net field change was usually from one-fifth to one-hundredth that of the R_b change. The hook process was found to occupy between 200 and 800 μsec, although longer values were observed. Longer hook duration was found to be associated with longer time intervals between the return stroke and the occurrence of the M-component.

Kitagawa *et al.* (1962) have found that M-field changes occur during both short- and long-continuing current. They have correlated M-field changes, which for distant strokes appear as unidirectional changes in the electric field, with increases in channel luminosity. Examples of these data are shown in Fig. 1.9. Kitagawa *et al.* (1962) present data showing that during the first

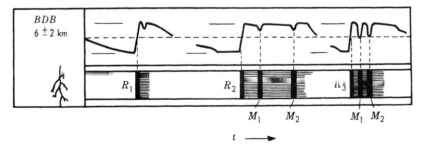

Fig. 9.3 Electric field changes correlated with channel luminosity for M-components observed near the lightning discharge. Adapted from Malan and Schonland (1947).

15 msec after a return stroke the time interval between M-components tends to increase rapidly with elapsed time, that between 15 and 40 msec this tendency decreases, and that after 40 msec the time interval between M-components shows no dependence on the elapsed time. Kitagawa *et al.* (1962) have shown that K-changes, the small, rapid electric field changes that occur in the intervals between and after the strokes of a multiple-stroke flash (Kitagawa, 1957; Kitagawa and Brook, 1960), are associated with the photographed M-component during periods of continuing channel luminosity. That is, the M-field changes observed at a distance during continuing luminosity are K-changes. Further, the most frequent time between all K-field changes and between field changes occurring during continuing luminosity was found to be about 6 msec. Kitagawa *et al.* (1962) suggest that there is no essential difference between K-changes associated with M-components and K-changes that occur during the nonluminous interstroke periods. Kitagawa *et al.* (1962) report that in the absence of continuing luminosity (and M-components), K-changes are sometimes associated either with luminosity in the cloud or with luminous waves (similar to the dart leader) that propagate down the channel but fail to reach the ground. On the basis of their investigations, Kitagawa *et al.* (1962) suggest that the K-change "is evidence of the movement of penetrative streamers into fresh regions of cloud, streamers whose occurrence must be determined wholly by conditions inside the cloud." Data regarding the occurrence and characteristics of K-changes in intracloud discharges are given in Chapter 13.

9.5   INITIATION, MAINTENANCE, AND DEMISE

Brook *et al.* (1962) show in their Fig. 2 that the charge transfer in return strokes that are followed by continuing current is small compared to the charge transfer in strokes that are not. Krehbiel *et al.* (1979) refer to this observation and add that the dart leaders preceding strokes that initiate continuing current are of relatively long (2 msec) duration. Livingston and Krider (1978) state that a large fraction of the slow field changes they associate with continuing current is initiated by abrupt field transitions that are smaller in amplitude and slower in risetime than those due to the preceding return stroke. In the absence of photographic measurement, Livingston and Krider (1978) could not determine whether these small field transitions were return strokes but state that, if they were, they did not transfer as much charge as strokes not followed by continuing current. It appears from the foregoing that there may well be something different about the leaders and return strokes, if indeed they are properly referred to as such, involved in initiating continuing current.

Kitagawa *et al.* (1962) present evidence to show that after an especially long continuing current of 100–500 msec duration has ended, the time interval preceding the next stroke, during which the channel is nonluminous, is comparable to the time interval between strokes without continuing current. For flashes without long-continuing current but containing short-continuing current having one to three M-components, these currents, after they could no longer be observed photographically, were followed by relatively long time intervals before a new stroke was initiated. Kitagawa *et al.* (1962) suggest that channel conductivity decays more slowly following a short-continuing current than after a stroke without continuing current, but it is not obvious why this effect was not observed with long-continuing current. Kitagawa *et al.* (1962) also report that after a channel loses its luminosity, a subsequent stroke can follow the same channel if dart leader initiation occurs within about 100 msec. Otherwise, if another stroke occurs, it takes a new channel forged by a new stepped leader.

From analyzing the multiple-station electric field measurements referred to in Section 9.3, Krehbiel *et al.* (1979) found that electric field changes in the interstroke interval associated with junction or J-processes (Section 10.3) did not serve to develop a source of charge for the continuing current flow, as suggested by Malan (1954), but rather than the two processes were independent and could occur simultaneously in different locations in the cloud. Krehbiel *et al.* (1979) conclude that the feeding process for the continuing current is initiated directly by the return stroke.

Observational evidence for the cutoff of channel current during an intermediate- or short-continuing current is provided from electric field measurements by Malan and Schonland (1947) and by Krehbiel *et al.* (1979). In this case, the R_c field change exhibits a fast recovery or abrupt flattening when observed near a ground discharge but continues to increase when observed far from the discharge. Such observations can be interpreted as due to charge moving down the channel but unable to reach its lower portion, implying that the channel cuts off first in the lower atmosphere where the pressure is highest and while current is still flowing into the top of the channel (Malan and Schonland, 1947).

An alternative view to that of the R_c field change being due to intermediate or continuing current flow from cloud charge source to earth has been given by Pierce and Wormell (1953). They have suggested that the slow R_c field change is caused by the collapse of that part of the leader charge still surrounding the channel after the return stroke phase is over. Pierce (1958) and Rao and Bhattacharya (1966) have estimated the time of this collapse to be of the order of a millisecond, comparable to the time of the R_c field change. Lin *et al.* (1980), on the other hand, have concluded from electric field measurements made on a microsecond time scale that the return stroke removes the bulk of the leader charge in a few tens of microseconds (Section 7.3), a conclusion consistent with the laboratory spark experiment of Wagner and Hileman (1961).

Latham (1980) has published results obtained from a computer model of an air arc approximating the lightning continuing current. As current increased from tens to hundreds of amperes, channel electric field decreased from about 2×10^4 to 1×10^4 V/m while channel axis temperature increased from 7,000 to 11,000 K. The channel radius at which a temperature of 2000 K was present was found to be about 0.5 cm, consistent with lightning channel radii determined in various ways (Sections 7.2.4 and 15.3.1).

There does not appear to be a good explanation for why some strokes are followed by little if any continuing current while others are followed by hundreds of milliseconds of continuing current. Malan (1954) argues that the continuing current is to be associated with specific levels of charge density located relatively high in the cloud, but the measurements of Krehbiel et al. (1979) do not support this view. Perhaps the continuing current is always present but to a varying degree depending on the location and amount of available cloud charge. On the other hand, the continuing current may well be due to a completely different type of physical process that perhaps involves larger regions of the cloud charge than the more localized areas of charge involved in strokes to ground without appreciable continuing current. Not unrelated questions are those of why positive flashes (Chapter 11) are almost always composed of a single stroke followed usually by a continuing current but hardly ever by a subsequent stroke, and why discrete strokes are even possible in negative flashes when a continuous discharge would appear to be sufficient to lower the available charge. Krehbiel et al. (1979) suggest that the difference between discrete strokes and continuous discharge may be partly associated with the physical characteristics of the channel to ground that cause the channel to cut off in its lower portions, as discussed earlier, when the available current falls below a certain level for a sufficient length of time, this cutoff being related to the availability of source current from the cloud. In this view, channel current will flow if the source can provide the proper level of current; otherwise a discrete stroke will occur even if the charge source has not yet been exhausted.

REFERENCES

Berger, K., and E. Vogelsanger: Messungen und Resultate der Blitzforschung der Jahre 1955–1963 auf dem Monte San Salvatore. *Bull. Schweiz. Elektrotech. Ver.*, **56**:2–22 (1965).

Brook, M., N. Kitagawa, and E. J. Workman: Quantitative Study of Strokes and Continuing Currents in Lightning Discharges to Ground. *J. Geophys. Res.*, **67**:649–659 (1962).

Fisher, F. A., and J. A. Plumer: Lightning Protection of Aircraft. NASA Reference Publication 1008, NASA Lewis Research Center, October, 1977.

Fuquay, D. M., R. G. Baughman, A. R. Taylor, and R. G. Hawe: Characteristics of Seven Lightning Discharges That Caused Forest Fires. *J. Geophys. Res.*, **72**:6371-6373 (1967).

Fuquay, D. M., A. R. Taylor, R. G. Hawe, and C. W. Schmid, Jr.: Lightning Discharges That Caused Forest Fires. *J. Geophys. Res.*, **77**:2156-2158 (1972).

Hagenguth, J. H., and J. G. Anderson: Lightning to the Empire State Building, Part III. *AIEE*, **71** (Pt. III):641-649 (1952).

Kitagawa, N.: On the Mechanism of Cloud Flash and Junction or Final Process in Flash to Ground. *Pap. Meteorol. Geophs. (Tokyo)*, **7**:415-424 (1957).

Kitagawa, N., and M. Brook: A Comparison of Intracloud and Cloud-to-Ground Lightning Discharges. *J. Geophys. Res.*, **65**:1185-1201 (1960).

Kitagawa, N., M. Brook, and E. J. Workman: The Role of Continuous Discharges in Cloud-to-Ground Lightning. *J. Geophys. Res.*, **65**:1965 (1960).

Kitagawa, N., M. Brook, and E. J. Workman: Continuing Currents in Cloud-to-Ground Lightning Discharges. *J. Geophys. Res.*, **67**:637-647 (1962).

Krehbiel, P. R., M. Brook, and R. A. McCrory: An Analysis of the Charge Structure of Lightning Discharges to Ground. *J. Geophys. Res.*, **84**:2432-2456 (1979).

Latham, D. J.: A Channel Model for Long Arcs in Air. *Phys. Fluids*, **23**:1710-1715 (1980).

Lin, Y. T., M. A. Uman, and R. B. Standler: Lightning Return Stroke Models. *J. Geophys. Res.*, **85**:1571-1583 (1980).

Livingston, J. M., and E. P. Krider: Electric Fields Produced by Florida Thunderstorms. *J. Geophys. Res.*, **83**:385-401 (1978).

Malan, D. J.: Les Descharges Orageuses Intermittentes et Continu de la Colonne de Charge Negative. *Ann. Geophys.*, **10**:271-281 (1954).

Malan, D. J.: La Distribution Verticale de la Charge Negative Orageuse. *Ann. Geophys.*, **11**:420-426 (1955).

Malan, D. J.: The Relation between the Number of Strokes, Stroke Interval, and the Total Durations of Lightning Discharges. *Pure Appl. Geophys.*, **34**:224-230 (1956).

Malan, D. J., and B. F. J. Schonland: Progressive Lightning, 7, Directly Correlated Photographic and Electrical Studies of Lightning from near Thunderstorms. *Proc. R. Soc. London Ser. A*, **191**:485-503 (1947).

Pierce, E. T.: Electrostatic Field Changes Due to Lightning Discharges. *Q. J. R. Meteorol. Soc.*, **81**:211-228 (1955).

Pierce, E. T.: Some Topics in Atmospheric Electricity. *In* "Recent Advances in Atmospheric Electricity" (L. G. Smith, ed.), pp. 5-16. Pergamon, Oxford, 1958.

Pierce, E. T., and T. W. Wormell: Field Changes Due to Lightning Discharges. *In* "Thunderstorm Electricity" (H. R. Byers, ed.), pp. 251-266. Univ. of Chicago Press, Chicago, Illinois, 1953.

Rao, M.: Corona Currents after the Return Stroke and the Emission of ELF Waves in a Lightning Flash to Earth. *Radio Sci.*, **2**:241-244 (1967). Correction, **2**:1394 (1967).

Rao, M., and H. Bhattacharya: Lateral Corona Currents from the Return Stroke Channel and the Slow Field Change after the Return Stroke in a Lightning Discharge. *J. Geophys. Res.*, **71**:2811-2814 (1966).

Rust, W. D., D. R. MacGorman, and W. L. Taylor: Photographic Verification of Continuing Current in Positive Cloud-to-Ground Flashes. *J. Geophys. Res.*, **90**:6144-6146 (1985).

Schonland, B. F. J.: The Lightning Discharge. *Handb. Phys.*, **22**:576-628 (1956).

Takeuti, T.: Studies on Thunderstorm Electricity, II, Ground Discharge, *J. Geomagn. Geoelectr.*, **18**:13-22 (1966).

Thomson, E. M.: Characteristics of Port Moresby Ground Flashes. *J. Geophys. Res.*, **85**:1025-1036 (1980).

Wagner, C. F., and A. R. Hileman: The Lightning Stroke, II. *Trans. Am. Inst. Electr. Eng.*, **80**(3):622–642 (1961).

Williams, D. P., and M. Brook: Fluxgate Magnetometer Measurements of Continuous Currents in Lightning. *J. Geophys. Res.*, **67**:1662 (1962).

Williams, D. P., and M. Brook: Magnetic Measurement of Thunderstorm Currents, 1. Continuing Currents in Lightning. *J. Geophys. Res.*, **68**:3243–3247 (1963).

Chapter 10 | J- and K-Processes in Discharges to Ground

10.1 INTRODUCTION

The J- or "junction"-process takes place in the cloud during the time interval between return strokes. It is identified with an electric field having a relatively steady change on a time scale of tens of milliseconds. The J-change can be either positive or negative, is generally smaller than the field change due to continuing current (Chapter 9), and, as differentiated from the continuing current field change, is not associated with a luminous channel between cloud and ground. Small, relatively rapid electric field variations termed K-changes also occur between strokes generally at intervals of 2–20 msec, and hence appear to be superimposed on the overall electric field change associated with the J-process.

There is controversy regarding the physical interpretation of the interstroke field change associated with the "junction"-process. As we shall discuss, most investigators have associated the J-field change with a cloud discharge that makes available charge to the top of the previous stroke channel so as to initiate a dart leader, while the recent work of Krehbiel *et al*. (1979) indicates that the J-process can be independent of the dart-leader charge source and related to the mopping up of charge from charge centers involved with previous strokes.

10.2 VISUAL AND TV OBSERVATIONS OF THE J-PROCESS

Brook and Vonnegut (1960) reported the following visual observation of what they interpret to be the J-process:

> The uppermost region from which the return stroke originated was observed to move (in steps reminiscent of darts) upward and outward, illuminating new regions of cloud, before a new section was added to the channel of the previous return stroke and a new stroke occurred. Some discharges to ground appeared to originate from a vertical column, but by

far the greater number were seen to progress horizontally or inclined to about 30° to the horizontal. .... The manner of horizontal spread of the junction streamer is also of interest. It appeared sometimes to progress horizontally in divergent directions, and, although the in-cloud paths were distinct and separate, and often quite long, the return strokes were seen to follow the same below-cloud path to ground. This observation is important because it shows that lightning strokes may often bridge individual convective cells when many thunderstorms are simultaneously active along a line.

Brantley *et al.* (1975) reproduce television frames illustrating the propagation of in-cloud horizontal channels similar to those described by Brook and Vonnegut (1960).

10.3   MEASUREMENTS OF THE ELECTRIC FIELDS OF THE J-PROCESS

The J-field change occurring between strokes of a ground discharge is almost always negative for flashes within a few kilometers and can be positive or negative for discharges beyond about 5 km (Malan and Schonland, 1951a,b; Malan, 1955, 1965). J-changes reported from the South African studies are shown in Fig. 10.1. Similar J-changes from measurements made in New Mexico are evident in Fig. 4.3. Malan and Schonland (1951a) report on J-changes occurring in 105 flashes containing 388 strokes. For 19 flashes within 5 km they reported 80 negative field changes, 8 zero field changes, and 2 positive field changes. For 34 distinct flashes at ranges between 12 and 30 km, they reported 7 negative field changes, 29 zero field changes, and 64 positive field changes. Malan and Schonland (1951a) argue that some of the zero change in the 12- to 30-km range "may well have been positive since the J-field changes at a considerable distance were rather small for accurate measurement." Inconsistent with this observation is the fact that

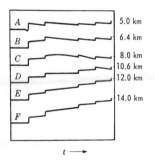

Fig. 10.1   J-field changes observed at various distances according to Schonland (1956).

Malan (1965) found that 44% of interstroke field changes beyond 25 km were negative. For 52 flashes in the intermediate range between 5 and 12 km, Malan and Schonland (1951a) report 71 negative field changes, 64 zero field changes, and 63 positive field changes, and Malan and Schonland (1951b), using the most reliable data of Malan and Schonland (1951a) and some additional measurements, found 43 negative field changes, 8 zero field changes, and 29 positive field changes for 18 flashes in the same distance range.

In addition to the J-field change reversing from primarily negative at close range to mixed polarity with increasing distance from the discharge, the J-change may reverse polarity within a single flash (see C in Fig. 10.1). That is, early interstroke intervals may have positive J-changes and later ones negative J-changes. This type of reversal occurs typically when observations are made at intermediate ranges (Malan and Schonland, 1951b).

The maximum electric dipole moment change (dipole moment change is defined and discussed in Section A.1.2) reported by Brook *et al.* (1962) for all measured J-changes (defined by them as interstroke field changes not associated with channel luminosity) was 1.6 C-km, about 10% of the value of the average moment change for strokes and about 1% of the average moment change for continuing currents. Brook *et al.* (1962) argue that a slow field change of 0.1 V/m, which is 10^{-3} times the ambient fair weather field, is the lower limit for a field to be detectable in the presence of ambient atmospheric noise and that, since a moment change of 1.6 C-km is equivalent to about 0.1 V/m at about 50 km, J-field changes should not be detectable beyond about that distance. Thus some of the positive interstroke field changes observed at relatively large distances and attributed to J-processes (e.g., Malan, 1955; Pierce, 1955) were probably due to continuing current, making the available statistics on the sign of distant J-changes of questionable validity (see also Section 9.2).

It is conceivable that low-level currents could flow to ground during the total interstroke period. These currents would have to produce insufficient channel luminosity to darken photographic film at the distances that streak cameras have been employed. Brook *et al.* (1962) have discussed this possibility. For example, if 10 A were to flow for 50 msec, a charge transfer of 0.5 C would result which, for a charge source height of 5 km, would yield a moment change of 5 C-km, considerably larger than the maximum moment change observed for the J process, 1.6 C-km. Current flow to earth yields a positive field change at all distances. Arguing against the existence of such "photographically dark" currents are the direct current measurements of McCann (1944), who found that in 21 of 24 multiple-stroke flashes the current fell below 0.1 A between strokes, and of Berger (1967), who reported that for lightning to the towers on Mt. San Salvatore the current decreased to less than 1 A between strokes.

10.4   INTERPRETATION OF J-PROCESS ELECTRIC FIELDS

Schonland (1938) proposed that during the time interval between strokes "stepped leaders" traveled from previously untapped negative charge centers in the cloud to the top of the previous return stroke channel at which time a dart leader would be initiated at the top of that channel. He referred to Y-shaped discharge channels with different upper branches of the Y being associated with different strokes. Schonland (1938), in his Fig. 1, depicts separate negative charge sources displaced horizontally in the cloud, each providing charge for a common channel to earth. Interestingly, Malan and Schonland (1951a,b) refuted this early picture on the basis of their single-station electric field measurements of the polarity change of the J-process as a function both of distance and of stroke order at intermediate distances and suggested that the stroke charge regions must be located in a primarily vertical column with the J-process involving the raising of positive charge from the top of a previous return stroke channel into a new region of negative charge above. If most distant J-changes were positive while most close ones were negative, as Malan and Schonland (1951a,b) apparently believed, there would be clear evidence for J-changes being due to the vertical motion of either negative charge downward or positive charge upward (Section A.1.3). Since distant J-changes are apparently a mixture of positive and negative (even if account is taken of the effects of charge relaxation in the atmosphere, e.g., Illingworth, 1971; Krehbiel et al., 1979), the single-station measurements do not represent a persuasive argument for a vertical charge model. Evidence that in-cloud channels of ground discharges are oriented more horizontally than vertically has been provided by a variety of investigators: the two-station electric field measurements of Ogawa and Brook (1969), the eight-station electric field measurements of Krehbiel et al. (1979), various photographic, TV, and visual observations (e.g., Malan, 1956; Brook and Vonnegut, 1960; Brantley et al., 1975), and the location of thunder sources (e.g., Teer and Few, 1974; Section 15.5).

Krehbiel et al. (1979) have deduced from their multiple-station electric field measurements (Figs. 4.2 and 4.3) that the J-process usually moved negative charge horizontally toward the top of a previous stroke in a manner that probably should not be referred to as a "junction"-process since this negative charge was not necessarily the same charge that was involved in the next stroke to ground. Thus the interstroke field changes observed by Krehbiel et al. (1979) were indicative of charge motion that did not always link successive stroke volumes. In fact, charge transport between strokes was observed to persist in the direction of an earlier stroke while subsequent strokes discharged more distant regions of the cloud. Krehbiel et al. (1979) found that the leader and return-stroke field changes appeared to be superimposed upon the interstroke field change, as if independent of it, implying that there was not a strong coupling between the field change associated with the J-process and the following leader–return stroke sequence. Further, Krehbiel *et al.* (1979) suggest that the single-station J-change measurements of Malan and Schonland (1951a,b) in South Africa are consistent with the New Mexico measurements and can be interpreted as due to predominantly horizontal channels rather than to the vertical ones assumed by Malan and Schonland (1951a,b).

10.5   THE K-PROCESS

The K-process is generally viewed as a "recoil streamer" or small return stroke that occurs when a propagating discharge within the cloud encounters a pocket of charge opposite to its own (Ogawa and Brook, 1964). In this view, the J-process represents a slowly propagating discharge which initiates the K-process. Ogawa and Brook (1964) have supplied evidence from electric field and streak-photographic measurement that this is the case for K-changes in cloud discharges (Section 13.5). It is reasonable to expect that cloud discharge K-changes are similar to K-changes in the in-cloud portion of ground discharges, particularly in view of the fact that the distribution of time intervals between K-changes was found by Kitagawa and Brook (1960) to be almost identical for cloud and for ground discharges (Section 13.5). Brook and Vonnegut (1960) and Kitagawa and Brook (1960) suggest that the slow J-field change can be interpreted as due to the instrumental time integration of a series of rapid K-field changes of duration less than 1 msec and moment change between a few hundredths and 1 C-km. In this view, the J-change is the smoothed trace of the electric field record which actually consists of a number of small K-change steps. Ogawa and Brook (1964) state that the steps associated with K-processes observed at close range constitute the major contribution to the overall field change associated with the last stage of a cloud discharge (Section 13.5). It is not clear why the recoil process should produce a significant field change while the slowly propagating discharge that initiates it does not, except perhaps that the K-change moves a relatively large charge a relatively long distance compared to the equivalent charge motion in the time between K-changes. It is interesting to note that apparently another view of the K-process has been suggested by Kitagawa *et al.* (1962) who argue that the K-change "is evidence of the movement of penetrative streamers into fresh regions of cloud, streamers whose occurrence must be determined wholly by conditions inside the cloud."

The detailed characteristics of the K-change currents are very much in doubt, although there is evidence that there may be one or more fast pulses and considerable VHF noise over an interval from less than 50 to about 1000 μsec (Kitagawa and Kobayashi, 1959; Brook and Kitagawa, 1964; Proctor, 1976; Hayenga, 1969; Rustan et al., 1980; Hayenga and Warwick, 1981). Rustan et al. (1980) measured for a three-stroke flash that struck the 150-m weather tower at the Kennedy Space Center, the location of four sets of VHF sources whose radiation was termed "solitary pulses" and was probably associated with K-changes and found them to propagate upward for a few kilometers at speeds between 1 and 4 × 10⁷ m/sec. Krehbiel et al. (1984) point out the possibility of location errors in the VHF time-of-arrival technique used (Section C.7.3) and suggest from their own multiple-station electric field measurements (Section C.7.1) on the same flash that the interstroke discharges were more horizontal than vertical. Apparently the same phenomena, observed by VHF interferometry (Section C.7.3) in New Mexico, was found to be oriented horizontally by Hayenga (1979, 1984) and Hayenga and Warwick (1981). They refer to these events as "fast bursts," which have a duration of about 20 μsec, exhibit a propagation speed of about 10⁷ m/sec, and occur on average every 100 μsec. Proctor (1976) in South Africa also finds similar horizontal discharges that he refers to as Q-noise. Speeds of propagation were between 2 × 10⁶ and 2 × 10⁷ m/sec with lengths less than 1 km. Similar propagating Q-noise was observed by Proctor (1981) in cloud discharges to be accompanied by K-electric field changes and was attributed to recoil streamers (Section 13.7).

Arnold and Pierce (1964) have reviewed the literature on the ratio of the peak electric field of K-changes to those of return strokes. They find a typical ratio of about 0.1 which implies that K-processes are associated with peak currents not much different than those of return strokes, since the K-processes apparently have velocities an order of magnitude lower than return strokes (see paragraph above and Section A.4.3). Arnold and Pierce (1964) have combined the data of Steptoe (1958) and Ishikawa (1961) to produce a model K-process current. This current has a rise to peak of 7500 A in 9 μsec and a fall to half value in tens of microseconds, the current propagating along a previous channel at an initial speed of 2 × 10⁷ m/sec. Kitagawa and Brook (1960) show that the distribution of time intervals between K-changes in the interstroke intervals of discharge to ground, the mean time interval being 8.5 msec, is essentially the same as the distribution of K-change intervals in the final portion, the so-called J-portion, of an intracloud discharge (Section 13.5) although the polarity of the K-changes observed far from the discharge is generally different in cloud and ground discharge. Ogawa and Brook (1964) found cloud discharge K-changes, which they attributed to mini-return strokes raising negative charge, to have positive field changes for close discharges (less than about 5 km), negative field changes beyond about 6 km, and initially negative changes followed by positive changes, although being overall negative, in the intermediate range (Section 13.5). Thomson (1980) reported that K-changes in ground flashes in the 6- to 40-km range usually had negative field changes, opposite in polarity to the field change of the ground strokes, but apparently the same polarity as the J-changes. Ogawa and Brook (1964) state that the K-changes in cloud discharges are an order of magnitude larger than K-changes in ground discharges (Section 13.5) while others (e.g., Ishikawa, 1961; Wadhera and Tantry, 1967a,b) claim they are of the same magnitude.

Kitagawa et al. (1962) suggest that there is no essential difference between the K-changes associated with M-components (Section 9.4) and K-changes that occur during the nonluminous interstroke periods. Kitagawa et al. (1962) report that in the absence of continuing luminosity (and M-components) K-changes are sometimes associated either with luminosity in the cloud or with a luminous wave similar to the dart leader, which propagates down the channel but fails to reach the ground.

REFERENCES

Arnold, H. R., and E. T. Pierce: Leader and Junction Processes in the Lightning Discharge as a Source of VLF Atmospherics. *Radio Sci.*, **68D**:771–776 (1964).

Berger, K.: Novel Observations on Lightning Discharges. *J. Franklin Inst.*, **283**:478–525 (1967).

Brantley, R. D., J. A. Tiller, and M. A. Uman: Lightning Properties in Florida Thunderstorms from Video Tape Records. *J. Geophys. Res.*, **80**:3402–3406 (1975).

Brook, M., and N. Kitagawa: Radiation from Lightning Discharges in the Frequency Range 400 to 1000 Mc/s. *J. Geophys. Res.*, **69**:2431–2434 (1964).

Brook, M., and B. Vonnegut: Visual Confirmation of the Junction Processes in Lightning Discharges. *J. Geophys. Res.*, **65**:1302–1303 (1960).

Brook, M., N. Kitagawa, and E. J. Workman: Quantitative Study of Strokes and Continuing Currents in Lightning Discharges to Ground. *J. Geophys. Res.*, **67**:649–659 (1962).

Funaki, K., K. Sakamoto, R. Tanaka, and N. Kitagawa: A Comparison of Cloud and Ground Lightning Discharges Observed in South-Kanto Summer Thunderstorms, 1980. *Res. Lett. Atmos. Electr.*, **1**:99–103 (1981).

Hayenga, C. O.: Positions and Movement of VHF Lightning Sources Determined with Microsecond Resolution by Interferometry. Ph.D. thesis, University of Colorado, Boulder, Colorado, 1979.

Hayenga, C. O.: Characteristics of Lightning VHF Radiation near the Time of Return Strokes. *J. Geophys. Res.*, **89**:1403–1410 (1984).

Hayenga, C. O., and J. W. Warwick: Two-Dimensional Interferometric Positions of VHF Lightning Sources. *J. Geophys. Res.*, **86**:7451–7462 (1981).

Hewitt, F. J.: Radar Echoes from Inter-Stroke Processes in Lightning. *Proc. Phys. Soc. London, Ser. B*, **70**:961–979 (1957).

Illingworth, A. J.: Electric Field Recovery after Lightning as the Response of the Conducting Atmosphere to a Field Change. *Q. J. R. Meteorol. Soc.*, **98**:604–616 (1972).

Ishikawa, H.: Nature of Lightning Discharges as Origins of Atmospherics. *Proc. Res. Inst. Atmos, Nagoya Univ.*, **8A**:1–273 (1961).

Kitagawa, M.: On the Mechanism of Cloud Flash and Junction or Final Process in Flash to Ground. *Pap. Meterol. Geophys. (Tokyo)*, **7**:415–424 (1957).

Kitagawa, N.: Types of Lightning. *In* "Problems of Atmospheric and Space Electricity" (S. C. Coroniti, ed.), pp. 337–348. American Elsevier, New York, 1965.

Kitagawa, N., and M. Brook: A Comparison of Intracloud and Cloud-to-Ground Lightning Discharges. *J. Geophys. Res.*, **65**:1189–1201 (1960).

Kitagawa, N., and M. Kobayashi: Field Changes and Variations of Luminosity Due to Lightning Flashes. *In* "Recent Advances in Atmospheric Electricity" (L. G. Smith, ed.), pp. 485–501. Pergamon, Oxford, 1959.

Kitagawa, N., M. Brook, and E. J. Workman: Continuing Currents in Cloud-to-Ground Lightning Discharges. *J. Geophys. Res.*, **67**:637–647 (1962).

Kobayashi, M., N. Kitagawa, T. Ikeda, and Y. Sato: Preliminary Studies of Variation of Luminosity and Field Change Due to Lightning Flashes. *Pap. Meteorol. Geophys. (Tokyo)*, **9**:29–34 (1958).

Krehbiel, P. R., M. Brook, and R. McCrory: Analysis of the Charge Structure of Lightning Discharges to Ground. *J. Geophys. Res.*, **84**:2432–2456 (1979).

Krehbiel, P. R., M. Brook, S. Khanna-Gupta, C. L. Lennon, and R. Lhermitte: Some Results Concerning VHF Lightning Radiation from the Real-Time LDAR System at KSC, Florida. *In* "Preprints, 17th Int. Conf. Atmos. Electr., June 3–8, 1984, Albany, New York" (R. E. Orville, ed.), pp. 388–393. Am. Meteorol. Soc., Boston, Massachusetts, 1984.

McCann, G. D.: The Measurement of Lightning Currents in Direct Strokes. *Trans. AIEE*, **63**:1157–1164 (1944).

Mackerras, D.: A Comparison of Discharge Processes in Cloud and Ground Lightning Flashes. *J. Geophys. Res.*, **73**:1175–1183 (1968).

Malan, D. J.: La Distribution Verticale de la Charge Negative Orageuse. *Ann. Geophys.*, **11**:420–426 (1955).

Malan, D. J.: Visible Electrical Discharges Inside Thunderclouds. *Geofis. Pura Appl.*, **34**:221–223 (1956).

Malan, D. J.: The Theory of Lightning. *In* "Problems of Atmospheric and Space Electricity" (S. C. Coroniti, ed.), pp. 323–331. American Elsevier, New York, 1965.

Malan, D. J., and B. F. J. Schonland: The Electrical Processes in the Intervals between the Stroke of a Lightning Discharge. *Proc. R. Soc. London Ser. A*, **206**:145–163 (1951a).

Malan, D. J., and B. F. J. Schonland: The Distribution of Electricity in Thunderclouds. *Proc. R. Soc. London Ser. A*, **209**:158–177 (1915b).

Ogawa, T., and M. Brook: The Mechanism of the Intracloud Lightning Discharge. *J. Geophys. Res.*, **69**:5141–5150 (1964).

Ogawa, T., and M. Brook: Charge Distribution in Thunderstorm Clouds. *Q. J. R. Meteorol. Soc.*, **95**:513–525 (1969).

Pathak, P. P., J. Rai, and N. C. Varshneya: VLF Radiation from Lightning. *Geophys. J.R. Astron. Soc.*, **69**:197–207 (1982).

Pierce, E. T.: Electrostatic Field Changes Due to Lightning Discharges. *Q. J. R. Meteorol. Soc.*, **81**:211–228 (1955).

Proctor, D. E.: A Radio Study of Lightning. Ph.D. thesis, University of Witwatersrand, Johannesburg, South Africa, 1976.

Proctor, D. E.: VHF Radio Pictures of Cloud Flashes. *J. Geophys. Res.*, **86**:4041–4071 (1981).

Rustan, P. L., Jr.: Properties of Lightning Derived from Time Series Analysis of VHF Radiation Data. Ph.D. thesis, University of Florida, Gainesville, Florida, 1979.

Rustan, P. L., M. A. Uman, D. G. Childers, W. H. Beasley, and C. L. Lennon: Lightning Source Locations from VHF Radiation Data for a Flash at Kennedy Space Center. *J. Geophys. Res.*, **85**:4893–4903 (1980).

Schonland, B. F. J.: Progressive Lightning, Pt. 4, The Discharge Mechanisms. *Proc. R. Soc. London Ser. A*, **164**:132–150 (1938).

Schonland, B. F. J.: The Lightning Discharge. *Handb. Phys.*, **22**:576–628 (1956).

Steptoe, B. J.: Some Observations on the Spectrum and Propagation of Atmospherics. Ph.D. thesis, University of London, London, 1958.

Takagi, M.: The Mechanism of Discharges in a Thundercloud. *Proc. Res. Inst. Atmos., Nagoya Univ.*, **8B**:1–105 (1961).

Teer, T. L., and A. A. Few: Horizontal Lightning. *J. Geophys. Res.*, **79**:3436–3441 (1974).

Thomson, E. M.: Characteristics of Port Moresby Ground Flashes. *J. Geophys. Res.*, **85**:1027–1036 (1980).

Wadhera, N. S., and B. A. P. Tantry: VLF Characteristics of K Changes in Lightning Discharges. *Indian J. Pure Appl. Phys.*, **5**:447–449 (1967a).

Wadhera, N. S., and B. A. P. Tantry: Audio Frequency Spectra of K Changes in a Lightning Discharge. *J. Geomagn. Geolectr.*, **19**:257–260 (1967b).

Wang, C. P.: Lightning Discharges in the Tropics-2. Component Ground Strokes and Cloud Dart Streamer Discharges. *J. Geophys. Res.*, **68**:1951–1958 (1963).

Chapter 11 | Positive Lightning

11.1 INTRODUCTION

Most cloud-to-ground lightning lowers negative charge to earth. Those flashes that lower positive charge are generally referred to as positive lightning. Although the rest of this book is primarily concerned with the more common and more extensively studied negative cloud-to-ground lightning, discussions of various aspects of positive lightning are found in Sections 1.4, 2.6, 3.2 (Fig. 3.3), 5.3, 7.2.2, 7.2.4, 9.1, 9.5, 12.2, and 12.3. Positive lightning is of particular interest because it is responsible for the largest of the recorded lightning currents, those in the 200- to 300-kA range, and for charge transfers to earth that are considerably larger than those of negative flashes.

11.2 OCCURRENCE STATISTICS AND GENERAL PROPERTIES

Bruce and Golde (1941) have reviewed the early evidence for the existence of positive flashes as inferred from the sign of measured electric and magnetic field changes and from direct current measurements made on grounded structures. In the 17 papers they reference, the percentage of positive flashes ranges from essentially 0 to about 30%.

From peak current measurements made on transmission line towers in the United States, Lewis and Foust (1945) found that 18% of 2721 currents recorded were due to positive ground flashes. They found a correlation between the altitude of the line and the percentage of flashes lowering positive charge, the range being from about 3% near sea level to about 30% in the mountains of Colorado at altitudes of 2–4 km above sea level.

Pierce (1955a), working in England, made electric field measurements with sufficient time resolution to identify rapid negative changes in the field (such as those shown in Figs. 1.9, 1.12, 4.3, and 9.1 but of opposite polarity) which he attributed to return strokes in positive flashes. For flashes that were too distant to be observed visually, he found 34 overall slow field changes that were negative and had rapid negative changes, of which six had more than one rapid field change per flash, while he observed 373 slow positive field changes containing rapid positive changes, which he attributed to normal negative strokes. Pierce (1955b) suggested that his observations of prestroke field magnitude and duration (Section 11.4) indicated that those flashes lowering positive charge were initiated in the P region of the cloud (Chapter 3 and Figs. 3.1, 3.2, and 3.4). As further evidence of this supposition, he observed that flashes lowering positive charge most frequently occurred in the later stages of a storm, when the P region could be expected to be relative less shielded from ground by the N region as a result of the lateral displacement of the main charges.

From measurements of the magnetic fields accompanying over 2000 visually observed ground strokes in 1000 flashes, Norinder (1956), working in Sweden, inferred that about 90% of the strokes and flashes lowered predominantly negative charge and about 3% lowered positive charge, the remainder lowering both negative and positive charge (or at least producing complex magnetic fields that were interpreted from theory as indicating the lowering of both charge signs).

Berger and Vogelsanger (1965) reported that 15% of all downward-moving leaders to the towers on Mt. San Salvatore in Switzerland (Section 7.2.2) were positively charged. This inference was made on the basis of percentage of impulsive currents in the towers that indicated positive charge was lowered. Berger (1972) and Berger et al. (1975), in summarizing all of the San Salvatore data, listed about 20% of the lightning as being initiated by downward-moving leaders lowering positive charge, all discharges being composed of single-current pulses (Table 7.2). More recently, however, Berger (1977, 1978) has suggested that these positive discharges are better described as being due to upward-moving negative leaders rather than to downward-moving positive leaders and that, in his opinion, positive downward leaders above mountainous terrain are almost always met by relatively long upward-moving negative leaders. Berger (1977, 1978) argues that in this type of discharge, a cloud flash causes rapid electric field changes at the ground initiating an upward-moving negative leader typically from a few hundred meters to a few kilometers in length which subsequently connects with the cloud flash. Apparently, there is no direct photographic evidence to indicate what type of discharge was actually met by the long upward-going leaders, whose length was estimated from photographically measured velocities near the tower and the measured time durations (generally 4–25 msec) of the tower currents (generally 100–1000 A).

Studying lightning to the Empire State Building, Hagenguth and Anderson (1952) found 2 of 84 recorded current peaks were of positive polarity. The maximum current recorded, 58 kA, and the largest charge transferred in the time to current half-value, 4.6 C, were due to a positive stroke.

Mackerras (1968), working in Australia, reported from electric field measurements that 8 of 100 ground flashes within 6 km lowered positive charge.

Fuquay (1982) using electric field and optical measurements found that about 3% (75 events) of the lightning studied in a mountainous region of Montana was positive. The measurements were made during three summers in which there were 48 days with lightning. From 1 to 11 positive flashes occurred on each of 16 of these days. All were single-stroke flashes, most, if not all, were initiated by downward-moving positive leaders, and all were followed by field changes that were interpreted as being due to continuing current. The positive flashes occurred at the ends of thunderstorms after negative lightning to ground had ceased. Intracloud discharges often were also present during these final storm periods. Positive discharges photo-graphed and observed visually were reported to exhibit long horizontal components through clear air.

Rust *et al.* (1981) using electric field and television measurements documented 31 positive flashes to earth in five severe spring thunderstorms in Oklahoma and show a microsecond-scale electric field attributed to a positive return stroke. The positive flashes were a small fraction of the total lightning observed. They generally came from high in the cloud, and none was associated with the precipitation core in which negative lightning was common. Most positive flashes were composed of a single stroke and about half of the positive flashes had electric fields indicative of a continuing current following the ground stroke. To support the view that these fields were indeed due to continuing currents and not to some process in the cloud, Rust *et al.* (1985) present simultaneous streak-camera, TV, and electric field records for four single-stroke positive flashes, proving that in those typical cases channel luminosity was correlated with an electric field change characteristic of continuing current.

Orville *et al.* (1983), using a magnetic direction-finding system (Sections C.2 and C.7.2) that covered the northeastern United States, found that for a severe fall storm (October 1982) that produced over 11,000 flashes to ground in about 22 hr, the percentage of positive ground flashes increased with the age of the storm, reaching 37% in the last hour of significant activity. The average positive flash percentage was 4. Orville *et al.* (1983) state that the percentage of positive flashes had been observed to increase toward the end of other storms as well. The same lightning detection system has been employed both in the United States (Orville *et al.*, 1987) and in Scandinavia (Huse and Olsen, 1984) to show that the percentage of positive ground flashes is smallest in summer and highest in winter. For example,

Orville *et al.* (1987) found that about 80% of the overall lightning in the eastern United States in February 1985 was positive, decreasing to less than 10% in April and to a few percent during the summer. The 1-year sample included about 720,000 negative flashes and about 17,700 positive flashes. In the winter, most flashes occurred over the Atlantic Ocean. Idone *et al.* (1984), using the same lightning detection system, describe a storm in July in New York that produced only one cloud-to-ground flash, that being positive. The discharge had an estimated peak current of 70 kA and did considerable damage to a residential structure.

Cooray and Lunquist (1982) and Cooray (1984), in Sweden, characterized electric field waveforms that they attribute to distant positive return strokes on the basis of the overall similarity of the fields to those of negative return strokes (except for polarity) and the dissimilarity of the fields to any known cloud discharge processes. Cooray (1986) shows that correlation of these positive waveforms with the onset and temporal characteristics of the associated 3-MHz radiation indicates that the positive waveforms are radiated from near the ground and hence are due to return strokes (see also the similar discussion for negative return strokes in Section 7.2.1). Hojo *et al.* (1985) provide a characterization of positive return strokes similar to that of Cooray and Lundquist (1982) and Cooray (1984), but for both winter and summer storms in Japan. Cooray and Lundquist (1982) state that in an unpublished study in Sweden it was found that 10–15% of the ground flashes lowered positive charge to earth. Beasley *et al.* (1983) in Florida have shown that microsecond-scale electric fields similar to those attributed to positive lightning by Rust *et al.* (1981), Cooray and Lundquist (1982), and Cooray (1984) are accompanied by microsecond-scale optical signals radiated from near ground and hence that there can be little doubt that the field waveshapes in question are associated with positive return strokes. During 15 min (one instrumentation tape) of a summer convective thunderstorm, Beasley *et al.* (1983) recorded the fields and light of 31 negative first strokes and 3 positives. Pictures of the channels of these 3 positive flashes were also recorded on video tape. In addition, three additional electric field waveforms indicative of positive return strokes were recorded out of the field-of-view of the optical detector.

Positive lightning to normal ground has been reported in winter thunderstorms in Japan (Takeuti *et al.*, 1973, 1976–1978, 1980; Brook *et al.*, 1982; Hojo *et al.*, 1985) and in Norway (Takeuti *et al.*, 1983, 1985). These studies used remote electric field measurements supplemented by photographic and TV recordings. A drawing of the postulated charge structure of a winter thunderstorm and of a positive lightning to ground is given in Fig. 3.3. Positive lightning in winter storms apparently is generated by clouds that have the normal dipole structure, positive charge higher than negative

(Takeuti *et al.*, 1978; Brook *et al.*, 1982), but with the dipole severely tilted (Brook *et al.*, 1982). The picture drawn (e.g., Brook *et al.*, 1982; Takagi *et al.*, 1986) is one in which the upper positive charge is horizontally separated from the lower negative so that positive downward leaders from the upper charge region are not intercepted by the lower negative charge. Experimental evidence for this view is found in the fact that the percentage of positive flashes increases with wind shear (Takeuti *et al.*, 1980; Brook *et al.*, 1982). Brook *et al.* (1982), on the basis of the data from Japanese winter storms, give a minimum wind shear of 1.5 m/sec/km to produce any positive lightning and a value of about 7 m/sec/km to produce only positive lightning, although Takeuti *et al.* (1985) present observations to show that this is not necessarily the case in Norwegian winter storms. Brook *et al.* (1982) report observing 26 positive and 37 negative ground flashes (out of 264 total flashes, both cloud discharges and ground discharges) in 8 Japanese winter thunderstorms, so that while winter thunderstorms may not produce large number of ground flashes, they apparently produce relative high percentages of positive flashes. In this regard, Takeuti *et al.* (1983) found five positive and no negative ground flashes out of eight total flashes in two Norwegian winter thunderstorms.

Although most of the positive cloud-to-ground lightning to normal ground in winter thunderstorms is thought to be initiated by downward-moving positive leaders (Takeuti *et al.*, 1978; Brook *et al.*, 1982), Takeuti *et al.* (1976) show an upward-branched channel, implying an upward-moving negative leader, and several of the multiple-station electric field records reproduced in the literature can be interpreted as due to upward-moving negative leaders (e.g., flash 10 and 20 of storm B, Takeuti *et al.*, 1978). In the study on Mt. San Salvatore, either downward-moving positive leaders or cloud discharge channels were often met by upward-moving negative leaders of up to kilometers in length. Thus, it is perhaps not unreasonable to expect positive flashes to nonmountainous terrain to be characterized by upward-moving negative leaders of length greater than the 10 to a few hundred meters commonly found for the upward-moving positive leaders associated with normal negative lightning striking either towers or the earth (see Chapter 6).

11.3 PHOTOGRAPHIC MEASUREMENTS

Television images or still photographs of positive flashes to ground have been published, for example, by Takeuti *et al.* (1976, 1978, 1980), Fuquay (1982), and Beasley *et al.* (1983). Some of these show the long horizontal channel sections often reported to be associated with positive flashes (e.g., Fuquay, 1982; Takeuti *et al.*, 1978).

Apparently the only streak photograph of a downward-moving positive stepped leader is due to Berger and Vogelsanger (1966) (also reproduced by Berger, 1967, 1977). That streak photograph and still photographs of the same flash are shown in Fig. 11.1. The leader was identified as positively charged solely on the basis of the similarity in luminous characteristics to

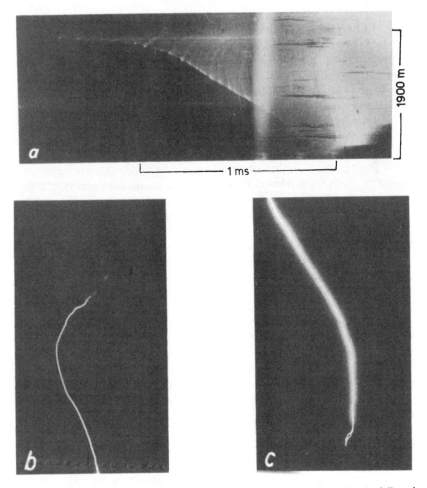

Fig. 11.1 (a) Streak photograph of the last millisecond of a positive stepped leader followed by a positive return stroke obtained from a range of 3.3 km, (b) a still camera view from the same location as a, (c) a different still camera view showing the strike point on Lake Lugano and the "loop" connecting upward- and downward-going leaders (Section 6.2). Previously published by Berger and Vogelsanger (1966) and Berger (1967, 1977). Courtesy, K. Berger and The Swiss High Voltage Research Committee (FKH), Zurich.

upward-moving positively charged leaders from the Mt. San Salvatore towers (Section 12.2, Fig. 12.5). Positive stepped leaders do not exhibit the clear pulsing on and off of luminosity characteristic of negative stepped leaders. Rather they show a luminosity that is more or less continuous but modulated in intensity.

11.4 ELECTRIC FIELDS AND OPTICAL PROPERTIES

Electrical fields on a millisecond time scale attributed to positive ground flashes have been published, for example, by Pierce (1955a), Fuquay (1982), Rust et al. (1981, 1985), Beasley et al. (1983), Beasley (1985), Takeuti et al. (1978), and Brook et al. (1982), the latter two papers describing lightning in winter thunderstorms in Japan. Millisecond-scale field waveforms preceding and following a single positive stroke are shown in Fig. 11.2, and statistics derived from waveforms on that time scale are found in Table 11.1. Electric fields from positive flashes observed on a slow time scale are apparently similar to fields from normal negative flashes except for polarity, the fact that only one return-stroke field change is generally observed, and possibly the fact that a slow field change, which can be interpreted as due to relatively large continuing current, commonly follows the ground stroke. Additionally, Pierce (1955a) states that the field changes he observed preceding positive strokes were of significantly longer duration than, and had nearly twice the magnitude of, those preceding the negatives observed in the same study (Table 11.1) leading him to postulate that the discharges were initiated high in the cloud (Section 11.2).

Electric fields of positive return strokes on a microsecond or submicrosecond time scale have been published by Rust et al. (1981, 1985), Cooray and Lundquist (1982), Beasley et al. (1983), Cooray (1984, 1986), and Hojo et al. (1985). An example of electric fields on that time scale are shown in Fig. 11.2. Also shown is the correlated luminosity emitted by about 100 m of channel at a height between 0.5 and 1 km above ground. Note that the light signal begins some microseconds after the electric field signal as would be expected if the return stroke luminosity propagated upward from near ground level at a speed of about 10^8 m/sec and the beginning of the electric field signal was generated when the return stroke luminosity was near the channel base. Note also the slow field increase following the return-stroke field change. This slow field is indicative of either continuing current to ground or of charge motion in the cloud, with photographic evidence (Rust et al., 1985) and evidence from tower current and multiple-station field measurements indicating the likelihood that the field is due to continuing current (Section 11.5).

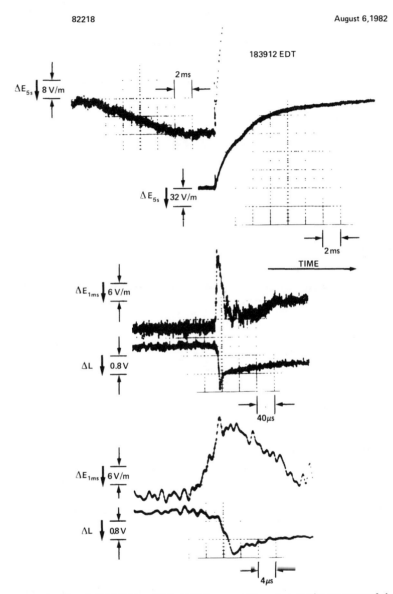

Fig. 11.2   Electric field (ΔE) and light (ΔL) from a positive return stroke at a range of about 20 km in Florida. The top two waveforms show the millisecond-scale electric field. The center and bottom pair of waveforms are time-correlated electric field and light near ground on faster time scales. Note the delay, 4 to 7 μsec, between electric field and light signals and the large, slowly varying electric field after the return stroke. Adapted from Beasley *et al.* (1983).

Table 11.1

Statistics on Vertical Electric Fields from Flashes Lowering Positive Charge to Earth

	Number of strokes	Mean	Standard deviation	Range of values
Duration of prestroke field change (msec)				
Pierce (1955a)				
L(β)[a]	15	310		
L(α)[a]	8	45		
Rust *et al.* (1981)	25	241		40–800
Fuquay (1982)	41	130	36	65–210
Duration of field change associated with stepped leader (msec)				
Fuquay (1982)	13	50		40–70
Fraction of positive strokes followed by continuing current				
Takeuti *et al.* (1978)	12	0.7		
Rust *et al.* (1981)	31	0.5		
Brook *et al.* (1982)	12	0.8		
Fuquay (1982)	75	1.0		
Beasley *et al.* (1983)	3	1.0		
Duration of field change associated with continuing current (msec)				
Rust *et al.* (1981)	14	121		30–240
Fuquay (1982)	58	50		5–160
Fraction of total electrostatic field change due to return stroke				
Rust *et al.* (1981)	20	0.12		0.01–0.54
Flash duration (msec)				
Rust *et al.* (1981)	31	520		100–1200

[a] Pierce labels those total prestroke electric field changes in which there is a flat or I portion (of a B, I, L-like waveform as discussed in Chapter 4) as L(β) and those without the I portion as L(α). For 332 negative flashes, of which 15% were L(β), the mean L(α) duration was 50 msec and the mean L(β) duration 175 msec.

Some statistics on positive return-stroke electric fields are given in Table 11.2. Comparison can be made with similar parameters of negative return strokes listed in Table 7.1. Perhaps the most meaningful comparisons of positive vs negative stroke properties are due to Cooray and Lundquist (1982) in Sweden who studied negative and positive return strokes in the same storm with the same equipment and to Hojo *et al.* (1985) in Japan who studied negative and positive strokes over the same time period and the same general area for both summer and winter storms. Comparison between these data and other studies probably should not be made because of the potentially different affect of propagation on the waveshapes in different studies (Sections 7.2.1 and 7.3). As can be seen by comparing data in Tables 7.1 and 11.2, Cooray and Lundquist (1982) found mean peak normalized (to 100 km) return-stroke fields from positive strokes to be about two times those from negatives and zero-to-peak risetimes and slow front durations for positives about twice as long as for negatives. On the other hand, the slow front

Table 11.2

Statistics on Return-Stroke Vertical Electric Fields from Strokes Lowering Positive Charge to Earth

	Number of strokes	Mean	Standard deviation	Range of values
Initial peak normalized to 100 km (V/m)				
Cooray and Lundquist (1982)	58	11.5	4.5	4.5–24.3
Zero-to-peak risetime (μsec)				
Rust et al. (1981)	15	6.9		4–10
Cooray and Lunquist (1982)	64	13	4	5–25
	52	12	3	5–25
Cooray (1986)	20	8.9	1.7	4–12
Hojo et al. (1985)	—	22.3		
10–90% risetime (μsec)				
Beasley et al. (1983)	6			1.6, 2.0, 4.5
				1.2, 2.8, 4.0
Hojo et al. (1985)				
Winter	32	8.7		
Summer	44	6.7		
Cooray (1986)	15	6.2	1.4	3–9
Slow front duration (μsec)				
Cooray and Lundquist (1982)	63	10	4	3–23
	33	9	3	3–19
Cooray (1986)	20	8.2	1.7	3–11
Hojo et al. (1985)	—	19.3		
Slow front amplitude as percentage of peak				
Cooray and Lundquist (1982)	57	39	11	10–70
	31	44	14	10–80
Cooray (1986)	20	45	7	30–60
Fast transition 10–90% risetime (nsec) for propagation over salt water				
Cooray (1986)	20	560	70	400–800

amplitude as a percentage of the initial peak was about the same for strokes of either polarity. Supporting the observation that initial peak fields from positive return strokes are larger than from negatives are the measurements of Orville *et al.* (1987) who reported that the median of about 17,700 peak magnetic fields from positive strokes exceeded the median of about 720,000 negative strokes by 50%. Cooray and Lundquist (1982) also compared the separation time between the stepped-leader pulses just preceding return strokes and found a mean of 26 μsec for positive strokes and 14 μsec for negatives. The latter value is in good agreement with the values given in Chapter 5. Apparently, however, most of the positive waveforms did not show stepped-leader pulses above the noise level, so one might question whether all positive pulses were being identified on those records that did show pulses. Cooray and Lundquist (1982) and Cooray (1984) state that electric radiation fields from both distant positive and negative strokes have similar behavior for 50–100 μsec following the peak field, after which time some of the positive waveforms reach a second peak at a few hundred microseconds, a waveform feature that was not present in the negative strokes observed. They suggest that the second field peak in positive strokes is due to an additional surge of current down the positive stroke channel, a feature observed on tower measurements (Section 11.5).

Hojo *et al.* (1985) found a mean first stroke 10–90% field rise of 8.7 μsec for positive strokes in winter storms and 3.9 μsec for negatives. For summer storms, the mean positive stroke field risetime was 6.7 μsec and the mean negative was 4.0 μsec. For winter storms, the mean zero-to-peak risetime for positive stroke fields was 22.3 μsec and for negatives was 7.5 μsec, the difference being due to the longer slow fronts associated with the positive strokes since the fast transitions for strokes closer than 50 km were in the range 0.6–1.0 μsec. The mean duration of the slow front for positives was 19.3 μsec and for negatives was 6.0 μsec. Thus the slow front duration observed by Hojo *et al.* (1985) in Japan is significantly longer than that observed by Cooray and Lundquist (1982). Only one-third to one-fourth of all the positive first return strokes studied by Hojo *et al.* (1985) had discernible leader pulses. For these, the mean time interval between leader pulses in the 100 μsec just prior to the return stroke was 17.4 μsec.

11.5 CURRENT AND CHARGE TRANSFER

Current and charge transfer in positive lightning have been determined directly from current measurements on towers and indirectly from analysis of electric field records. We consider the former first.

Berger *et al.* (1975) give statistics on the waveshapes of 26 positive current impulses measured on towers on Mt. San Salvatore, as well as presenting some typical waveforms. Some of these statistics are found in Table 7.2 and Fig. 7.8. Tracings of typical return-stroke current waveforms are shown in Fig. 11.3, and a specific waveform is shown in Fig. 11.4 on a time scale so that the current attributable to the negatively charged upward-going leader is included. From the available data it is evident that the measured positive currents have slower current risetimes and much slower current falltimes than do negatives. The median positive stroke front duration of 22 μsec with a 5% value of 200 μsec is about 4 times as long as that of the comparable negative first stroke and is probably to be associated with the effects of the long upward-going leader on an impulse current waveform that starts at the top of that leader but is measured at the bottom (Section 11.2). In this regard,

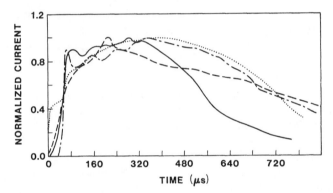

Fig. 11.3 Four typical current waveforms from positive return strokes. Each waveform is normalized so that its peak value is unity. Adapted from Berger *et al.* (1975).

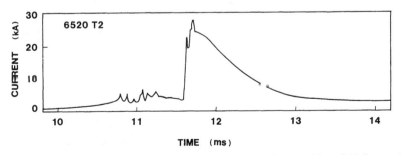

Fig. 11.4 Current in tower on Mt. San. Salvatore during latter portion of 11.6 msec of upward-going negative leader and following positive return stroke. Adapted from Berger (1977).

Berger (1967) presents data to show that the maximum rate of change of current in the rise to peak is inversely related to the length of the upward-going leader. The slower positive current decay after peak is reflected in Table 7.2 in that (1) the median flash duration is about 7 times larger for positive than for negative single-stroke flashes, (2) the median action integral ($\int i^2\,dt$) is an order of magnitude larger for positive than for negative first strokes, (3) the median impulse charge is a factor of 3 larger for positive than for negative single-stroke flashes, and (4) the median total flash charge is over an order of magnitude larger for positive than for negative single-stroke flashes. Median first-stroke peak currents are about the same for positive and negative strokes, 35 kA for positives and 30 kA for negatives, but the positives contain a higher percentage of relatively large current peaks, as illustrated in Fig. 7.8. In fact, the largest reliably measured currents are from positive strokes (6 above 100 kA of which one is above 200 kA and one near 300 kA, Berger, 1972). The median charge transfer for positive impulse current was 16 C with 150 C at the 5% level, while total positive flash charge, due to the single impulsive current plus all following lower level current was 80 C with 350 C at the 5% level. Clearly, positive flashes as measured on towers lower the bulk of their total charge via continuing currents which flow after the primary impulse.

Berger (1978), in a paper primarily on upward-initiated lightning, has included statistics for 35 positive return strokes occurring between 1963 and 1973, two additional years beyond the work of Berger et al. (1975). He found a median current peak of 36.5 kA with the 10 and 90% values of 127 and 10.4 kA, respectively. The median charge transfer for these flashes was 84.2 C, with 348 and 20.4 C at the 10 and 90% levels, respectively. For the median action integral and its 10 and 90% values, Berger (1978) reported 6.6×10^5, 9×10^6, and 5×10^4 A^2·sec, for front time 39, 340, and 4.5 μsec, for current rate of rise 1.9, 12.2, and 0.28 kA/μsec, and for flash duration 68, 240, and 19 msec.

Analysis of simultaneous multiple ground-based electric field measurements allows the determination of initial charge location and charge transferred to ground, as well as an estimate of the current flowing in the channel to ground on a millisecond time scale (Krehbiel et al., 1979). Brook et al. (1982) used this technique for winter thunderstorms in Japan and reported that discharges lowering positive charge to earth often exhibit continuing currents greater than 10^4 A for periods up to 10 msec, with 1 of 12 discharges studied in detail having had a current of the order of 10^5 A at about 2 msec and a charge transfer to earth that exceeded 300 C after 4 msec, at which time the current was about 10^4 A. The continuing currents following positive strokes were about an order of magnitude greater than those for the negative discharges in the same winter storms and more than an order of magnitude greater than continuing currents usually associated with negative discharges to earth (see Section 9.3). Brook *et al.* (1982) found that 10 of 12 positive strokes recorded were followed by continuing current and Takeuti *et al.* (1978) report 8 of 12, as indicated in Table 11.1. The positive charges contributing to positive flashes to earth were located higher than the negative charge contributing to the negative flashes, with wind shear suggested as the mechanism by which the two main charge centers were separated horizontally so that the positive downward-moving leader could bypass the negative charge region (see also Section 11.2).

In addition to the statistics on the occurrence of continuing current in positive discharges in winter thunderstorms given above, Fuquay (1982) reports that all of the 75 positive flashes he studied had field changes that could be interpreted as due to continuing current; Rust *et al.* (1981, 1985) indicate that about half of the positive cloud-to-ground flashes in their two studies had such field changes. All three field changes from positive ground flashes published by Beasley *et al.* (1983) show field changes that could be interpreted as due to continuing current. For two of these flashes Beasley *et al.* (1983) estimated a continuing current of the order of 10^3 A averaged over the first 2 msec falling to a few hundred amperes in about 20 msec. The current values are susceptible of large errors, however, if the range to the lightning was significantly different from the 20-km estimate.

From the available statistics, shown in Table 11.1, it appears that the duration of continuing current for positive lightning is not substantially different from that for negative lightning (Section 9.3). For positive lightning, the bulk of the charge transfer, as determined from both field change (Table 11.1) and tower measurements (earlier in this section), is due to the continuing current. This may well also be the case for negative single-stroke flashes that may have about the same probability of being followed by continuing current as do positives, although there is controversy about this probability for negative single-stroke flashes (Section 9.2).

As noted in the previous section, Cooray and Lundquist (1982) and Cooray (1984) found that in some positive strokes the field waveshapes could be interpreted to indicate that there was a second channel current maximum a few hundred microseconds after the start of the stroke field. Positive stroke current waveshape measured on towers by Berger *et al.* (1975), examples of which are shown in Figs. 11.3 and 11.4, typically have a broad current peak of roughly the same magnitude as the initial peak current at a time of 200–500 μsec. In fact, it is often difficult to identify separate initial and later peaks in the waveforms. It is interesting to note that Mackerras (1973) (Section 7.2.4) reported frequently observing a second peak in optical luminosity a few hundred microseconds after the initial peak in negative strokes to earth, whereas Cooray and Lundquist (1982) and Cooray (1984) do not see evidence of a current responsible for such peaks in their electric field waveforms from distant negative return strokes but do see such evidence for positive return strokes.

REFERENCES

Beasley, W. H.: Positive Cloud-to-Ground Lightning Observations. *J. Geophys. Res.*, **90**:6131–6138 (1985).

Beasley, W. H., M. A. Uman, D. M. Jordan, and C. Ganesh: Positive Cloud to Ground Lightning Return Strokes. *J. Geophys. Res.*, **88**:8475–8482 (1983).

Berger, K.: Novel Observations on Lightning Discharges: Results of Research on Mount San Salvatore. *J. Franklin Inst.*, **283**:478–525 (1967).

Berger, K.: Methoden und Resultate der Blitzforshung auf dem Monte San Salvatore bei Lugano in den Jahren 1963–1971. *Bull. Schweiz. Elektrotech. Ver.*, **63**:1403–1422 (1972).

Berger, K.: Oszillographische Messungen des Feldverlaufs in der Nähe der Blitzeinschläge auf dem Monte San Salvatore. *Bull. Schweiz. Elektrotech. Ver.*, **64**:120–136 (1973).

Berger, K.: The Earth Flash. *In* "Lightning, Vol. 1, Physics of Lightning" (R. H. Golde, ed.), pp. 119–190. Academic Press, New York, 1977.

Berger, K.: Blitzstrom-Parameter von Aufwartsblitzen. *Bull. Schweiz. Elektrotech. Ver.*, **69**:353–360 (1978).

Berger, K., and E. Vogelsanger: Messungen und Resultate der Blitzforschung der Jahre 1955–1963 auf dem Monte San Salvatore. *Bull. Schweiz. Elektrotech. Ver.*, **56**:2–22 (1965).

Berger, K., and E. Vogelsanger: Photographische Blitzuntersuchungen der Jahre 1955–1965 auf dem Monte San Salvatore. *Bull. Schweiz. Elektrotech. Ver.*, **57**:599–620 (1966).

Berger, K., R. B. Anderson, and H. Kroninger: Parameters of Lightning Flashes. *Electra*, **80**:23–37 (1975).

Brook, M., N. Kitagawa, and E. J. Workman: Quantitative Study of Strokes and Continuing Currents in Lightning Discharges to Ground. *J. Geophys. Res.*, **67**:649–659 (1962).

Brook, M., M. Nakano, P. Krehbiel, and T. Takeuti: The Electrical Structure of the Hokuriku Winter Thunderstorms. *J. Geophys. Res.*, **87**:1207–1215 (1982).

Brook, M., P. Krehbiel, D. MacLaughlin, T. Takeuti, and M. Nakano: Positive Ground Stroke Observations in Japanese and Florida Storms. *In* "Proceedings in Atmospheric Electricity" (L. H. Ruhnke and J. Latham, eds.), pp. 365–369. Deepak, Hampton, Virginia, 1983.

Bruce, C. E. R., and R. H. Golde: The Lightning Discharge. *J. Inst. Electr. Eng. (London)*, **88** (Pt. 2):487–524 (1941).

Cooray, V.: Further Characteristics of Positive Radiation Fields from Lightning in Sweden. *J. Geophys. Res.*, **89**:11,807–11,815 (1984).

Cooray, V.: A Novel Method to Identify the Radiation Fields Produced by Positive Return Strokes and Their Submicrosecond Structure. *J. Geophys. Res.*, **91**:7907–7911, 13,318 (1986).

Cooray, V., and S. Lundquist: On the Characteristics of Some Radiation Fields from Lightning and Their Possible Origin in Positive Ground Flashes. *J. Geophys. Res.*, **87**:11,203–11,214 (1982).

Fuquay, D. M.: Positive Cloud-to-Ground Lightning in Summer Thunderstorms. *J. Geophys. Res.*, **87**:7131–7140 (1982).

Garbagnati, E., E. Giudice, and G. B. Lo Piparo: Measurement of Lightning Currents in Italy—Results of a Statistical Evaluation. *Elektrotech. Z. etz-a*, **99**:664–668 (1978).

Hagenguth, J. H., and J. G. Anderson: Lightning to the Empire State Building. *Trans. AIEE*, **71** (Pt. 3):641–649 (1952).

Halliday, E. C.: The Polarity of Thunderclouds. *Proc. R. Soc. London Ser. A*, **138**:205–229 (1932).

Hojo, J., M. Ishii, T. Kawamura, F. Suzuki, and R. Funayama: The Fine Structure in the Field Change Produced by Positive Ground Strokes. *J. Geophys. Res.*, **90**:6139–6143 (1985).

Huse, J., and K. Olsen: Some Characteristics of Lightning Ground Flashes Observed in Norway. "Lightning and Power Systems," pp. 72–76. IEE Conference Publication Number 236. Inst. Electr. Eng., London and New York, 1984.

Idone, V. P., and R. E. Orville: Lightning Return Stroke Velocities in the Thunderstorm Research International Program (TRIP). *J. Geophys. Res.*, **87**:4903–4916 (1982).

Idone, V. P., R. E. Orville, and R. W. Henderson: Ground Truth: A Positive Cloud-to-Ground Lightning Flash. *J. Climate Appl. Meteorol.*, **23**:1148–1151 (1984).

Jensen, J. C.: The Relation of Branching of Lightning Discharges to Changes in the Electrical Field of Thunderstorms. *Phys. Rev.*, **40**:1012–1014 (1932).

Jensen, J. C.: The Branching of Lightning and the Polarity of Thunderclouds. *J. Franklin Inst.*, **216**:707–747 (1933).

Jordan, D. M., and M. A. Uman: Variations in Light Intensity with Height and Time from Subsequent Lightning Return Strokes. *J. Geophys. Res.*, **88**:6555–6562 (1983).

Krehbiel, P. R., M. Brook, and R. A. McCrory: An Analysis of the Charge Structure of Lightning Discharges to Ground. *J. Geophys. Res.*, **84**:2432–2456 (1979).

Lewis, W. W., and C. M. Foust: Lightning Investigation on Transmission Lines, Pt. 7. *Trans. AIEE*, **64**:107–115 (1945).

Mackerras, D.: A Comparison of Discharge Processes in Cloud and Ground Lightning Flashes. *J. Geophys. Res.*, **73**:1175–1183 (1968).

Mackerras, D.: Photoelectric Observations of the Light Emitted by Lightning Flashes. *J. Atmos. Terr. Phys.*, **35**:521–535 (1973).

Nakano, M.: The Cloud Discharge in Winter Thunderstorms of the Hokuriku Coast. *J. Meteorol. Soc. Japan*, **57**:444–445 (1979a).

Nakano, M.: Initial Streamer of the Cloud Discharge in Winter Thunderstorms of the Hokuriku Coast. *J. Meteorol. Soc. Japan*, **57**:452–457 (1979b).

Norinder, H.: Magnetic Field Variations from Lightning Strokes in Vicinity Thunderstorms. *Ark. Geofys.*, **2**:423–451 (1956).

Orville, R. E., R. W. Henderson, and L. F. Bosart: An East Coast Lightning Detection Network. *Bull. Am. Meteorol. Soc.*, **64**:1029–1037 (1983).

Orville, R. E., R. A. Weisman, R. B. Pyle, R. W. Henderson, and R. E. Orville, Jr.: Cloud-to-Ground Lightning Flash Characteristics from June 1984 through May 1985. *J. Geophys. Res.*, **92**, in press, 1987.

Pierce, E. T.: Electrostatic Field Changes Due to Lightning Discharges. *Q. J. R. Meteorol. Soc.*, **81**:211–228 (1955a).

Pierce, E. T.: The Development of Lightning Discharges. *Q. J. R. Meteorol. Soc.*, **81**:229–240 (1955b).

Rust, W. D., D. R. MacGorman, and R. T. Arnold: Positive Cloud to Ground Lightning Flashes in Severe Storms. *Geophys. Res. Lett.*, **8**:791–794 (1981).

Rust, W. D., D. R. MacGorman, and W. L. Taylor: Photographic Verification of Continuing Current in Positive Cloud-to-Ground Flashes. *J. Geophys. Res.*, **90**:0144–0140 (1330).

Schonland, B. F. J., and T. E. Allibone: Branching of Lightning. *Nature (London)*, **128**:794–795 (1931).

Takagi, N., T. Takeuti, and T. Nakai: On the Occurrence of Positive Ground Flashes. *J. Geophys. Res.*, **91**:9905–9909 (1986).

Takeuti, T., M. Nakano, M. Nagatani, and H. Nakada: On Lightning Discharges in Winter Thunderstorms. *J. Meteorol. Soc. Japan*, **51**:494–496 (1973).

Takeuti, T., M. Nakano, and Y. Yamamoto: Remarkable Characteristics of Cloud-to-Ground Discharge Observed in Winter Thunderstorms in Hokuriku Area, Japan. *J. Meteorol. Soc. Japan*, **54**:436–439 (1976).

Takeuti, T., M. Nakano, H. Ishikawa, and S. Israelsson: On the Two Types of Thunderstorms Deduced from Cloud-to-Ground Discharges Observed in Sweden and Japan. *J. Meteorol. Soc. Japan*, **55–56**:613–616 (1977).

Takeuti, T., M. Nakano, M. Brook, D. J. Raymond, and P. Krehbiel: The Anomalous Winter Thunderstorms of the Hokuriku Coast. *J. Geophys. Res.*, **83**:2385–2394 (1978).

Takeuti, T., S. Israelsson, M. Nakano, H. Ishikawa, D. Lundquist, and E. Astrom: On Thunderstorms Producing Positive Ground Flashes. *Proc. Res. Inst. Atmos., Nagoya Univ., Japan*, **27-A**:1–17 (1980).

Takeuti, T., K. Funaki, and N. Kitagawa: A Preliminary Report on the Norwegian Winter Thunderstorm Observation. *Res. Lett. Atmos. Electr., Japan*, **3**:69–72 (1983).

Takeuti, T., Z. Kawasaki, K. Funaki, N. Kitagawa, and J. Huse: Notes and Correspondence on the Thundercloud Producing the Positive Ground Flashes. *J. Meteorol. Soc. Japan*, **63**:353–358 (1985).

Uman, M. A.: "Lightning." McGraw-Hill, New York, 1969. See also, Dover, New York, 1984.

Chapter 12 | Upward Lightning and the Artificial Initiation of Lightning

12.1 INTRODUCTION

We define artificially initiated lightning as a discharge that occurs because of the presence of a man-made structure or event. Such lightning is characterized by an initial upward-moving leader. Discharges initiated by upward-moving leaders also occur naturally, for example, from mountain tops (e.g., Berger and Vogelsanger, 1966), and are expected to be similar to those that are initiated artificially. It is the artificially initiated lightning that has been primarily studied. Upward-initiated lightning has no "first return stroke" of the type always observed in normal downward-initiated lightning. Rather its place is taken by an upward-moving leader and any continuous current that may follow when that leader reaches the cloud. This initial upward discharge is often, however, followed by combinations of downward-moving dart or dart-stepped leaders and upward-moving subsequent return strokes that appear to be very similar to subsequent strokes in normal cloud-to-ground flashes. Occasionally, the initial upward-moving leader initiates a downward-moving cloud-to-ground return stroke, apparently to date observed only for the case of negatively charge leaders (e.g., Berger, 1978).

The most commonly occurring artificially initiated lightning is that generated by leaders moving upward from the tops of tall man-made structures such as the Empire State Building in New York City (McEachron, 1939, 1941; Hagenguth and Anderson, 1952), the towers on Mt. San Salvatore in Lugano, Switzerland (e.g., Berger and Vogelsanger, 1965, 1966; Berger, 1967, 1972, 1978), and the CN Tower in Toronto, Canada (e.g., Chang et al., 1985). These upward-moving leaders are often, if not always, immediately preceded by cloud discharges that provide, in a fraction of a second, the high electric fields at structure level that serve to initiate the leaders (Berger, 1977).

Lightning can also be artificially initiated by rapidly introducing conductors into strong static electric fields. Brook et al. (1961) found that small balloons flown on metal wires of several kilometers length in thunderstorms in New Mexico did not get struck, even during periods of active lightning. However, they showed that, in the laboratory, artificial initiation could be simulated by the rapid introduction of a conductor into an electric field when its steady presence did not result in a discharge. They suggested that corona acts to shield a stationary conductor so that the high fields necessary to initiate lightning are not obtained, whereas the field distortion due to the rapid introduction of a conductor is not significantly reduced by corona since there is insufficient time for its production.

Newman *et al.* (1967) first demonstrated that artificial initiation of lightning was possible by launching small rockets with trailing ground wires from a ship at sea. More recently, lightning has been artificially initiated over land in a similar manner (e.g., Fieux *et al.*, 1975, 1978; Fieux and Hubert, 1976; Horii, 1982; Hubert *et al.*, 1984; Hubert, 1984; Laroche *et al.*, 1985; Kito *et al.*, 1985), and extensive measurements have been made on the resultant discharges. In a related, but accidental, example of artificial initiation, a plume of water from an exploding depth charge in Chesapeake Bay initiated a three-stroke lightning flash that decayed to bead lightning (Brook *et al.*, 1961; Young, 1962; Uman, 1971; Barry, 1980). A photograph is shown in Fig. 1.14. Interestingly, a similar beaded decay is observed in rocket-initiated lightning with long-duration continuing current (Fieux *et al.*, 1975; Hubert, 1985).

Long, conducting objects not connected to earth can also initiate lightning. Illustrations of this are provided by the two lightning strikes to the Apollo 12 spacecraft (Godfrey *et al.*, 1970; Krider *et al.*, 1974), by lightning initiation via small rockets trailing 100–200 m of ungrounded conducting wire, the lower portions of which were 50–150 m above ground at initiation (Hubert *et al.*, 1984; Hubert, 1984), and by accounts of lightning initiated by aircraft flying in clouds that were otherwise not producing lightning (Fitzgerald, 1967; Cobb and Holitza, 1968; Pierce, 1972; Clifford and Kasemir, 1982).

Finally, in a pathological vein, ground-level thermonuclear explosions rapidly deposit enough charge in the atmosphere to induce upward-going lightning from small structures on the Earth's surface (e.g., Uman *et al.*, 1972; Hill, 1973; Grover, 1981; Gardner *et al.*, 1984), as illustrated in Fig. 1.17.

12.2   UPWARD-INITIATED LIGHTNING FROM FIXED STRUCTURES

There is no reason to believe that naturally occurring upward-initiated lightning should be significantly different from upward lightning occurring from tall stationary man-made structures. Measurements of current have obviously been made only on instrumented structures although photographic data have been obtained for both cases. Photographs of upward lightning are shown in Figs. 12.1, 12.4, and 12.5. As illustrated in these figures, upward lightning is upward branched, the channel shape being determined by the propagation and branching of the initial upward leader.

The discovery of upward-initiated lightning is generally attributed to McEachron (1939) who made photographic and current measurements on the Empire State Building (a steel frame building 410 m in height above the New York City street level) beginning in 1935. Most of the lightning discharges observed at the Empire State Building were found to be initiated by an upward-moving stepped leader whose point of origin was the top of the building. The leader current merged smoothly into a continuous current flow between cloud and building without the occurrence of a return stroke. In about half of these discharges subsequent return-stroke current peaks initiated by downward-moving dart leaders followed the initial discharge stage. There was an average of 2.3 subsequent strokes per discharge of this type. A drawing of a streak-camera photograph of upward lightning at the

Fig. 12.1 Four upward lightning flashes initiated concurrently, by visual observation, from four 300-m-tall television transmission towers during a frontal thunderstorm in Kansas City. The TV towers are located along a line 10 km long. Courtesy, C. Gill Kitterman.

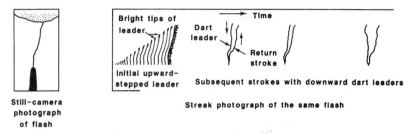

Fig. 12.2   Drawing of streak-camera photograph illustrating usual lightning from the Empire State Building. Adapted from McEachron (1939).

Empire State Building is shown in Fig. 12.2. Hagenguth and Anderson (1952) reported that the average maximum continuous current flow was about 250 A and that the maximum observed was 1450 A. The duration of the continuous current was typically tenths of a second, half the flashes having a duration in excess of 0.27 sec. The maximum flash duration was 1.5 sec.

McEachron (1939) found that the upward-moving stepped leaders had an average length of 8.2 m with a range from 6.2 to 23 m. The mean time interval between steps was reported to be 30 μsec with a range from 20 to 100 μsec. For 20 upward-moving stepped leaders, the mean upward speed was 2.6×10^5 m/sec with a range from 4.7×10^4 to 6.4×10^5 m/sec. Thus stepped-leader speeds and time intervals were observed to be similar to those parameters for downward stepped leaders. Apparently the polarities of all upward-initiated discharges at the Empire State Building were the same and were such as to lower negative charge, that is, initiated by upward-going positively charged leaders, but this fact is not specifically addressed in the papers by McEachron (1939, 1941) and Hagenguth and Anderson (1952).

About 85% of the discharges observed on Mount San Salvatore were initiated by upward-moving stepped leaders (Berger, 1972). Upward-moving leaders of both polarities were observed, about 90% of these total being positively charged and thus resulting in the lowering of negative charge (Berger and Vogelsanger, 1965; Berger, 1972). Current traces for representative discharges are given by Berger and Vogelsanger (1965) and Berger (1967, 1977) and examples are reproduced in Fig. 12.3. The continuous part of the current curves are characterized by amplitudes from 20 to several hundred amperes and times to maximum current of from hundredths to tenths of seconds. The current curves were found to be relatively smooth as the leaders stepped upward. The total continuous current duration was tenths of a second. Continuous currents were often followed by current pauses during which time dart leader–return stroke sequences occurred. Berger and Vogelsanger (1965) reported that current peaks which occurred during continuous current flow were relatively small compared to those during periods of no current flow.

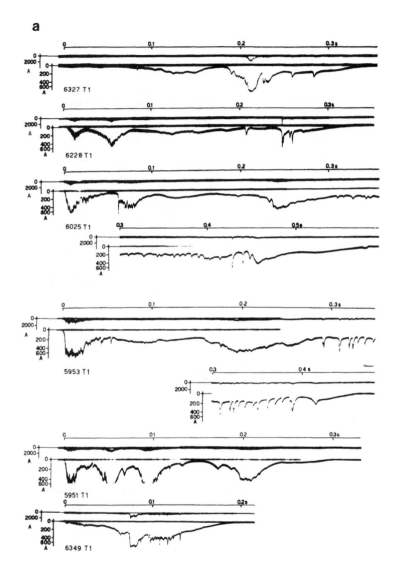

Fig. 12.3 Current traces for upward-initiated lightning from Mt. San Salvatore. (a) Continuous current without appreciable impulses, (b) continuous current with superimposed impulses, (c) continuous current with impulses attributed to subsequent strokes during periods of no continuous current. Adapted from Berger (1967).

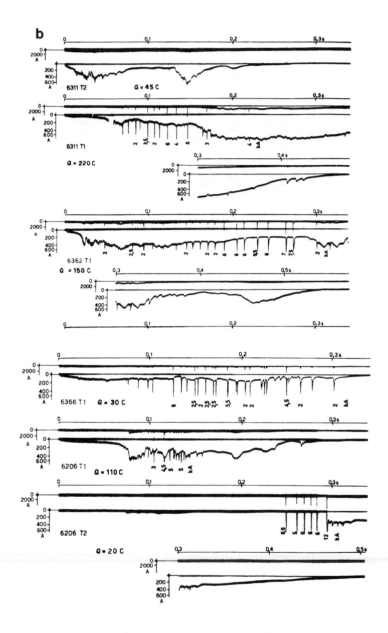

Fig. 12.3b (See legend on p. 209.)

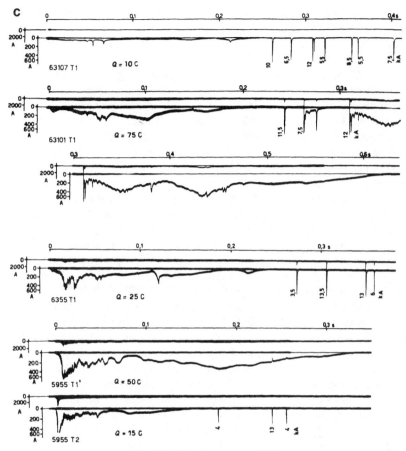

Fig. 12.3c (See legend on p. 209.)

As noted in Section 11.2, upward-going negative leaders of relatively long duration, 4–25 msec, and hence relatively long length, of the order of a kilometer or more, may initiate return strokes that traverse the leader channel downward and thus could be classified as cloud-to-ground return strokes (Berger, 1977, 1978). About 20% of the observed upward negative leaders initiated these positive return strokes (Berger, 1978). Upward-going positive leaders have not been observed to result in downward return strokes, although such leaders can, of course, be followed by downward dart leader–upward return stroke combinations. Reporting on data obtained between 1955 and 1963, Berger and Vogelsanger (1965) state that the average positive charge lowered in 28 discharges initiated by upward-going negatively charged leaders was 64 C, with a maximum of 310 C. Berger (1978) reported for data taken between 1963 and 1973 a median of 26 C for 137 cases in which there was no downward positive return stroke and 84 C for 35 cases with downward positive return strokes. In this latter category 10% of the charge transfers were over 348 C. For upward positive leaders, Berger and Vogelsanger (1965) found an average negative charge lowered of 22 C for 329 discharges, with a maximum charge of 220 C. Berger (1978), for 172 discharges, found a median charge of 23 C, with 10 and 90% values of 100 and 5.4 C, respectively.

Berger and Vogelsanger (1966) have used streak photography to identify the properties of and the differences between positively charged and negatively charged upward-moving stepped leaders. Streak photographs of each type of leader and still photographs of the resultant discharges are shown in Figs. 12.4 and 12.5. Negatively charged stepped leaders, whether upward or downward moving, exhibited regular distinct steps throughout the length of the leader. Positively charged stepped leaders (all observed positive stepped leaders were upward going except the single downward example discussed in Chapter 11 and shown in Fig. 11.1) exhibited steps only over a small portion of the channel length. The remainder was occupied by an apparently continuous leader that exhibited periodic or irregular intensity variations. Positive leaders emitted considerably less light than did negative leaders. Leaders could be identified in the streak photographs of only 7 of the 46 upward positive discharges for which currents and still photographs were available. In these 7 cases the positive leader was not visible until the channel attained a length of at least 40 m at which time it had a current of several hundred amperes. Stepping occurred for a distance of 60–150 m above the tower top. Above about 200 m the leader assumed a more continuous form. When stepping could be measured, the step lengths were under 10 m with 50–100 μsec between steps with an upward speed between 0.4 and 0.8 $\times$ 10^5 m/sec. During the more continuous portion of the leader, speeds were between 1 and 10 $\times$ 10^5 m/sec; times between light intensity maxima were in the range 40–115 μsec, and vertical lengths between intensity maxima were in the range 12–40 m. For upward-moving negatively charged stepped leaders, step lengths were between 3 and 18 m, time between steps between 4 and 50 μsec, and leader average speeds between 1 and 12 $\times$ 10^5 m/sec. Negatively charged steps observed within about 1 km of the tower top had lengths in agreement with the Empire State Building measurements.

Berger and Vogelsanger (1966) show streak photographs of two upward-moving negatively charged stepped leaders that show a faint corona discharge in advance of the bright leader tip. This corona extends upward about one step length and occurs at the time of the bright leader step under it rather than continuously between two steps. Apparently, the corona eventually becomes the next leader step.

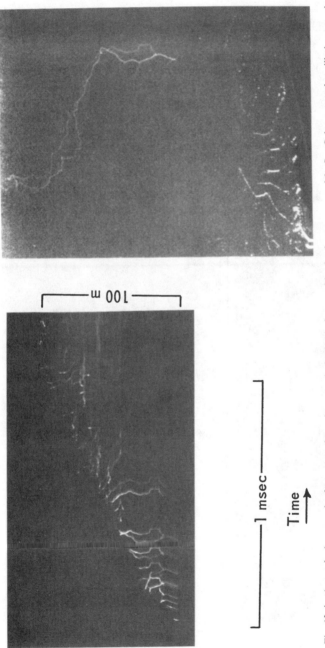

Fig. 12.4 A streak photograph of an upward-moving negatively charged stepped leader from tower on Mt. San Salvatore and a still photograph of the resultant flash. Originally published by Berger and Vogelsanger (1966). Courtesy, K. Berger and the Swiss High Voltage Research Committee (FKH), Zurich.

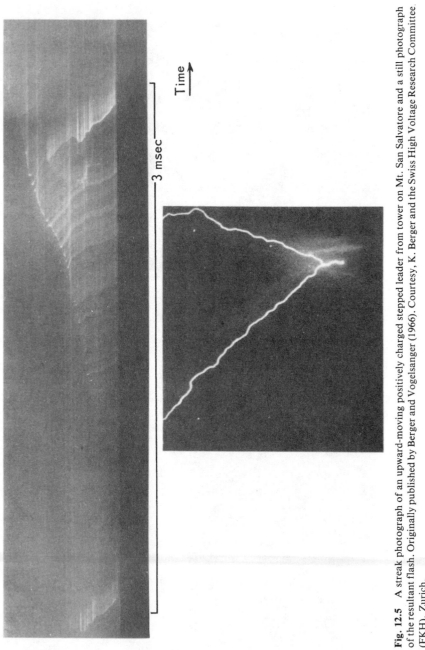

Fig. 12.5 A streak photograph of an upward-moving positively charged stepped leader from tower on Mt. San Salvatore and a still photograph of the resultant flash. Originally published by Berger and Vogelsanger (1966). Courtesy, K. Berger and the Swiss High Voltage Research Committee. (FKH), Zurich.

Time →

3 msec

Berger (1977) shows, in his Fig. 13a and b, a cloud discharge field changes of the order 100 kV/m immediately preceding and apparently leading to the initiation of upward-moving negatively charged leaders. He does not specifically state whether similar field changes are necessary to initiate all upward leaders or whether some upward leaders can be initiated in the presence of a large ambient static or slowly varying field.

12.3 ARTIFICIAL INITIATION BY SMALL ROCKETS

12.3.1 MECHANICS OF ROCKET INITIATION

The technique of artificially initiating lightning via small rockets with trailing ground wires has been described by Fieux *et al.* (1978), Hubert *et al.* (1984), Hubert (1984), Horii (1982), Kito *et al.* (1985), and Laroche *et al.* (1985). A photograph of rocket-intitiated lightning is found in Fig. 12.6. Basically, a small rocket is fired upward at a speed of near 200 m/sec and a grounded wire is unspooled either from the earth or from the rocket. Wire used to date has been about 0.2 mm in diameter and either steel or Kevlar-covered copper. The upward leader is initiated from the rocket tip when the

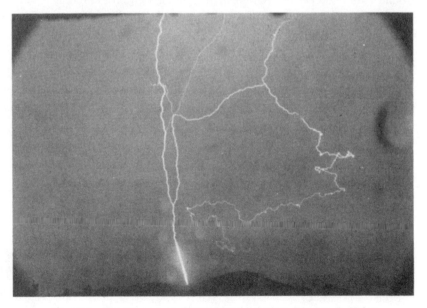

Fig. 12.6 A photograph of rocket-initiated lightning. Courtesy, P. Hubert.

rocket is at an altitude of typically 200–300 m. A high probability of success is assured if the static field at ground level is near 10 kV/m. In general, the higher the value of the field at ground, the lower the height of the rocket at which the upward leader is initiated. Interestingly, because of the possibility of the decrease in the ambient field due to nearby lightning in the time period between the observation of a high-static field at ground and the firing of a rocket, initiation is more successful under conditions of a relatively low lightning flashing rate than when there is a relatively high lightning activity (Hubert, 1984). This fact may provide biases in the data obtained in that more rocket initiations are likely to be achieved near the end of a storm.

Rocket initiation of lightning in winter has taken place from sea level in the Hokuriku area of central Japan (Horii, 1982; Kito et al., 1985; Akiyama et al., 1985; Horii and Sakurano, 1985). Rocket initiation in summer has been achieved from an altitude of about 1100 m in St. Privat d'Allier, France (Fieux et al., 1978; St. Privat d'Allier Research Group, 1982; Hubert, 1984), from an altitude of about 3330 m in Socorro, New Mexico (Hubert et al., 1984), and from sea level at and near the Kennedy Space Center, Florida (Laroche et al., 1985). Rocket heights and static field values at the time of initiation are summarized in Table 12.1.

In the Japanese winter studies, initiation occurred for rocket heights between about 40 and 330 m with a median height of 142 m. The fields at ground prior to rocket launch were in the 5-kV/m range for initiating rocket heights above 100 m and in the 10-kV/m range for heights below. The minimum, maximum, and median values for successful lightning initiation were 4.0, 11.5, and 7.4 kV/m, respectively. Nearly equal numbers of positive and negative flashes were initiated.

At St. Privat d'Allier, rocket heights at the time of initiation of the upward-going leader were between 50 and 530 m with a mean of 210 m and with half of the initiations occurring between 50 and 210 m. The initiation

Table 12.1

Mean Rocket Height and Mean Static Electric Field at Leader Initiation

	Hokuriku area of Japan	St. Privat d'Allier, France	Langmuir Laboratory, near Socorro, New Mexico	Near Melbourne, Florida
H (m)	142[a]	210	216	380
E (kV/m)	7.4[a]	10[b]	8.8[c]	6.3

[a] Median given instead of mean.
[b] Negative lightning only.
[c] Calculated from Eq. (12.1).

success rate over an 8-year period (1973–1980) was 73%, the total initiated discharges being 94. The static field at ground prior to launch was generally in the range 6–15 kV/m with an average of 10 kV/m for lightning bringing negative charge to earth and higher for positive lightning.

For New Mexico, where an initiation efficiency of 62% was achieved for 56 rocket firings and all flashes were initiated in the field of negative overhead charge, Hubert *et al.* (1984) give the following relation between the rocket height H at initiation and the field E in kV/m just before rocket launch:

$$H = 3900E^{-1.33} \quad \text{m} \tag{12.1}$$

This expression fits the measured data with a correlation coefficient of -0.82. Lightning was initiated in fields between about 5 and 13 kV/m at heights in the range from about 100 to 600 m. The mean rocket height at initiation was 216 m with a standard deviation of 103 m.

In Florida, where the cloud charge height above local terrain is greater than that of the mountainous regions of France and New Mexico, Laroche *et al.* (1985) report 20 initiation heights in the 210- to 740-m range with a mean of 380 m for static fields at ground between 2.6 and 8.1 kV/m with a mean of 6.3 kV/m. The static fields were all from negative charge overhead. An initiation efficiency of 66% was reported, the total initiated discharges being 28. Apparently, the fields at ground level necessary to initiate lightning are lower and the rocket heights higher at sea level in Florida than in mountainous terrain. The rockets used south of Melbourne, Florida in 1983 and at the Kennedy Space Center in 1984 and 1985 had Kevlar-covered copper wire that spooled off the rocket, whereas essentially all previous studies had the spool at the ground and employed steel wire. Because of these differences, the Florida rocket had higher speed, up to 250 m/sec, and the wires had different electrical properties than the previously used rockets, but Laroche *et al.* (1985) argue that these differences do not affect the overall physical properties of the triggered lightning.

12.3.2 TYPES OF ROCKET-INITIATED LIGHTNING

Hubert *et al.* (1984) and Hubert (1984) have divided rocket-initiated lightning into various categories by virtue of differing physical characteristics. The two basic categories are "classical" and "anomalous." Laroche *et al.* (1985) label these categories "A" and "B," respectively, while in earlier papers from the French groups they are assigned different designations. In the "classical" case, the upward-going leader initiates a current flow of a few hundred amperes lasting of the order of several hundred milliseconds. After the cessation of this continuous current, "normal" dart leader–return stroke sequences occur accompanied by current peaks in the kiloampere to tens of kiloampere range. The classical case is similar to light-ning initiated by a tall structure. A subcategory of the classical case is termed the "slow" case. In slow flashes there are no subsequent strokes or associated current peaks, only the upward leader and any following continuous current. In the "anomalous" case, the current at ground is cut off at about the time that the rocket wire melts, some tens of milliseconds after the upward leader is launched. The current is interrupted when it is typically 40 A. Apparently, the melting wire acts as a fuse to stop the current flow temporarily. Negligible current flows for a period of a few tens of milliseconds, after which a down-ward leader traverses the previous upward leader channel until the downward leader reaches the wire at which point it no longer follows the path of the wire but becomes a stepped leader in virgin air. A "first" return stroke and then "normal" subsequent strokes follow. If a flash has the same current charac-teristics as an anomalous flash but the downward leader appears to follow the wire, it is termed "pseudoclassical," more or less a subcategory of the anomalous. The distinction between anomalous and pseudoclassical is not clear-cut according to Hubert *et al.* (1984). Examples of currents from clas-sical, slow, anomalous, and psuedoclassical flashes are shown in Fig. 12.7.

Hubert *et al.* (1984) found that the 14 classical flashes they observed were triggered at a mean height of 158 m and the 3 slow flashes at 161 m, somewhat lower than the mean heights for the 7 anomalous and 7 pseudoclassical flashes, 215 and 330 m, respectively. They cite this observation to support their view that there is a significant physical difference between the classical and anomalous cases. The St. Privat d'Allier Research Group (1982) found 48 classical cases, 31 slow cases, and 15 anomalous cases in 132 rocket firings over a period of 8 years (1973–1978). While classical and anomalous trig-gerings occurred with not vastly differing probabilities in the mountainous regions of France and New Mexico, at sea level in Florida all triggered events were of the classical or of the slow type (Laroche *et al.*, 1985). The reason for this may be associated with the facts that (1) the Kevlar-covered copper wire, because of its higher conductivity, does not melt until the upward leader has traveled further upward than in the case of the steel wire and (2) the copper wire is observed to carry a sufficiently high current, over 100 A, so that, when it does melt, the current can be maintained in air as in the case of a normal continuing current. In this latter view, if the steel wire melts with relatively low currents, such currents cannot be maintained in air.

12.3.3 CURRENTS, CHARGES, FIELDS, AND VELOCITIES

We consider now some of the information derived about the physics of artificially initiated lightning from the French, Japanese, and United States

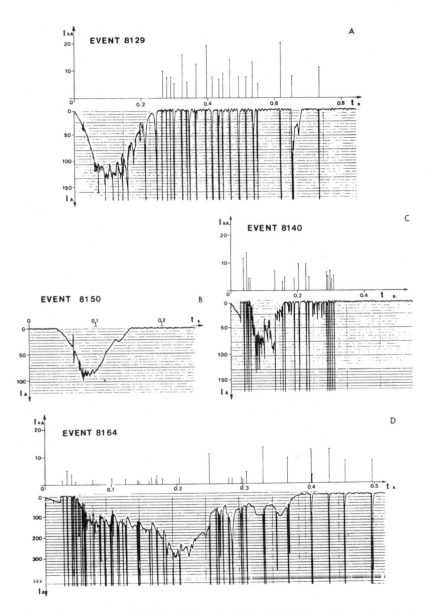

Fig. 12.7   Currents in rocket-initiated lightning: (A) classical, (B) slow, (C) anomalous, and (D) pseudoclassical. Wire melting is evident in all four waveforms in the current discontinuity at a time of the order of 30 msec. The vertical bars indicate the time and magnitude of the main current pulses. Adapted from Hubert *et al.* (1984).

studies (e.g., Fieux *et al.*, 1978; Waldteufel *et al.*, 1980; Hubert and Mouget, 1981; Djebari *et al.*, 1981; Horii, 1982; St. Privat d'Allier Research Group, 1982; Hubert *et al.*, 1984; Idone *et al.*, 1984; Idone and Orville, 1984, 1985; Kito *et al.*, 1985). Among the diagnostic techniques employed in these experiments were direct current measurement, static to millisecond-scale electric field measurement, submicrosecond-scale electric and magnetic field measurements, still and streak photography and photoelectric measurements, and acoustic measurements.

12.3.3.1 France

Fieux *et al.* (1975, 1978), the St. Privat d'Allier Research Group (1982), Hubert and Mouget (1981), and Hubert (1984) reported on the physical parameters of triggered lightning observed in France. The triggered flashes were initiated at the rocket by an upward-going leader of speed between 2×10^4 and some 10^5 m/sec, which displayed considerable upward branching. Dart leader–return stroke combinations generally occurred after an initial upward-going discharge. The dart leaders had speeds of 2–10×10^6 m/sec. The triggered flashes were characterized by continuous current components of the order of a few hundred amperes with durations of hundreds of milliseconds and large numbers of fast current peaks (53 in one case) of magnitude and risetime similar to normal subsequent strokes. For 94 flashes, the median peak current for the largest current of each flash was 12 kA, and the maximum value was 42 kA. The median total charge transferred per flash was 50 C with 10% of the events transferring at least 100 C. The median flash duration was 350 msec, with 10% above 850 msec and a maximum of 1.35 sec. For 110 current peaks over 2 kA, the median risetime was 0.3μsec, with 10% less than 0.1 μsec and 10% greater than 1.0 μsec.

From the measured fields and currents in subsequent strokes of triggered events, Fieux *et al.* (1978) and Djebari *et al.* (1981) have calculated return stroke speeds. The results of this study are discussed in Section 7.3 where they are used as an argument for the validity of Eq. (7.5).

Acoustic measurements have shown that fast current peaks are necessary to produce strong acoustic signals (Fieux *et al.*, 1978). Apparently, purely upward-going discharges lacking subsequent strokes do not make the sounds we regard as thunder, as first suggested in the Empire State Building study (McEachron, 1939). On the other hand, Hubert (1985) argues, from personal observations, that rocket-triggered lightning makes a surprisingly small noise when observed at about 100 m even if there are subsequent return strokes and that for observations beyond 1 km the noise is similar to normal thunder even if there are no return strokes.

Waldteufel *et al.* (1980) studied an "anomalous" multiple-stroke flash having 32 current pulses above 1 kA. The locations of the charge sources of 10 of the strokes were found from a field mill network to be in clear air at the top of the photographed channels. This observation apparently provides evidence for the existence of "bolts from the blue," references to visual observations of this phenomenon being given in Section 1.7.1.

Hubert and Mouget (1981) measured return stroke speed in two triggered lightning flashes with photoelectric detectors and correlated these data with peak current and current risetime information. For nine subsequent strokes with risetimes shorter than 1.5 μsec, the data points on a speed vs peak current plot were reasonably well fit by the Lundholm (1957) expression given in Section 7.2.3 with $W = 40\,kA$, as Idone *et al.* (1984) also found to be the case for 56 subsequent strokes from the New Mexico study. For two anomalous first strokes and two subsequent strokes with long risetimes, 15 and 20 μsec, the data of Hubert and Mouget (1980) fall considerably below the Lundholm (1957) curve, that is, speeds were lower than expected for a given peak current.

12.3.3.2 New Mexico

Analyses of the 1981 New Mexico triggering experiments have been described by Hubert (1984), Hubert *et al.* (1984), Idone *et al.* (1984), and Idone and Orville (1984, 1985). The paper by Hubert *et al.* (1984) is primarily concerned with presenting data regarding the currents in the 35 rocket-initiated flashes. As in the work in France, the rocket-initiated lightning exhibited a relatively large number of current pulses, a median value of 15 above 2 kA, 10 above 3 kA, and 2 above 10 kA. Anomalous and pseudo-classical flashes produced a median value of 22 current peaks greater than 2 kA, while classical flashes produced only 9. The highest current peak was 40 kA. Pseudoclassical events had first-stroke peak currents near 2 kA, typically an order of magnitude smaller than the current in anomalous events in which the downward leader left the upward leader channel above the wire. In anomalous event, the current was determined from close magnetic field measurements after the technique was calibrated on classical events whose currents were directly measured by a resistive shunt. The median value of charge transferred for total flashes was 35 C, most of the charge transported being associated with the continuous current. The median charge per current pulse was about 0.35 C. Current pulses occurring during continuous current seldom had an identifiable dart leader–return stroke sequence, and, when such did exist, the return stroke speed was smaller than usual. Pulses occurring during continuous current also had relatively slow risetimes, with a median value of about 20 μsec. Twenty-two percent of all pulses had

10–90% risetimes less than 0.3 μsec, the instrumentation limit. Subsequent stroke risetimes not occurring during continuing current had a median risetime of about 0.5 μsec. Typical values for the time to half value for subsequent pulses were 20–50 μsec, with a median value of about 35 μsec. The median risetime for first strokes in anomalous events was 14 μsec.

Idone *et al.* (1984), in addition to relating return stroke speed to peak current for 56 subsequent strokes, as discussed above and in Section 7.2.3, measured a mean three-dimensional return stroke speed of 1.2×10^8 m/sec with a range of 0.67×10^8 to 1.7×10^8 m/sec. They also found for 32 dart leaders a mean three-dimensional speed of 2.0×10^7 m/sec with a range of 0.95×10^7 to 4.3×10^7 m/sec. Linear correlation coefficients were evaluated for dart leader speed and return stroke current (0.84), stroke peak current and interstroke interval (0.57), stroke speed and interstroke interval (0.49), and dart leader speed and interstroke interval (0.43), the latter value being more strongly correlated for natural lightning. Additional discussion is found in Sections 7.2.3 and 8.2.

Idone and Orville (1984) have photographed and analyzed the transformation of dart leaders to stepped leaders and the reflections of dart leader luminosity at the top of the rocket-borne wire.

Idone and Orville (1985) found a strong correlation between the peak light intensity and peak subsequent stroke current (range 1.6–21 kA) in each of two triggered flashes containing 19 and 20 subsequent strokes, respectively. They also measured the ratio of dart-leader to return-stroke light intensity near ground and found a mean of 0.1 with a range from 0.22 to 0.23. They estimated individual dart leader currents near ground to be in the range from 0.1 to 6 kA with a mean of about 1.7 kA. Additional discussion is found in Sections 7.2.4 and 8.2.

12.3.3.3 Japan

Horii (1982) and Kito *et al.* (1985) have described the primary physics in the Japanese experiments in triggering 39 flashes in the winter at Hokuriku. Kito *et al.* (1985) discuss and categorize the luminous properties of the initial leaders and following events while Horii *et al.* (1983) consider all aspects of the triggered flashes. Supporting data are found in Horii and Sakurano (1985), Akiyama *et al.* (1985), and Nakano *et al.* (1983). Positive leaders, associated with negative charge overhead, were observed to leave the rocket tip with a speed of about 0.1×10^5 m/sec and exhibited obvious branching, while negative leaders started at speeds of $0.5–1.0 \times 10^5$ m/sec and showed few branches. For both polarities, speed tended to increase as the leader progressed upward. Current pulses in the tens of ampere range were superimposed on a background current of several milliamperes for 200–300 msec prior to the initiation of an upward leader and melting of the wire which is associated with a current of the order of 100 A. Four flashes having both positive and negative current pulses associated with subsequent return strokes were observed. The median peak pulse current per flash was 15 kA and the median duration was 420 msec.

12.3.3.4 Florida

Laroche *et al.* (1985) have given a preliminary report on the Florida studies carried out during the summers of 1983 and 1984. Experiments were also performed in 1985 (Section 7.3) and 1986. During the ascent of the Kevlar-covered copper wire, isolated current pulses were observed with amplitudes less than 1 A and widths less than 1 msec. According to Hubert (1985), similar isolated pulses but with currents near 50 A were observed during the experiments in France, leading him to question whether the bandwidth of the recorder used in Florida was sufficient to resolve the true pulse peaks. The time interval between the isolated pulses in the Florida studies ranged between 10 and 100 ms. These current pulses generally increased with time, and the last pulse did not decay to zero but rather merged into the continuous current of the upward leader. The continuous current ranged from a few hundred to a few thousand amperes. All flashes triggered were of the classical type and lowered negative charge. The initial rate of change of current, which has a duration of the order of 10 msec and is to be associated with the upward leader, had a median value of about 10^2 kA/sec, whereas the median values of this parameter at St. Privat and Socorro were 2 to 5 kA/sec. The relatively high value of initial current rate of change is associated by Laroche *et al.* (1985) with the launch from sea level rather than with the properties of the rocket or wire, and unpublished results from rocket-initiated lightning at sea level in Le Barp, France using the rockets with steel wires that were employed in St. Privat D'Allier and Socorro are cited in support of this claim.

12.4 COMPARISON OF ROCKET-INITIATED LIGHTNING WITH THAT FROM FIXED STRUCTURES

Rocket-initiated lightning has much in common with upward lightning from fixed structures. There may also be significant differences, but these are difficult to identify given the various biases in the different data sets, perhaps the greatest of which is associated with the time chosen to launch a rocket, generally a time when there is a relatively high ambient electric field magnitude and a relative absence of field variation. Median values of pertinent parameters discussed below for rocket-initiated and for structure-initiated lightning are summarized in Table 12.2.

In rocket-initiated lightning, the field necessary to produce an upward leader is generated by rapidly introducing a grounded conductor several hundred meters in length into a static field of roughly 5–10 kV/m (Section 12.2.1, Table 12.1). For upward lightning from a fixed structure, the necessary field magnitude is probably produced by an overhead cloud discharge or perhaps a nearby ground discharge (Section 12.2), but may possibly, in the case of very tall structures, be initiated in the presence of the static field alone.

For rocket-initiated lightning, the discharges with only continuous current, the so-called slow discharges, occur with a frequency of 20–40% (Hubert *et al.*, 1984). At the Empire State Building (Hagenguth and Anderson, 1952), about half of the upward-initiated leaders produced discharges with only continuous current while the other half started flashes containing dart leader and subsequent strokes. On Mt. San Salvatore, Berger (1978) found that about 80% of the upward flashes contained only continuous current.

The median duration of 94 rocket-initiated flashes in France was 350 msec (850 msec at the 10% level, 70 msec at the 90%) and of 35 rocket-initiated flashes in New Mexico was 470 msec (940 msec at the 10% level, 250 msec at the 90%) (Hubert, 1984). For 112 upward flashes at Mt. San Salvatore the median duration was 338 msec (790 msec at the 10% level, 144 msec at the 90%) (Berger, 1978), and for 73 flashes at the Empire State Building, the great majority of which were upward initiated and lowered negative charge, the median duration was 270 msec (Hagenguth and Anderson, 1952).

The median charge lowered in the 94 rocket-initiated flashes in France was 50 C (100 C at the 10% level, 4 C at the 90%) (St. Privat d'Allier Research Group, 1982; Hubert, 1984) and in the 35 rocket-initiated flashes in New Mexico was 35 C (175 C at the 10% level, 6 C at the 90%) (Hubert *et al.*, 1984; Hubert, 1984). These values are to be compared with the median of 23 C (100 C at the 10% level, 3.7 C at the 90%) reported by Berger (1978) for 172 upward flashes bringing negative charge to earth at Mt. San Salvatore and the median of 19 C reported by Hagenguth and Anderson (1952) for the 73 flashes at the Empire State Building.

For 94 rocket-initiated events, the median peak current for the largest current of each flash was 12 kA (29 kA at the 10% level, 2 kA at the 90%) in France (St. Privat d'Allier Research Group, 1982; Hubert, 1984) and for 35 events was 18 kA (30 kA at the 10% level, 4 kA at the 90%) in New Mexico (Hubert *et al.*, 1984; Hubert, 1984). For 176 subsequent strokes in upward lightning at Mt. San Salvatore, Berger (1978) found a median value of 10 kA (25 kA at the 10% level, 4.2 kA at the 90%, and the median value at the

Table 12.2

Median Values for Parameters of Rocket-Initiated, Structure-Initiated, and Natural Lightning[a]

	Natural lightning		Rocket-initiated lightning		Structure-initiated lightning	
	Near Socorro, New Mexico	Mt. San Salvatore, Switzerland	St. Privat D'Allier, France	Langmuir Laboratory, near Socorro, New Mexico	Empire State Building, New York City	Mt. San Salvatore, Switzerland
Upward flashes with continuous current only (%)				20–30	50	80
Flash duration (msec)	103 (single-stroke flashes without cc[b]) 370 (without cc[b]) 550 (with cc[b])	13 (all flashes) 180 (multiple-stroke flashes)	350	470	270	338
Flash charge (C)	19 (without cc[b]) 34 (with cc[b])	7.5	50	35	19	23
First stroke charge (C)	6–7	4.5		0.8[c]		
Subsequent-stroke charge (C)	1.3–1.4	0.95 (impulse) 1.4 (total)		0.35	0.15	0.77
Subsequent-stroke peak current (kA)		12		5	10	10
Number of return strokes or current peaks	6 4[d]		11	15	2–3	4

[a] Sources of values given are found in Sections 12.4 and 12.5.
[b] cc, Continuing current; about half of the flashes studied contained at least one cc between two strokes (see Chapter 10).
[c] Anomalous flashes with stepped leaders from the wire top to ground.
[d] Schonland (1956) in South Africa.

Empire State Building was also about 10 kA for a data set composing 82 negative current peaks and 2 positive current peaks (Hagenguth and Anderson, 1952).

The median subsequent stroke charge in rocket-initiated lightning was 0.35 C (Hubert *et al.*, 1984), while the comparable value at the Empire State Building was 0.15 C (Hagenguth and Anderson, 1952) and at Mt. San Salvatore was 0.77 C (Berger, 1978).

For rocket-initiated lightning, there was a median of 11 current pulses above 2 kA per flash in France (St. Privat d'Allier Research Group, 1982) and 15 above 2 kA in New Mexico (Hubert *et al.*, 1984). Berger (1978) reported that for both polarities of upward-going discharges on Mt. San Salvatore there was a median of 3 subsequent strokes per flash with 10% of the flashes having over 12 subsequent strokes and 19% having no subsequent strokes. On the Empire State Building there was a median of two to three current peaks per flash with a maximum of 14 (Hagenguth and Anderson, 1952).

In view of the above, it is difficult to identify any significant differences between rocket-initiated and structure-initiated lightning other than the number of current pulses per discharge and the percentage of discharges exhibiting subsequent strokes.

12.5 COMPARISON OF ROCKET-INITIATED AND NATURAL LIGHTNING

Rocket-initiated lightning does not, in general, have a first stroke to compare to the first stroke of natural lightning. The "first stroke" of an anomalous rocket-initiated flash does, however, provide some analog to a natural first stroke in that the last few hundred meters above ground, the length of the wire, is traversed by a downward-moving stepped leader. The anomalous first stroke, however, transfers considerably less charge, a median value of 0.8 C, according to Hubert *et al.* (1984), than does a normal first stroke, a median value of 6 to 7 C according to Brook *et al.* (1962) and 4.5 C according to Berger *et al.* (1975). Subsequent-stroke median charge in natural lightning is near 1 C (Brook *et al.*, 1962; Berger *et al.*, 1975) and in rocket-triggered lightning is 0.35 C (Hubert *et al.*, 1984). Median total flash charge for natural negative lightning without continuing current is 19 C and for flashes with continuing current is 34 C from remote field measurements in New Mexico (Brook *et al.*, 1962), while Berger *et al.* (1975) report only 7.5 C associated with negative downward lightning to Mt. San Salvatore. The comparable rocket-initiated median values are 50 C in France and 35 C in New Mexico (Hubert *et al.*, 1984). Median flash durations for natural flashes with continuing current in New Mexico were 550 msec and without 370 msec, while Berger *et al.* (1975) report 180 msec for multiple-stroke negative flashes in Switzerland. The comparable rocket-initiated values are 350 msec in France and 470 msec in New Mexico (Hubert *et al.*, 1984).

Rocket-initiated lightning is characterized by a relative continuous current flow interrupted by a median of 10 or 11 current pulses above 2 or 3 kA (Hubert *et al.*, 1984; St. Privat d'Allier Research Group, 1982), whereas natural lightning has a median of 4 strokes (Schonland, 1956) with about one-half of all lightning flashes to ground containing at least one long continuing current interval (Section 9.2). The median value of peak subsequent stroke current per rocket-initiated lightning in New Mexico is near 5 kA (Hubert *et al.*, 1984, Table 4) and is appreciably lower than the value for downward lightning to Mt. San Salvatore, 12 kA for 135 negative strokes over 2 kA (Berger *et al.*, 1975, Table 7.2). Hubert *et al.* (1984) conclude from a comparison of their data with the literature on natural lightning that "there is a great similarity in shape, size, and number between the subsequent strokes of natural flashes and the largest among the pulses in the triggered events."

Idone *et al.* (1984) have found dart leaders in rocket-initiated flashes to be faster, a mean of 20×10^6 m/sec for 32 leaders, than for their own measurements on natural lightning, a mean of 11×10^6 m/sec for 21 leaders, and for the previous measurements of others as listed in Table 8.1.

Djebari *et al.* (1981) have compared subsequent return-stroke electric and magnetic fields in rocket-initiated flashes with those of natural lightning to ground near the facility at which the rockets were launched and found that the rocket-initiated and natural stroke fields were similar except that the rise to peak of the rocket-initiated strokes was faster. For 44 natural subsequent strokes at 3 km, the median 30–90% risetime was about 0.8 μsec. The difference in the risetimes for rocket-initiated and natural stroke fields is apparently due to the absence in the rocket-initiated fields of the slow front (Section 7.2.1, Fig. 7.4) in the initial return-stroke field variation. Djebari *et al.* (1981) suggest that the few hundred meters of vaporized wire is responsible for the decreased risetime for the artificially initiated events. The effect of having a good grounding medium, the tower from which the rockets were launched, at the bottom of the wire could also be important in this regard.

12.6 COMPARISON OF STRUCTURE-INITIATED AND NATURAL LIGHTNING

The properties of subsequent strokes in structure-initiated and natural lightning are reasonably similar. The best comparison can be made from data taken at Mt. San Salvatore where the same equipment was used for both types of measurements. Statistical information for natural lightning to the towers on Mt. San Salvatore is summarized in Table 7.2. Parameters for upward lightning from the same towers are presented in Section 12.2 and extensive statistical data are found in Berger (1978). Median values for subsequent-stroke peak current and charge transfer are given in Table 12.2: 12 kA (30 kA at the 5% level, 4.6 kA at the 95%) and 0.95 C (4.0 C at the 5% level, 0.22 C at the 95%) for natural subsequents and 10 kA and 0.77 C (4.1 C at the 10% level, 0.14 at the 90%) for structure-initiated subsequents.

Subsequent-stroke maximum rate of rise of current for natural lightning to Mt. San Salvatore had a median value of 40 kA/μsec with 120 kA/μsec at the 5% level and 12 kA/μsec at the 95% level. For upward lightning, the median was 25 kA/μsec with 120 kA/μsec at the 10% level and 6 kA/μsec at the 90% (Berger, 1978).

Upward flashes from the towers on Mt. San Salvatore had about 3 times the duration and lowered about 3 times the charge as downward flashes, as is evident from Table 12.2.

REFERENCES

Akiyama, H., K. Ichino, and K. Horii: Channel Reconstruction of Triggered Lightning Flashes with Bipolar Currents from Thunder Measurements. *J. Geophys. Res.*, **90**:10,674–10,680 (1985).

Barry, J. D.: "Ball Lightning and Bead Lightning." Plenum, New York, 1980 (Fig. 2.10).

Berger, K.: Novel Observations on Lightning Discharges: Results of Research on Mount San Salvatore. *J. Franklin Inst.*, **283**:478–525 (1967).

Berger, K.: Methoden und Resultate der Blitzforschung auf dem Monte San Salvatore bei Lugano in den Jahren 1963–1971. *Bull. Schweiz. Elektrotech. Ver.*, **63**:1403–1422 (1972).

Berger, K.: The Earth Flash. *In* "Lightning, Vol. 1, Physics of Lightning" (R. H. Golde, ed.), pp. 119–190. Academic Press, New York, 1977.

Berger, K.: Blitzstrom—Parameter von Aufwärtsblitzen. *Bull. Schweiz. Elektrotech. Ver.*, **69**:353–360 (1978).

Berger, K., and E. Vogelsanger: Messungen und Resultate der Blitzforschung der Jahre 1955–1963 auf dem Monte San Salvatore. *Bull. Schweiz. Elektrotech. Ver.*, **56**:2–22 (1965).

Berger, K., and E. Vogelsanger: Photographische Blitzuntersuchungen der Jahre 1955–1965 auf dem Monte San Salvatore. *Bull. Schweiz. Elektrotech. Ver.*, **57**:599–620 (1966).

Brook, M., G. Armstrong, R. P. H. Winder, B. Vonnegut, and C. B. Moore: Artificial Initiation of Lightning Discharges. *J. Geophys. Res.*, **66**:3967–3969 (1961).

Brook, M., N. Kitagawa, and E. J. Workman: Quantitative Study of Strokes and Continuing Currents in Lightning Discharges to Ground. *J. Geophys. Res.*, **67**:649–659 (1962).

Chang, J. S., T. G. Beuthe, G. G. Hu, G. Stamoulis, and W. Janischewskyj: Thundercloud Electric Field Measurements in the 553-m CN Tower during 1978–1983. *J. Geophys. Res.*, **90**:6087–6090 (1985).

Clifford, D. W., and H. W. Kasemir: Triggered Lightning. *IEEE Trans. Electromagn. Compat.*, **EMC-24**:112–122 (1982).

Cobb, W. E., and F. J. Holitza: A Note on Lightning Strikes to Aircraft. *Mon. Weather Rev.*, **96**:807–808 (1968).

Djebari, B., J. Hamelin, C. Lateinturier, and J. Fontaine: Comparison between Experimental Measurements of the Electromagnetic Field Emitted by Lightning and Different Theoretical Models-Influence of the Upward Velocity of the Return Stroke. *Electromagn. Compat. Symp., 4th Zurich, March 1981*. Available from T. Dvorak, ETH Zentrum-IKT, CH-8092 Zurich, Switzerland.

Fieux, R., and P. Hubert: Triggered Lightning Hazards. *Nature (London)*, **260**:188 (1976).

Fieux, R. P., C. Gary, and P. Hubert: Artificially Triggered Lightning above Land. *Nature (London)*, **257**:212–214 (1975).

Fieux, R. P., C. H. Gary, B. P. Hutzler, A. R. Eybert-Berard, P. L. Hubert, A. C. Meesters, P. H. Perroud, J. H. Hamelin, and J. M. Person: Research on Artificially Triggered Lightning in France. *IEEE Trans. Power Appar. Syst.*, **PAS-97**:725–733 (1978).

Fitzgerald, D. R.: Probable Aircraft "Triggering" of Lightning in Certain Thunderstorms. *Mon. Weather Rev.*, **95**:835–842 (1967).

Gardner, R. L., M. H. Frese, J. L. Gilbert, and C. L. Longmire: A Physical Model of Nuclear Lightning. *Phys. Fluids*, **27**:2694–2698 (1984).

Gary, C., A. Cimador, and R. Fieux: La Foudre: Étude du Phénomene. Application à la Protection des Lignes de Transport. *Rev. Gen. Electr.*, **84**:24–62 (1975).

Godfrey, R., E. R. Mathews, and J. A. McDivitt: Analysis of Apollo 12 Lightning Incident. NASA MSC-01540, January, 1970.

Grover, M. K.: Some Analytical Models for Quasi-Static Source Region EMP: Application to Nuclear Lightning. *IEEE Trans. Nucl. Sci.*, **NS-28**:990–994 (1981).

Hagenguth, J. H., and J. G. Anderson: Lightning to the Empire State Building. *Trans. AIEE* (Pt. 3):641–649 (1952).

Hamelin, J., J. F. Karczewsky, and F. X. Sene: Sonde de Mesure du Champ Magnétique due à une Décharge Orageuse. *Ann. Telecommun.*, **33**:198–205 (1978).

Hill, R. D.: Lightning Induced by Nuclear Bursts. *J. Geophys. Res.*, **78**:6355–6358 (1973).

Horii, K.: Experiment of Artificial Lightning Triggered with Rocket. *Mem. Fac. Eng., Nagoya Univ., Japan*, **34**:77–112 (1982).

Horii, K., and H. Sakurano: Observation on Final Jump of the Discharge in the Experiment of Artificially Triggered Lightning. *IEEE Trans.*, **PAS-104**:2910–2917 (1985).

Hubert, P.: Triggered Lightning in France and New Mexico. *Endeavour*, **8**:85–89 (1984).

Hubert, P.: Personal communication, 1985.

Hubert, P., and G. Mouget: Return Stroke Velocity Measurements in Two Triggered Lightning Flashes. *J. Geophys. Res.*, **86**:5253–5261 (1981).

Hubert, P., P. Laroche, A. Eybert-Berard, and L. Barret: Triggered Lightning in New Mexico. *J. Geophys. Res.*, **89**:2511–2521 (1984).

Idone, V. P., and R. E. Orville: Three Unusual Strokes in a Triggered Lightning Flash. *J. Geophys. Res.*, **89**:7311–7316 (1984).

Idone, V. P., and R. E. Orville: Correlated Peak Relative Light Intensity and Peak Current in Triggered Lightning Subsequent Return Strokes. *J. Geophys. Res.*, **90**:6159–6164 (1985).

Idone, V. P., R. E. Orville, P. Hubert, L. Barret, and A. Eybert-Berard: Correlated Observations of Three Triggered Lightning Flashes. *J. Geophys. Res.*, **89**:1385–1394 (1984).

Kitagawa, N., M. Brook, and E. J. Workman. Continuing Currents in Cloud to Ground Lightning Discharges. *J. Geophys. Res.*, **67**:637–647 (1962).

Kito, Y., K. Horii, Y. Higashiyama, and K. Nakamura: Optical Aspects of Winter Lightning Discharge Triggered by Rocket-Wire Technique in Hokuriku District of Japan. *J. Geophys. Res.*, **90**:6147–6157 (1985).

Kitterman, C. G.: Concurrent Lightning Flashes on Two Television Transmission Towers. *J. Geophys. Res.*, **86**:5378–5380 (1981).

Krider, E. P., R. C. Noggle, M. A. Uman, and R. E. Orville: Lightning and the Apollo 17/ Saturn V Exhaust Plume. *J. Spacecraft Rockets*, **11**:72–75 (1974).

Laroche, P., A. Eybert-Bérard, and L. Barret: Triggered Lightning Flash Characteristics. "10th International Aerospace and Ground Conference on Lightning and Static Electricity, Paris, June 1985," pp. 231–239. Available from Lés Éditions de Physique, BP112, 91944 Les Ulis Cedex, France.

Lhermitte, R.: Doppler Radar Observations of Triggered Lightning. *Geophys. Res. Lett.*, **9**:712–715 (1982).

Lundholm, R.: Induced Overvoltage Surges on Transmission Lines. *Chalmers Tek. Hoegsk. Handl.*, **188**:1–117 (1957).

McEachron, K. B.: Lightning to the Empire State Building. *J. Franklin Inst.*, **227**:147–217 (1939).

McEachron, K. B.: Lightning to the Empire State Building. *Trans. AIEE*, **60**:885–889 (1941).

Nakano, M., N. Takagi, Z. Kawasaki, and T. Takeuti: Return Strokes of Triggered Lightning Flashes. *Res. Lett. Atmos. Electr.*, **3**:73–78 (1983).

Newman, M. M., J. R. Stahmann, J. D. Robb, E. A. Lewis, S. G. Martin, and S. V. Zinn: Triggered Lightning Strokes at Very Close Range. *J. Geophys. Res.*, **72**:4761–4764 (1967).

Pierce, E. T.: Triggered Lightning and Some Unsuspected Lightning Hazards. *Naval Res. Rev.*, **25**:14–28 (March, 1972).

Ruhling, F.: Gezielte Blitzentladung Mittels Raketen. *Umschau Wiss. Tech.*, **74**:520–521 (1974a).

Ruhling, F.: Raketen-Getriggerter Blitz im Dienste des Freileitungsschutzes vor Gewetterüberspannung. *Bull. Schweiz. Elektrotech. Ver.*, **65**:1893–1898 (1976).

Saint Privat d'Allier Research Group: Eight Years of Lightning Experiments at Saint Privat d'Allier. Extrait de la Revue Générale de l'Electricité, Paris. September, 1982.

Schonland, B. F. J.: The Lightning Discharge. *Handb. Phys.*, **22**:576–628 (1956).

Uman, M. A.: "Understanding Lightning." BEK Technical Publication, Pittsburgh, Pennsylvania, 1971 (Fig. 16.4).

Uman, M. A., D. F. Seacord, G. H. Price, and E. T. Pierce: Lightning Induced by Thermonuclear Detonations. *J. Geophys. Res.*, **77**:1591–1596 (1972).

Waldteufel, P., P. Metzger, J. L. Boulay, P. Laroche, and P. Hubert: Triggered Lightning Strokes Originating in Clear Air. *J. Geophys. Res.*, **85**:2861–2868 (1980).

Young, G. A.: A Lightning Strike of an Underwater Explosion Plume. U.S. Naval Ordnance Laboratory, NOLTR 61-43, March, 1962.

Chapter 13 | Cloud Discharges

13.1 INTRODUCTION

Although the majority of lightning discharges occurs within the confines of a thundercloud (Section 2.4), these cloud flashes have been the subject of relatively little research. This is apparently the case because cloud flashes are not of great practical interest (they do not cause power lines outages, set forest fires, or produce the variety of other deleterious effects attributable to cloud-to-ground lightning) and because cloud flashes are not easily photographed (still and streak photography being one of the primary research tools used in the study of cloud-to-ground discharges). While the characteristics of cloud discharges are of practical interest primarily to those concerned with the flight of airborne vehicles that pass through clouds, an understanding of cloud and ground lightning is of equal importance to the scientist attempting to learn about thunderstorm charge generation and disposition.

As we shall see in this chapter, different investigators have obtained conflicting results relative to the properties of cloud discharges. To complicate the situation further or perhaps to help explain it, the term "cloud discharge" may well refer to a variety of different phenomona, and different investigators may well have studied different types of cloud discharges. The distinction between the terms intracloud lightning, intercloud lightning, and cloud-to air lightning is rarely discussed in the literature. In one of the few instances that it is, Ogawa and Brook (1964) argue that intracloud and cloud-to-air lightning discharges are similar on the basis of the observed similarity of the electric fields from the two types of discharges

Recent advances in the techniques of data acquisition and processing now make practical the relatively accurate location of in-cloud electrical activity via two different approaches: (1) multiple-station electric field networks (e.g., Krehbiel et al., 1979; Liu and Krehbiel, 1985; Section C.7.1) and multiple-station VHF source locating networks of both the time-of-arrival and interferometric type (e.g., Proctor, 1971, 1981; Hayenga, 1984; Richard and Auffrey, 1985; Section C.7.3). Research using these location techniques shows great promise for effecting a better understanding both of cloud discharges and of the in-cloud portion of ground discharges.

13.2   SIMPLE MODELS FOR THE CHARGE TRANSFER OF THE OVERALL FLASH

As discussed in Section 3.2 and illustrated in Figs. 3.1 and 3.2, the idealized configuration of the primary charge structure in a thundercloud is that of a dipole: a lower negative charge Q_N and an upper positive charge Q_P. Limitations of this type of model are also discussed in Section 3.2. If the two main charge regions in the cloud charge model can be represented as roughly spherical with centers at heights H_N and H_P and horizontal distances from the observer D_N and D_P, respectively, then the total electric field at ground, which is vertical, can be found from Eq. (A.3):

$$E = \frac{1}{4\pi\varepsilon_0}\left[\frac{2Q_P H_P}{(D_P^2 + H_P^2)^{3/2}} - \frac{2Q_N H_N}{(D_N^2 + H_N^2)^{3/2}}\right] \qquad (13.1)$$

where Q_N in Eq. (13.1) and below is treated as an absolute value, its negative sign appearing in front of the term containing it.

In order to illustrate the general features of the electrostatic field from such a cloud charge distribution, we simplify the dipole model whose electric field is given by Eq. (13.1) by considering the dipole charges to be equal and the dipole to be vertical, that is $Q_P = Q_N = Q$, $D_P = D_N = D$.

$$E = \frac{1}{4\pi\varepsilon_0}\left[\frac{2QH_P}{(D^2 + H_P^2)^{3/2}} - \frac{2QH_N}{(D^2 + H_N^2)^{3/2}}\right] \qquad (13.2)$$

The electric field of Eq. (13.2) is plotted as a function of D in Fig. 3.2. Also plotted is the field for the case that there is present a smaller positive charge Q_p beneath the primary negative charge Q_N. The effect of this lower positive charge is primarily local, that is, its effect is evident primarily under the thunderstorm, as illustrated in Fig. 3.2. The physical origins of all three charge regions are discussed in Section 3.3. Perhaps the most interesting feature of the field of Eq. (13.2) is that it reverses sign with range: it is negative at close distance and positive far away. The distance D_0 at which the field passes through zero, the so-called reversal distance, is found by setting the electric field in Eq. (13.2) equal to zero,

$$D_0 = [(H_P H_N)^{2/3}(H_P^{2/3} + H_N^{2/3})]^{1/2} \qquad (13.3)$$

To illustrate the concept of a field reversal distance, data from Ogawa and Brook (1964) in New Mexico showing the electrostatic field from a thundercloud as it approached to within a few kilometers of the observing station and then receded are shown in Fig. 13.1. The observed reversal distance from the data in Fig. 13.1 according to Ogawa and Brook (1964) is about 7.5 km.

The electrostatic field change measured on the ground at D due to an intracloud discharge destroying a portion of a vertically oriented positive dipole can be found from Eq. (13.2):

$$\Delta E = -\frac{1}{4\pi\varepsilon_0}\left[\frac{2\Delta Q H_P}{(H_P^2 + D^2)^{3/2}} - \frac{2\Delta Q H_N}{(H_N^2 + D^2)^{3/2}}\right] \tag{13.4}$$

where the charge ΔQ disappears from both the Q_N and Q_P charge regions. The electric field change given in Eq. (13.4) also suffers a reversal in sign with increasing D: at small D the field change is positive; at large D the field change is negative. That this is the case can also be seen from Fig. 3.2. If a

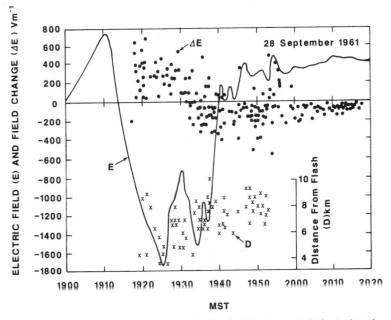

Fig. 13.1 Electric field E (solid curve) and electric field change ΔE (dots) plotted as a function of time for the isolated thunderstorm of 28 September 1961, observed in Socorro, New Mexico. The approximate distance to the storm at any given time is shown by the crosses which are primarily distances determined by thunder ranging on cloud-to-ground flashes. Adapted from Ogawa and Brook (1964).

portion of the dipole is destroyed, the cloud field at all ranges will become smaller: the positive field at large distances will decrease, a negative field change, while the negative field at close distances will also decrease, a positive field change. If the magnitudes of the original charges Q_N and Q_P that support the discharge are not equal, the field change reversal distance will differ from the cloud field reversal distance given in Eq. (13.3). The reversal distances will also differ if the positions of the upper and lower ends of the discharge are different from the effective centers of the two charge regions supporting the discharge. As an illustration of these effects, Ogawa and Brook (1964) found from the field change data shown in Fig. 13.1 that the cloud-flash reversal distance was 6.5 km, about 1 km smaller than the cloud field reversal distance, consistent with their physical model in which the effective end points of the cloud flash were closer together than the centers of the static cloud charges and were between them. From Fig. 13.1 it is also evident that the cloud field values are, except near the reversal distances, larger than the field change values, consistent both with the above discussion and with ΔQ being smaller than Q.

If $D \gg H_P$, the electric field change given in Eq. (13.4) can be written

$$\Delta E = -\frac{2}{4\pi\varepsilon_0} \frac{\Delta Q(H_P - H_N)}{D^3} \tag{13.5}$$

and hence the electric dipole moment change (Section A.1.2) can be determined directly from the electric field change and the distance to the flash

$$\Delta M = -2\Delta Q(H_P - H_N) = 4\pi\varepsilon_0 \Delta E D^3 \tag{13.6}$$

Note that the dipole moment change is approximately the same as given in Eq. (13.6) even if the two charges being neutralized are separated horizontally so long as the horizontal separation is small compared to D. For the data illustrated in Fig. 13.1, Brook and Ogawa (1977) report an average dipole moment change of 110 C-km which they attribute to an average charge transfer of about 28 C between heights of about 4 and 6 km.

If individual cloud discharges are to be properly studied, a nonvertical dipole model is more appropriate than the vertical dipole model just discussed. From measured electric field changes from cloud flashes and use of the inclined dipole model of Eq. (13.1), the three coordinates of each of the two discharge end points and the charge ΔQ can, in principle, be determined. Since there are seven unknown parameters, a minimum of seven simultaneous and spatially separated electric field measurements is necessary to determine the unknowns. If the cloud discharge is vertical, only four simultaneous measurements are necessary, but one can never be certain, a priori, that this is the case. Jacobson and Krider (1976) and Krehbiel et al.

(1979) discuss the mathematical techniques available to determine efficiently the unknown parameters once the multiple-station field change data have been obtained (Sections C.7.1 and B.5). Cloud-flash charge location and magnitude determined from both single- and multiple-station measurements are given in the next section.

13.3 FLASH CHARACTERISTICS

While lightning flashes to ground are characterized by rapid return-stroke field changes occurring every 50 msec or so and lasting for times of the order of 1 msec as illustrated in Figs. 1.9, 1.12, and 4.3, cloud discharges produce slow, relatively smooth field changes as illustrated in Figs. 1.12, 13.2, and 13.3. Cloud and cloud-to-ground discharges have about the same total time duration, generally a fraction of a second. The average cloud discharge duration reported by Mackerras (1968) is 480 msec, by Takagi (1961) is 300 msec, by Ishikawa (1961) is 420 msec, and by Ogawa and Brook (1964) is 500 msec. Dipole moment changes of the order of 100 C-km associated with cloud-flash field changes have been reported, for example, by Mackerras (1968) in Australia who found a median of 100 C-km, by Pierce (1955a) in England who found a typical value of about 100 C-km, by Wang (1963a,b) in Singapore who found a mean of 200 C-km, by Brook and Ogawa (1977) in New Mexico who found a mean of 110 C-km (see previous section), while values an order of magnitude smaller have occasionally been measured (e.g., Reynolds and Neill, 1955; Tamura *et al.*, 1958), apparently associated with smaller discharges or with horizontally oriented discharges (see this section, below).

As noted in Section 13.2, if electric field change measurements are made at at least seven separate stations, the resultant data are sufficient to completely specify the seven discharge parameters. Workman and Holzer (1942) made a set of electric field measurements with eight stations in New Mexico, during the summer of 1939. They analyzed field changes from 16 cloud flashes and found an average height of 4.7 km above local terrain (which was 2.3 km above sea level) for the lower negative charge and 5.8 km for the upper positive. The average electric dipole moment change was 70 C-km, consistent with a mean vertical discharge length of about 1 km and a mean charge transfer of about 35 C. Workman *et al.* (1942) constructed an improved system for electric field change measurement during the summer of 1940 and were able to analyze over 100 cloud flashes. They reported charge transfers of from 0.3 to 100 C. The average vertical separation between the charges was 0.6 km, the lower negative being at a height of 5.2 km, and the upper positive being at 5.8 km, on the average. The horizontal separation between the charges varied between 1 and 10 km, the average separation being 3 km. Thus the intracloud discharges were found to be more nearly horizontal than vertical.

Reynolds and Neill (1955) made simultaneous field change observations at 12 stations, also in New Mexico, using a technique similar to the one developed by Workman *et al.* (1942). Reynolds and Neill (1955) reported data from 35 cloud discharges, 28 of which were found to have the positive charge center above the negative. The average negative charge center was at 5.1 km and the average positive charge center was at 5.5 km above local terrain. The average vertical separation of the charges was about 0.5 km, while the horizontal separation ranged from essentially negligible to about 0.5 km. The charges neutralized ranged from about 1 to 60 C with a mean value of about 20 C. The average electric dipole moment change for a cloud flash was about 20 C-km.

Pierce (1955a), working in England, measured the time duration of 685 slow, negative field changes and 143 slow, positive field changes, all observed at a single station far from the discharges. Pierce (1955a) attributed the slow, negative field changes to "air or cloud discharges which either lower positive charge, or more probably, raise negative charge"; the slow, positive field changes to "flashes which do not reach the earth, and which probably involve the downward movement of negative charge." Pierce (1955a) found a mean duration of 245 msec for the negative field changes and 145 msec for the positive field changes. Both the positive and the negative slow field changes were found to yield a typical moment change of about 100 C-km.

Kitagawa and Brook (1960) obtained electric field records for about 1400 cloud discharges. Typical data are shown in Fig. 13.2. Kitagawa and Brook (1960) divided the cloud-discharge field change into 3 portions: (1) an initial portion, (2) a very active portion, and (3) a later or J (junction)-type portion.

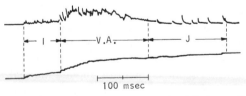

Fig. 13.2 Diagram of the typical field change due to a cloud discharge. The upper and lower traces represent simultaneous data recorded by an electric field system of high sensitivity and 70 μsec decay constant and by an electric field system of lower sensitivity and enough low-frequency response to accurately reproduce the electrostatic field change, respectively (Section C.1; Figs. C.2 and 1.9). I, Initial portion; V.A., very active portion; J, J-type portion. Adapted from Kitagawa and Brook (1960).

We review now their characterization of these three portions. (1) The initial portion is characterized by pulsations of relatively small amplitude with a mean pulse interval of about 680 μsec. The duration of the initial portion of the cloud discharge ranges from 50 to 300 msec. Thus significant differences are reported between the initial stages of cloud and of cloud-to-ground discharges: the time interval between pulses in the initial portion and the duration of the initial portion of the cloud discharge is reported to be significantly longer than the stepped-leader interpulse interval and time duration. In contrast, Schonland *et al.* (1938) and Schonland (1956) report that the pulsations superimposed on the slow electric field change due to a cloud discharge exhibit the same interpulse time intervals as do the pulsations from the stepped leader; and Proctor (1981) finds that the character of the VHF noise and the motion of the sources for stepped leaders in ground discharges and for the initial motion of leaders in cloud discharges differ only in their destination. Kitagawa and Brook (1960) report that they find the initial field change characteristics of cloud and of cloud-to-ground discharges so different that from the first 10 msec of the field records they can predict with over 95% certainty whether a discharge will reach ground or will remain within the cloud. (2) The very active portion of the cloud-discharge field change exhibits initially a relatively rapid overall field change and large field pulses. However, the transition between the initial and the very active portions is not always abrupt or distinct. (3) The later or J-type portion of the cloud-discharge field change is similar to the J-portion of the ground discharge (Chapter 10). K-changes occur at intervals from 2 to 20 msec (see Section 13.5). The field change in the J-type portion is not as rapid as in the preceding very active stage. The J-type portion of a cloud discharge is very distinct in character from the initial and very active portions.

Kitagawa and Brook (1960) report that out of about 1400 cloud discharges studied, 50% contained all three of the above-discussed portions, 40% consisted of very active and J-type portions, and the remaining 10% lacked the J-type portion and consisted of either the initial or very active portion, or both.

13.4   INITIATION

While the magnitude of the cloud-flash overall electric field change can be used to determine the charge values and locations associated with the flash, in order to gain information about the formation of the discharge it is necessary to analyze the detailed waveshape of that field change. The question of whether typical cloud flashes are initiated by negative discharges moving upward from the N region or by positive discharges moving downward from the P region, the two models most assumed and employed

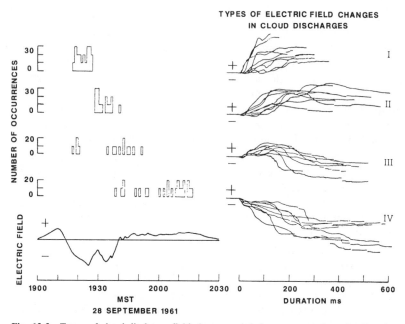

Fig. 13.3 Types of cloud-discharge field change and their occurrence throughout various stages of the storm shown in Fig. 13.1. The electric field curve is reproduced from Fig. 13.1 and correlated with the time of occurrence of each type of field change. Adapted from Ogawa and Brook (1964).

to analyze data to date, or whether some other model for typical discharges is more appropriate, such as one in which the discharges are more nearly horizontal, is perhaps the fundamental unanswered question in cloud flash research.

Waveforms of cloud-flash field changes from Ogawa and Brook (1964) are shown in Fig. 13.3 for the same storm whose field and field change data are given in Fig. 13.1. The waveforms in Fig. 13.3 have been divided by Ogawa and Brook (1964) into four types. Type 1 has field change values and slopes that are positive and were usually observed for flashes within 6 km. The waveforms were of relatively short duration and occurred only at the beginning of the storm. In type 2 the slope is first positive, then negative, while the field remains positive. The polarity of the steplike K-changes at the end of the discharges is positive for type 1 and negative for type 2. Type 2 discharges were observed for cloud flashes at ranges between 4 and 10 km as were type 3 which were similar to type 2 except that the final field became negative. In type 4, observed for flashes at distances greater than 8 km, both the field change slopes and values are negative.

Ogawa and Brook (1964) assumed that the initial and very active stages of the discharge, about the first half or more of the field change, were due to the propagation of a vertically oriented leader either lowering positive charge or raising negative charge. Theoretically expected waveshapes for their model are derived in Section A.1.3 and shown in Fig. A.4. By matching the shapes of the measured fields from a single station with the theoretical curves, Ogawa and Brook (1964) concluded that positive charge was being lowered from the upper positive charge center. For a leader length of a few kilometers and duration of a few hundred milliseconds, the downward leader speed was found to be about 10^4 m/sec. For a charge transfer of a few tens of coulombs in a time of a few hundred milliseconds, the average leader current was inferred to be of the order of 100 A.

Takagi (1961), on the basis of a statistical study of cloud-flash electric field waveshapes made from a single station in Japan, had earlier arrived at conclusions essentially identical to those of Ogawa and Brook (1964), whose research was performed without knowledge of Takagi's work.

Contrary to the conclusions of Ogawa and Brook (1964) and Takagi (1961), Smith (1957) and Nakano (1979a,b) found that the cloud discharges they studied were initiated by ascending negative leaders. They used a physical model similar to that of Ogawa and Brook (1964) and Takagi (1961) in which the vertical leader extension takes place in a time of the order of a hundred milliseconds. Smith (1957) used 2 stations separated by 13.2 km to study summer lightning in Florida while Nakano (1979a,b) employed 3 to 7 stations to study winter lightning in Japan.

Smith (1957) analyzed electric field change data from 54 discharges, each of which produced a field change containing a maximum or minimum at one or both of the recording stations. These data plus a knowledge of which station was closer to a given discharge allowed a determination of the sign and direction of motion of the charge within the cloud. Smith (1957) used as discharge models a vertical positive dipole and a vertical negative dipole in which one point charge was allowed to move. He found that in 39 of the 54 observations the dipole was positive. In the 39 positive dipole cases, negative charge was raised 30 times and positive charge lowered 9 times. In the 15 negative dipole cases, positive charge was raised 9 times and negative lowered 6 times. Smith (1957) infers from his data that the discharge of positive dipoles takes place at greater altitude than does that of negative dipoles. Smith (1957) reports that about 80% of the slow field changes measured can be explained satisfactorily by the uniform vertical movement of a single charge (point charge model) within the cloud.

Nakano (1979a,b) studied 221 cloud discharges in seven winter storms. He found that the negative cloud charge source for the upward discharge originated in a region having temperature between -6 and $-10°$C, similar to summer storms, and that the charge carried upward had an average value of 63 C. The leader orientation was within 30° of the vertical and its speed was of the order of 10^4 m/sec. The upward propagation of the leader had a duration of about 50 msec. Nakano (1979a,b) states that it is well established, although this is not the case in view of later research, that summer cloud flashes develop from descending positive discharges and thus his results show that winter cloud flashes are different from those occurring in summer.

Takeuti (1965), working in Japan, used 3 stations to measure cloud-flash field changes. He found both vertical and horizontal discharges. For the former, the field change data indicated 24 descending positive leader and 14 ascending negative for the 90% of the discharges involving a positive dipole. Vertical discharges had path lengths of 2 km or less and formed with a leader speed of $1-2 \times 10^4$ m/sec.

Weber et al. (1982) used in-cloud electric field measurements and thunder source reconstructions (Sections 15.5) to estimate the physical properties of seven cloud discharges whose orientation was more horizontal than vertical. Average propagation speed was about 5×10^4 m/sec for an average distance of 5 km. The average current was 390 A. Discharges of both polarities were observed.

Liu and Krehbiel (1985) argue that all of the studies discussed above have employed the wrong model. Liu and Krehbiel (1985) find from simultaneous 7 to 8 station electric field measurements made in Florida that the cloud discharge is initiated by a negative leader propagating upward at a speed of $1-3 \times 10^5$ m/sec for a duration of 10–50 msec. In their view, the final length of the intracloud channel in the direction of propagation is attained in a time which is an order of magnitude less than assumed in the previous models, and the overall shape of the field change is therefore not related to a relatively slow leader propagation, as previously assumed. In support of their results, discussed in more detail below, they note that Proctor (1976, 1981) found, via a VHF source location technique (Section C.7.3), that 4 of 5 intracloud discharges were initiated by negative leaders propagating at speeds between 1 and 2×10^5 m/sec. The fifth discharge was initiated by a positive leader with a higher initial speed. Three of four of the negative leaders were horizontal; the fourth was vertical. Proctor (1983) found essentially similar results in a study of 26 consecutive flashes in a dissipating storm. Proctor's results will be discussed later in this section and in Sections 13.5 and 13.6.

Liu and Krehbiel (1985) modeled the initial part of the intracloud leader as an arbitrarily oriented uniformly charged line extending from a spherical charge source. This model is similar to that used by Huzita and Ogawa (1976b). However, Liu and Krehbiel (1985) establish the leader characteristics for a fixed leader orientation for only the first 5–10 msec period. After that time, they allow a new leader section to leave from the end of the old section at an arbitrary angle. Thus the model is composed of piece-wise linear segments of variable direction, length, and charge. Liu and Krehbiel (1985) were able to successfully apply the model to only the first two or three time periods of 5-10 msec. The results of the analysis of four intracloud flashes using the above model was that the discharges were initiated by the upward motion of negative charge. The height of initiation was between 7.3 and 8.3 km above both ground and sea level for three flashes in a small storm and 10 km for one flash in a large storm. The initiation heights correspond to environmental temperatures of −17 to −24 and −38°C, respectively. The discharges progressed nearly vertically upward with speeds between 1.1 and 3.4×10^5 m/sec for the initial 15-30 msec of the discharge after which time any further development could not be determined with the model used. The negative charge density on the initial discharges was 1-4 C/km over several kilometers of path length.

Liu and Krehbiel (1985) argue that small growing thunderstorm cells will produce vertical intracloud discharges while large mature or dissipating storms will allow a channel orientation which may be more horizontal as observed by Proctor (1976, 1981, 1983), but, independent of channel orientation, negative charge is carried away from the lower negative charge center. Proctor (1976, 1981, 1983) calculated negative line charge densities for cloud discharge very similar to those found by Liu and Krehbiel (1985).

Krehbiel (1981) studied the electric field of 24 intracloud flashes in Florida. All but four of the discharges were vertical. Three of the four horizontal discharges occurred during the dissipating stages of a large storm system and the fourth was the tenth flash in a 15 flash sequence produced by a small storm. In most cases the direction of propagation and sign of the charge of the initial discharge propagation could not be determined with the dipole models [different from the model of Liu and Krehbiel (1985) discussed above] that were used. In several cases, however, the center of the charge transfer increased slightly in altitude during the first 15-30 msec, indicative of negative upward propagation at speeds of 10^5 m/sec or faster.

Proctor (1976, 1981), using a VHF source location technique in conjunction with a ground-based electric field measurement, found that most cloud flashes in South Africa were best characterized as horizontal and were initiated by several leaders moving negative charge away from a common origin. Detailed data are given on three cloud discharges that moved negative charge nearly horizontally at altitudes of 2.9, 4.5, and 6.5 km above ground level (which is 1.4 km above sea level), one that moved negative charge nearly vertically between 4.8 and 11.5 km, and one that moved positive charge horizontally at an altitude of 4 km, about 1 km above the freezing level. Proctor (1983) reported on the VHF source locations for 26 consecutive flashes in one thunderstorm. Of these, 19 were cloud flashes. Apparently, all flashes moved negative charge out of a region between 3 and 5 km above ground, and the cloud flashes followed horizontal paths influenced by a positive charge region on the rear flank of the storm. None of the cloud discharges discussed by Proctor (1976, 1981, 1983) is consistent with the previous pictures of a typical cloud discharge based solely on electric field measurements.

13.5   K-CHANGES

The millisecond-scale electric field changes that occur in cloud discharges are generally called K-field changes. The K-changes are evident in the waveforms of Figs. 1.12, 13.2, and 13.3. Kitagawa and Kobayashi (1958), who named the K-changes (according to Brook and Ogawa, 1977, as short for the German "Kleine Veränderungen"), observed that the short-duration field changes with accompanying pulses of luminosity that occurred in cloud flashes were similar to those they had previously observed between strokes of ground discharges (Section 10.4). Since the character of the cloud-flash field changes was very similar to the K-changes in the ground discharges, they were also so designated. Kitagawa and Kobayashi (1958) assumed the total discharge process of the cloud discharge to be similar to that of the J-process between strokes in the ground discharge and viewed K-changes as occurring throughout the discharge as the propagating positive leader encountered concentrated regions of negative charge. Ogawa and Brook (1964), on the other hand, argue that K-changes do not occur during the initial or very active portions of the cloud discharge and interpret K-changes as being due to upward-propagating negative recoil streamers or mini-return strokes that are initiated only when a downward-moving positive leader finally contacts negative charge during the latter or J-portion of the discharge. Recoil streamers in cloud-to-air discharges have been photographed by Ogawa and Brook (1964) and by Sourdillon (1952). Ogawa and Brook (1964) interpret their streak photographs of cloud-to-air discharge channels as indicating that continuous current flows for a time equivalent to the duration of the initial and very active stages, while the K-change recoil streamers occur only in the final stage of the discharge. It was assumed, on the basis of similar electric field records, that the air discharge photographed was essentially the same as an intracloud discharge. Many investigators have labeled any large rapid field change in a cloud discharge as a K-change (e.g., Malan, 1958; see Fig. 1.12; Kitagawa and Kobayashi, 1958) while others (e.g., Ogawa and Brook, 1964) reserve this designation only for those field changes occurring in the latter portion of the cloud discharge.

Kitagawa and Brook (1960) have plotted the frequency distribution of K-change intervals for the J-portion of 1318 cloud discharges and for the inter-stroke times of 671 ground discharges. The distribution of K-change intervals for the two types of discharges is almost identical, strongly suggesting that the in-cloud discharge mechanisms in the time between strokes of a ground discharge and the later portion of a cloud discharge are very similar. On the other hand, K-changes in cloud and ground discharges are usually of opposite polarity and, according to Ogawa and Brook (1964), the mean moment change associated with a cloud K-change, 8 C-km, is considerably larger than the largest moment change, 2 C-km, associated with K-changes in ground flashes. Kitagawa and Brook (1960) report that usually no field change is noticeable in the intervals between cloud flash K-changes. Thus, the net field change during the final part of the discharge is effectively the sum of the individual K-changes, and the slope of the J-type part is determined by the polarity of the K-changes. Ogawa and Brook (1964) suggest that the physical process associated with the negative cloud K-changes (or recoil streamers) has the following properties: time duration 1–3 msec, channel length 1–3 km, velocity 2×10^6 m/sec, charge neutralized 3.5 C, and average current 1–4 kA. They also report that during the J-period there are most frequently six of these K-processes.

The values of recoil-streamer charge, current channel length, propagation velocity, and frequency of occurrence reported by Takagi (1961) are all within a factor of 2 or 3 of the values obtained by Ogawa and Brook (1964). Takagi (1961) suggests that the intracloud discharge and the in-cloud inter-stroke processes of a cloud-to-ground flash are similar, except for the direction of discharge propagation. Takagi (1961) considers the J-process in a ground discharge to be an upward positive discharge down which negative recoil streamers propagate. Takagi (1961) suggests that a recoil streamer of large intensity may become a dart leader propagating to ground down the previous return stroke channel.

Photoelectric measurements of cloud luminosity correlated with electric field change data have been obtained by Takagi (1961). Strong luminous pulses were found to coincide with the K-field changes. The pulse luminosity rose to peak in about 0.5 msec and fell to half value in about the same time. The complete luminous cloud discharge was found to have a duration of 0.2–0.5 sec. The slow field change associated with the total discharge, which is attributed by Takagi (1961) to a downward-moving positive leader, was accompanied by weak continuous luminosity. The continuous luminosity fluctuated considerably, but on the average rose to a peak value at a time from initiation equal to about one-third of the total luminous duration. Similar measurements of the luminosity of cloud discharges have been made by Malan (1955b) and by Brook and Kitagawa (1960).

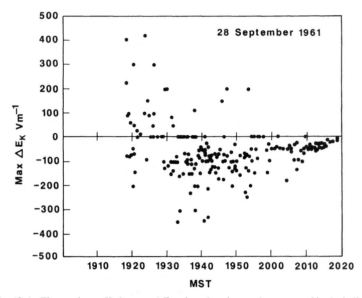

Fig. 13.4   The maximum K-changes, ΔE_k, plotted vs time as they occurred in the individual flashes in the storm shown in Fig. 13.1. The weaker K-changes (on the average there are about five per flash) have been omitted for clarity. Adapted from Ogawa and Brook (1964).

Ogawa and Brook (1964) measured the electric field change ΔE_k associated with K-processes in the cloud discharges of the storm whose field and field change data are shown in Figs. 13.1 and 13.3. The magnitude and the polarity of the largest of the observed K-changes of each cloud flash are plotted in Fig. 13.4 as a function of time. A single cloud discharge was found to produce as many as 20 detectable K-changes, the most frequent number being 6 per flash. These smaller K-change values fall on a line under the largest one, hence, for clarity in Fig. 13.4 only the largest K-change in each flash is plotted. The polarity and shape of the K-field change as a function of the distance of the discharge were used by Ogawa and Brook (1964) as evidence that K-changes was caused by recoil streamers or small return strokes of negative polarity that propagated upward. It is evident from a comparison of Fig. 13.4 with Fig. 13.1 that the sign of the field change produced by the K-processes changed from positive to negative somewhat earlier in time (and hence also in distance) than did both the electric field E and the field change ΔE. The ΔE reversal occurred when the storm was about 6.5 km from the station, the E reversal at about 7.5 km (Section 13.2).

Analyzing ΔE_k and ΔE data in a manner similar to the analysis of E and ΔE given in Section 13.2, Ogawa and Brook (1964) and Brook and Ogawa

(1977) suggest a similar constraint upon the effective end points of the K-discharge relative to the total cloud discharge. They conclude that the K-discharge is shorter in length than the total discharge. A calculation by Brook and Ogawa (1977) shows that the upper end of the recoil streamer associated with the K-change is at 5.3 km, assuming a starting point for the recoil streamer at 4.0 km, i.e., at the end point of the initial streamer. Thus the K-process channel length is estimated by Brook and Ogawa to be about 1.3 km in length, which corresponds to about 65% of the total cloud-discharge channel length. They calculate that a charge of 1.4 C is involved in the K-process for a measured field change value of $\Delta E_k = 10$ V m^{-1} at a distance of 8 km. The corresponding moment change is 3.6 C-km. With the charge magnitude and a duration of the order of 1 msec, Brook and Ogawa (1977) estimate the average current involved in the K-process to be about 1400 A. From the 1.3 km length and the 1 msec duration, Brook and Ogawa (1977) determine a recoil-streamer speed of approximately 1.3×10^6 m sec^{-1}, which is near the most frequently measured value of the dart leader speed in strokes to ground (see Section 8.2). This speed estimate is in fair agreement with the values given by Ishikawa (1961) and Takagi (1961) who estimated speeds of $(3-4) \times 10^6$ m sec^{-1} and of the order of 10^6 m sec^{-1}, respectively.

Proctor (1976, 1981) has used a VHF source location technique to follow the propagation of noise sources apparently associated with cloud flash K-changes. He studied 19 events that radiated VHF noise characteristic of K-changes, but sometimes occurring without a K-change, and found velocities from 2.5×10^6 to 4.4×10^7 m/sec, with a mean of 2.5×10^7 m/sec. Path lengths were from 136 m to 4.4 km. All but one of these events transferred positive charge, consistent with Proctor's (1976, 1981) observations that the cloud discharges initially moved negative charge out of the negative charge region of the cloud and the K-processes were recoil streamers back along these negative channels.

13.6 PULSE WAVESHAPES

Three types of radiation field pulses have been identified with wideband antenna systems as occurring in cloud discharges: (1) trains of unipolar pulses, (2) bipolar pulses with structure on the rise to peak, and (3) bipolar pulses with smooth rise to peak.

1. Krider et al. (1975) have recorded regular sequences of primarily unipolar electric and magnetic field pulses which occur at 5-μsec intervals and which have a total sequence duration of 100–400 μsec. The unipolar pulses have risetimes of 0.2 μsec or less and a full width at half maximum of about

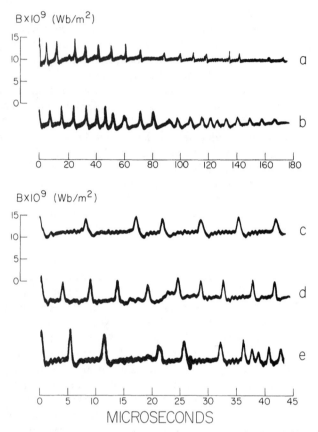

Fig. 13.5 Trains of unipolar magnetic field pulses (a–e) produced by five different intracloud lightning discharges within 50 km in Arizona. Adapted from Krider *et al.* (1975).

0.75 μsec. Examples of these pulses are given in Fig. 13.5. Krider *et al.* (1975) note the similarity of these pulse sequences to those radiated by the dart-stepped leaders that sometimes precede subsequent strokes in ground discharges (see Sections 8.2 and 8.3, and Fig. 7.4b) and suggest that a similar dart-stepping process may occur in the cloud discharge. If each of the 20–80 pulses occurring at 5-μsec intervals has a length of 10 m (a value similar to dart-stepped leader) the total channel length would be 200–800 m and the propagation velocity would be 2×10^6 m/sec. These values are similar to the description of a K-change given by Ogawa and Brook (1964). Krider *et al.* (1975) find that the sequences of regular pulses occur throughout the cloud discharge but state that there is a tendency for the sequences to occur at the end of the discharge. This implies that some may be associated with K-changes.

2. Weidman and Krider (1979) have identified large amplitude, bipolar pulses similar to those associated with the preliminary breakdown process in ground flashes (Section 4.3, Fig. 4.5) but with opposite polarity and larger time interval between pulses. Examples of these waveforms are shown in Figs. 13.6 and 13.7. The frequency spectrum of six pulses is given by Weidman *et al.* (1981). Weidman and Krider (1979) have found that these large bipolar pulses typically have several fast unipolar pulses superimposed on the initial rise to peak. The shape of the fast impulses is similar to individual pulses in the regular sequences just discussed. The large pulses have a mean full width of 63 μsec and a mean pulse interval of 780 μsec, a value which is similar to the 680-μsec mean pulse interval reported by Kitagawa and Brook (1960) for the initial portion of a cloud discharge. On the other hand, the bipolar pulses are among the largest events in the cloud discharge and thus perhaps should be associated with the very active portion of the discharge. Weidman and Krider (1979) suggest that the fast pulses on the initial rise are due to a steplike breakdown current and that the subsequent large bipolar field is due to a slower current surge flowing in the channel established by the steps. Krider *et al.* (1979) have shown that RF

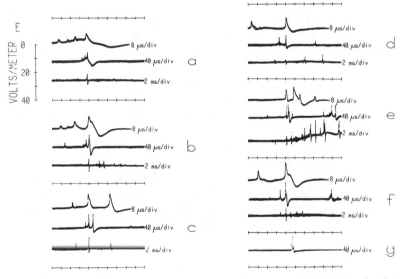

Fig. 13.6 Large bipolar electric fields (a–g) with structure on their fronts radiated by intracloud discharges at distances of 15–30 km. Each event is shown on time scales of 2 msec/division, 40 μsec/division, and 8 μsec/division. The polarity of the fields is consistent with the raising of negative charge or the lowering of positive charge. Adapted from Weidman and Krider (1979).

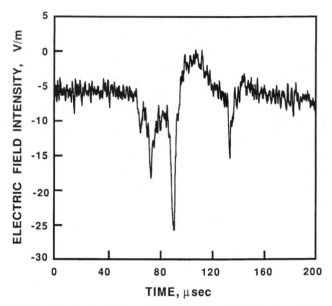

Fig. 13.7   A large bipolar cloud pulse with front structure of the type shown in Fig. 13.6 for a cloud flash in the 5- to 15-km range. The polarity of the field is consistent with the raising of negative charge or the lowering of positive charge. The high-frequency noise on the waveform is due to the tape recorder on which the fields were recorded. Data from the University of Florida group records.

radiation at frequencies between 3 and 295 MHz is coincident with each bipolar pulse, which implies that the process which generates the pulse probably involves the breakdown of virgin air rather than propagation along an already existing channel.

3. LeVine (1980) found that the strongest RF signals in both cloud and ground discharges were apparently from cloud discharges and that the bursts of RF noise had associated with them large isolated bipolar electric fields of 10–20 μsec duration and initial negative polarity. An example is shown in Fig. 13.8. The pulses described by LeVine (1980) have a initial smooth rise to peak as opposed to the bipolar pulses with structure on their fronts discussed in (2) above and illustrated in Figs. 13.6 and 13.7. LeVine (1980) attributes the smooth-fronted pulses to K-changes. Cooray and Lundquist (1985) have analyzed 26 pulses of the type described by LeVine (1980). Cooray and Lundquist (1985), however, triggered their recording equipment directly on the electric field pulses themselves whereas LeVine (1980) triggered on the RF noise burst associated with the pulses. Cooray and Lundquist (1985) found a mean zero-to-peak risetime of 4 μsec, a mean zero-crossing time of 13 μsec,

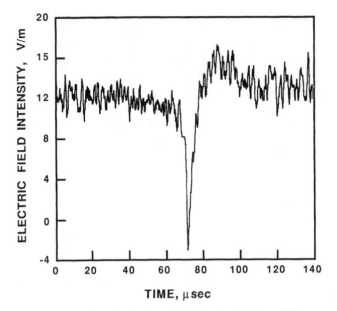

Fig. 13.8 A large bipolar electric field without structure on its front of the type studied by LeVine (1980) and Cooray and Lundquist (1985). The polarity of the field is consistent with the raising of negative charge or the lowering of positive. The cloud flash was in the 5- to 15-km range. The high-frequency noise on the waveform is due to the tape recorder on which the fields were recorded. Data from the University of Florida group records.

and a mean total duration of 75 μsec. The initial negative peak was about 3 times the value of the positive overshoot and the signals had 0.2–0.5 the amplitude of return strokes occurring at about the same time. The only significant difference between the observations of LeVine (1980) and those of Cooray and Lundquist (1985) would appear to be in the duration of the positive overshoot observed by Cooray and Lundquist, some tens of microseconds longer than that reported by LeVine (1980). It is certainly possible that the bipolar pulses discussed in (2) above are composed of sequences of overlapping smooth-fronted pulses of the type described by LeVine (1980) and by Cooray and Lundquist (1985).

13.7 NARROW-BAND RADIATION

Malan (1958) studied the relationship between electrostatic field changes and radiation in narrow frequency bands ranging from 3 kHz to 12 MHz for both cloud and ground discharge. His results are summarized in Fig. 1.12.

He found that, from 3 to 10 kHz, there were usually only a few cloud-flash narrow-band pulses, usually associated with the largest K-field changes. At higher frequencies, up to 2 MHz, more and more narrow-band pulses appeared, but those associated with the K-changes were the largest. From 4 to 12 MHz the radiation becomes more or less continuous, and the narrow-band radiation associated with the K-processes could no longer be distinguished. Clegg and Thomson (1979) have identified gaps or quiet periods that occur in the more or less continuous 10 MHz RF noise radiated during cloud flashes. The quiet periods have a typical duration of a few milliseconds and there are tens of these gaps per cloud discharge.

Brook and Kitagawa (1964) report that cloud discharges appear to be as strong or stronger microwave radiators at 420 and 850 MHz than ground discharges. The initial portion of a cloud discharge has continuous microwave radiation as well as strong pulse emission. The very active part produces large and frequent radiation pulses. The J-portion gives rise to K-change radiation pulses of the same magnitude as the K-change microwave radiation characteristic of the ground discharge.

Cloud flashes were divided by Proctor (1981) into two types according to the rates at which they emitted pulses in the VHF part of the radio spectrum. One type radiated pulses at rates that approximated 10^3 pulses per second. These pulses were often nearly rectangular in form, lasted approximately 1 μsec on average, and occurred simultaneously with pulses received at HF and at UHF. The second type emitted much shorter pulses (median durations of 0.2–0.4 μsec) at much higher rates, typically 10^5 pulses per second, and these pulses were generally not simultaneous with pulses at other radio frequencies. Diameters of the channels occupied by located radio sources varied from 100 m to several hundred meters. For low-pulse-frequency cloud flashes, average source sizes for individual pulses were near 300 m. For high-pulse-frequency sources, average sizes were near 60 m. It was found that pulses originated in regions near the tips of propagating discharges, and hence it was inferred that the pulses were associated with ionization of virgin air. The low-pulse-frequency cloud flashes showed marked stepping with channel extension speeds of approximately 6×10^4 m/sec, while the high-pulse-frequency flashes did not exhibit marked stepping and their channels extended at similar speeds, typically 10^5 m/sec.

Proctor (1981) also found that trains of band-limited noise, termed Q-noise, which lasted for times that ranged from about 10 μsec to about 1.5 msec, were emitted by lightning flashes of all kinds, but in cloud flashes this type of noise, which was distinctly different from the pulsed emission discussed above, occurred more frequently and with longer durations during the J-portion or final stages of cloud flashes. The band-limited noise was associated with propagation at speeds near 10^7 m/sec and often accompanied

K-electric field changes. Hence, the Q-noise was associated by Proctor (1981) with recoil streamers. The K-field changes were usually delayed by tens of microseconds after the start of this noise. The discharges that caused the noise transferred positive charge, with a single exception which was found to have been due to negative charge motion that occurred near the origin of an extensive, positive cloud flash.

REFERENCES

Aina, J. I.: Lightning Discharges Studies in a Tropical Area. II. Discharges Which Do Not Reach the Ground. *J. Geomagn. Geoelectr.*, **23**:359-368 (1971).

Aina, J. I.: Lightning Discharges Studies in a Tropical Area. III. The Profile of the Electrostatic Field Changes Due to Non-Ground Discharges. *J. Geomagn. Geoelectr.*, **24**:369-380 (1972).

Brook, M., and N. Kitagawa: Electric-Field Changes and Design of Lightning-Flash Counters. *J. Geophys. Res.*, **65**:1927-1931 (1960).

Brook, M., and N. Kitagawa: Radiation from Lightning Discharges in the Frequency Range 400 to 1,000 Mc/s. *J. Geophys. Res.*, **69**:2431-2434 (1964).

Brook, M., and T. Ogawa: The Cloud Discharge. *In* "Lightning, Vol. 1, Physics of Lightning" (R. H. Golde, ed.), pp. 191-230. Academic Press, New York, 1977.

Clegg, R. J., and E. M. Thomson: Some Properties of EM Radiation from Lightning. *J. Geophys. Res.*, **84**:719-724 (1979).

Cooray, V., and S. Lundquist: Characteristics of the Radiation Fields from Lightning in Sri Lanka in the Tropics. *J. Geophys. Res.*, **90**:6099-6109 (1985).

Funaki, K., K. Sakamoto, R. Tanaka, and N. Kitagawa: A Comparison of Cloud and Ground Lightning Discharges Observed in South-Kanto Summer Thunderstorms, 1980. *Res. Lett. Atmos. Electr.*, **1**:99-103 (1981).

Hayenga, C. O.: Characteristics of Lightning VHF Radiation near the Time of Return Strokes. *J. Geophys. Res.*, **89**:1403-1410 (1984).

Huzita, A., and T. Ogawa: Charge Distribution in the Average Thunderstorm Cloud. *J. Meteorol. Soc. Japan*, **54**:285-288 (1976a).

Huzita, A., and T. Ogawa: Electric Field Changes Due to Tilted Streamers in the Cloud Discharge. *J. Meteorol. Soc. Japan*, **54**:289-293 (1976b).

Ishikawa, H.: Nature of Lightning Discharges as Origins of Atmospherics. *Proc. Res. Inst. Atmos., Nagoya Univ.*, **8A**:1-273 (1961).

Ishikawa, H., and T. Takeuchi: Field Changes Due to Lightning Discharge. *Proc. Res. Inst. Atmos., Nagoya Univ.*, **13**:59-61 (1966).

Jacobson, E. A., and E. P. Krider: Electrostatic Field Changes Produced by Florida Lightning. *J. Atmos. Sci.*, **33**:103-119 (1976).

Khastgir, S. R., and S. K. Saha: On Intracloud Discharges and Their Accompanying Electric Field Changes. *J. Atmos. Terr. Phys.*, **34**:773-786 (1972).

Kitagawa, N.: On the Mechanism of Cloud Flash and Junction or Final Process in a Flash to Ground. *Pap. Meteorol. Geophys.*, **7**:415-424 (1957).

Kitagawa, N., and M. Brook: A Comparison of Intracloud and Cloud-to-Ground Lightning Discharges. *J. Geophys. Res.*, **65**:1189-1201 (1960).

Kitagawa, N., and M. Kobayashi: Field Changes and Variations of Luminosity due to Lightning Flashes. *In* "Recent Advances in Atmospheric Electricity" (L. G. Smith, ed.), pp. 485-501. Pergamon, Oxford, 1959.

Kobayashi, M., N. Kitagawa, T. Ikeda, and Y. Sato: Preliminary Studies of Variation of Luminosity and Field Change Due to Lightning Flashes. *Pap. Meteorol. Geophys.*, 9:29-34 (1958).

Krehbiel, P. R.: An Analysis of the Electric Field Change Produced by Lightning. Ph.D. thesis, University of Manchester Institute of Science and Technology, Manchester, England, 1981. Available as Report T-11, Geophysics Research Center, New Mexico Institute of Mining and Technology, Socorro, New Mexico 87801.

Krehbiel, P. R., M. Brook, and R. A. McCrory: An Analysis of the Charge Structure of Lightning Discharges to the Ground. *J. Geophys. Res.*, 84:2432-2456 (1979).

Krider, E. P., G. J. Radda, and R. C. Noggle: Regular Radiation Field Pulses Produced by Intracloud Discharges. *J. Geophys. Res.*, 80:3801-3804 (1975).

Krider, E. P., C. D. Weidman, and D. M. LeVine: The Temporal Structure of the HF and VHF Radiation Produced by Intracloud Lightning Discharges. *J. Geophys. Res.*, 74:5760-5762 (1979).

LeVine, D. M.: Sources of the Strongest RF Radiation from Lightning. *J. Geophys. Res.*, 85:4091-4095 (1980).

Liu, X., and P. R. Krehbiel: The Initial Streamer of Intracloud Lightning Flashes. *J. Geophys. Res.*, 90:6211-6218 (1985).

Mackerras, D.: A Comparison of Discharge Processes in Cloud and Ground Lightning Flashes. *J. Geophys. Res.*, 73:1175-1183 (1968).

Malan, D. J.: La Distribution Verticale de la Charge Negative Orageuse. *Ann. Geophys.*, 11:420-426 (1955a).

Malan, D. J.: Les Décharges Lumineuses dans les Nuages Orageux. *Ann. Geophys.*, 11:427-435 (1955b).

Malan, D. J.: Radiation from Lightning Discharges and Its Relation to the Discharge Process. *In* "Recent Advances in Atmospheric Electricity" (L. G. Smith, ed.), pp. 557-563. Pergamon, Oxford, 1958.

Nakano, M.: The Cloud Discharge in Winter Thunderstorms of the Hokuriku Coast. *J. Meteorol. Soc. Japan*, 57:444-445 (1979a).

Nakano, M.: Initial Streamer of the Cloud Discharge in Winter Thunderstorms of the Hokuriku Coast. *J. Meteorol. Soc. Japan*, 57:452-458 (1979b).

Ogawa, T., and M. Brook: The Mechanism of the Intracloud Lightning Discharge. *J. Geophys. Res.*, 69:514-519 (1964).

Petterson, B. J., and W. R. Wood: Measurements of Lightning Stroke to Aircraft. Report SC-M-67-549 and DS-68-1 on Project 520-002-03X to Dept. of Transportation, Federal Aviation Administration. Sandia Laboratory, Alburquerque, New Mexico, January, 1968.

Pierce, E. T.: Electrostatic Field-Changes Due to Lightning Discharges. *Q. J. R. Meteorol. Soc.*, 81:211-228 (1955a).

Pierce, E. T.: The Development of Lightning Discharges. *Q. J. R. Meteorol. Soc.*, 81:229-240 (1955b).

Prentice, S. A., and D. Mackerras: The Ratio of Cloud to Cloud-to-Ground Lightning Flashes in Thunderstorms. *J. Appl. Meteorol.*, 16:545-550 (1977).

Proctor, D. E.: A Hyperbolic System for Obtaining VHF Radio Pictures of Lightning. *J. Geophys. Res.*, 76:1478-1489 (1971).

Proctor, D. E.: Sources of Cloud-Flash Sferics. CSIR Special Report No. TEL 118, Pretoria, South Africa, 1974a.

Proctor, D. E.: VHF Radio Pictures of Lightning. CSIR Special Report No. TEL 120, Pretoria, South Africa, 1974b.

Proctor, D. E.: A Radio Study of Lightning. Ph.D. thesis, University of Witwatersrand, Johannesburg, South Africa, 1976.

Proctor, D. E.: VHF Radio Pictures of Cloud Flashes. *J. Geophys. Res.*, **86**:4041–4071 (1981).

Proctor, D. E.: Lightning and Precipitation in a Small Multicellular Thunderstorm. *J. Geophys. Res.*, **88**:5421–5440 (1983).

Proctor, D. E.: Correction to "Lightning and Precipitation in a Small Multicellular Thunderstorm." *J. Geophys. Res.*, **89**:11,826 (1984).

Rao, M., S. R. Khastgir, and H. Bhattacharya: Electric Field Changes. *J. Atmos. Terr. Phys.*, **24**:989–990 (1962).

Reynolds, S. E., and H. W. Neill: The Distribution and Discharge of Thunderstorm Charge-Centers. *J. Meteorol.*, **12**:1–12 (1955).

Richard, P., and G. Auffray: VHF–UHF Interferometric Measurements, Applications to Lightning Discharge Mapping. *Radio Sci.*, **20**:171–192 (1985).

Schonland, B. F. J.: The Lightning Discharge. *Handb. Phys.*, **22**:576–628 (1956).

Schonland, B. F. J., D. B. Hodges, and H. Collens: Progressive Lightning, Pt. 5, A Comparison of Photographic and Electrical Studies of the Discharge Process. *Proc. R. Soc. London Ser. A*, **166**:56–75 (1938).

Smith, L. G.: Intracloud Lightning Discharges. *Q. J. R. Meteorol. Soc.*, **83**:103–111 (1957).

Sourdillon, M.: Étude à la Chambre de Boys de "l'Eclair dans l'Air" et du "Coup de Foudre a Cime Horizontale." *Ann. Geophys.*, **8**:349–354 (1952).

Takagi, M.: The Mechanism of Discharges in a Thundercloud. *Proc. Res. Inst. Atmos., Nagoya Univ.*, **8B**:1–105 (1961).

Takeuti, T.: Studies on Thunderstorms Electricity, 1, Cloud Discharges. *J. Geomagn. Geoelectr.*, **17**:59–68 (1965).

Tamura, Y., T. Ogawa, and A. Okawati: The Electrical Structure of Thunderstorms. *J. Geomagn. Geoelectr.*, **10**:20–27 (1958).

Tepley, L. R.: Sferics from Intracloud Lightning Strokes. *J. Geophys. Res.*, **66**:111–123 (1961).

Wadhera, N. S., and B. A. P. Tantry: VLF Characteristics of K Changes in Lightning Discharges. *Indian J. Pure Appl. Phys.*, **5**:447–449 (1967).

Wang, C. P.: Lightning Discharges in the Tropics, (1) Whole Discharges. *J. Geophys. Res.*, **68**:1943–1949 (1963a).

Wang, C. P.: Lightning Discharges in the Tropics, (2) Component Ground Strokes and Cloud Dart Streamer Discharges. *J. Geophys. Res.*, **68**:1951–1958 (1963b).

Weber, M. E., H. J. Christian, A. A. Few, and M. F. Stewart: A Thundercloud Electric Field Sounding: Charge Distribution and Lightning. *J. Geophys. Res.*, **87**:7158–7169 (1982).

Weidman, C. D., and E. P. Krider: The Radiation Fields Wave Forms Produced by Intracloud Lightning Discharge Processes. *J. Geophys. Res.*, **84**:3159–3164 (1979).

Weidman, C. D., E. P. Krider, and M. A. Uman: Lightning Amplitude Spectra in the Interval from 100 kHz to 20 MHz. *Geophys. Res. Lett.*, **8**:931–934 (1981).

Wong, C. M., and K. K. Lim: The Inclination of Intracloud Lightning Discharge. *J. Geophys. Res.*, **83**:1905–1912 (1978).

Workman, E. J., and R. E. Holzer: A Preliminary Investigation of the Electrical Structure of Thunderstorms. *Tech. Notes Natl. Adv. Comm. Aeronaut.*, No. 850 (1942).

Workman, E. J., R. F. Holzer, and G. T. Pelsor: The Electrical Structure of Thunderstorms. *Tech. Notes Natl. Adv. Comm. Aeronaut.*, No. 864 (1942).

Chapter 14 | Lightning on Other Planets

14.1 INTRODUCTION

In order to discuss the possibilities of lightning on other planets and their satellites, it is useful to review briefly the circumstance under which lightning occurs on Earth. Lightning on Earth is almost always produced by convective, precipitating cumulonimbus clouds that contain supercooled water and ice. A discussion of the electrical properties of these clouds and of the processes that may produce the regions of cloud charge responsible for lightning is found in Chapter 3. In addition to being generated by thunderstorms, lightning (or long lightning-like electrical sparks) on Earth is sometimes produced in the ejected material above active volcanoes or between the ejected material and the volcano itself or the nearby earth, as illustrated in Fig. 1.16 (e.g., Anderson et al., 1965; Brook et al., 1974; Pounder, 1980), in sandstorms (Kamra, 1972), and apparently in clouds that have no portion higher than the freezing level and hence do not contain ice (e.g., Foster, 1950; Moore et al., 1960; Pietrowski, 1960; Michnowski, 1963). Transient luminous phenomena that may be associated with electrical discharges have been observed during earthquakes and may be due to electric fields generated by seismic strain (Finkelstein and Powell, 1970). Electrical sparks of short length occur in a variety of turbulent particulate media such as the material in grain elevators and the mixture of water and oil present during the water-jet cleaning of oil tanker holds (Pierce, 1974). The length of the electric spark generated under the circumstances indicated above depends on the scale of the separation of the charge sources. For example, Kamra (1972) observed sparks of a few meters length in New Mexico gypsum sandstorms, while Anderson et al. (1965) observed sparks of the order of 500 m at the volcano Surtsey near Iceland and Brook et al. (1974) interpreted electric field records from discharges at a volcano on the Westmann island of Heimaey, Iceland as indicating a discharge length of 200–500 m.

Based on the available evidence regarding the possibilities for the production of lightning and lightning-like electrical sparks on Earth, the probable requirements for the production of lightning on other planets are the following: (1) particulate material of at least two different types or particulate material of one type but with different properties such as size and temperature must interact to produce local charging so that different signs of charge are transferred to the different classes of particles, and (2) the different classes of particles of different charge sign must then be separated in space a distance of the order of the length of the resultant lightning. For thunderstorms on Earth this distance is measured in kilometers (Section 3.2).

A drawing of the major objects that comprise our solar system is found in Fig. 14.1. The eight planets of the solar system nearest to the sun can be divided into two groups. (1) The inner or terrestrial planets, from the sun outward, are Mercury, Venus, Earth, and Mars. These planets have well-defined surfaces but considerably different atmospheric conditions. Mercury and Mars have relatively thin atmospheres and no present volcanic activity and hence are not expected to produce lightning. The severe dust storms that occasionally occur on Mars could however produce electrical discharges (e.g., Eden and Vonnegut, 1973; Briggs *et al.*, 1977). (2) The outer, or giant planets, in order from the sun are Jupiter, Saturn, Uranus, and Neptune. These planets are relatively large and are thought to lack solid surfaces. Their atmospheres are composed primarily of hydrogen which increases in temperature and pressure with depth until it becomes liquid. The giant planets are shrouded with layers of clouds. The outermost cloud layers on Jupiter and Saturn are thought to be ammonia ice and on Neptune and Uranus are thought to be methane, with ammonia layers beneath (Weidenschilling and Lewis, 1973). Calculations further suggest that there are liquid water and water ice clouds in the lower atmosphere (Weidenschilling and Lewis, 1973). These water–ice clouds are expected to be capable of producing lightning, similar to thunderstorms on Earth. Any lightning will

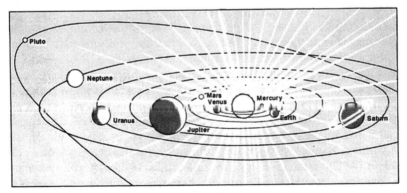

Fig. 14.1 The solar system.

be of the cloud discharge variety due to the relatively strongly downward-increasing atmospheric pressure and the lack of a nearby liquid or solid surface beneath the lowest cloud layer. Uranus and Neptune are thought to have more stable atmospheres than Jupiter or Saturn (Stone, 1973) and hence are less likely candidates for lightning. Completing the list of the known planets is Pluto, which orbits the Sun in a trajectory that is sometimes beyond and sometimes within that of Neptune, and about which relatively little is known.

Two planetary satellites have atmospheres in which lightning could potentially be generated. Io, one of 13 satellites of Jupiter and the closest to Jupiter of the 4 moon-sized, so-called Galilean satellites, has a relatively thin atmosphere, primarily of sulfur dioxide, and exhibits observed volcanic activity. Titan, the largest satellite in a complex system of at least 17 satellites and thousands of rings orbiting Saturn, is covered with thick clouds, primarily nitrogen and methane. Present evidence does not indicate the presence of lightning on either Io or Titan (Rinnert, 1985).

Optical and radio noise data obtained from a variety of planetary probes have been interpreted to indicate that there is lightning or some related form of electrical discharge on four planets: Venus, Jupiter, Saturn, and Uranus. The observations of Jovian lightning are the most convincing.

Review papers on planetary lightning have been published by Williams *et al.* (1983), Levin *et al.* (1983), and Rinnert (1982, 1985). In addition to discussing the evidence for planetary lightning, these reviews also contain considerable detail on the properties of the individual planetary atmospheres and the potential electrification mechanisms that could be operative in those atmospheres.

14.2 TECHNIQUES FOR DETECTION

According to Rinnert (1985), there is little possibility of successfully studying planetary lightning from Earth for the following two reasons: (1) optical observations are difficult at the ranges involved for the expected lightning light levels in the presence of reflected sunlight, particularly for the outer planets whose disks appear from Earth as fully sunlit as illustrated in Fig. 14.1; and (2) planets with substantial atmospheres necessarily have ionospheres that reflect back atmospheric radio emissions below a frequency determined by the maximum electron density in the particular ionosphere. This critical frequency is typically of the order of a megahertz and hence the bulk of the lightning RF energy, if it is earthlike lightning, will be trapped in the atmosphere. In addition to the ionospheric containment of electromagnetic waves from lightning, planetary magnetic fields (known to be present on Venus, Jupiter, Saturn, and Uranus) and their trapped electrons serve to guide the lower portion of the lightning spectrum along the magnetic field lines in the so-called "whistler" mode (e.g., Helliwell, 1965; Park, 1982), thus constraining that evidence of lightning to the planet's magnetosphere. Whistlers can propagate only at frequencies just below the frequency at which electrons orbit the magnetic field, the so-called electron gyrofrequency, which is typically in the audio range. Indeed whistlers received their names because they exhibit a dispersion in velocity with frequency so that if the electromagnetic wave is directly converted to an audible sound, that signal is a whistlelike noise caused by the fact that the higher frequencies arrive before the lower.

In view of the above, it is evident that the most efficient method of detecting and studying the properties of planetary lightning is *in situ* from specially instrumented spacecraft. Instrumented probes have entered the atmospheres of Venus and Mars and have measured the properties of those atmospheres. Probes have also landed on the surfaces of Venus and Mars and have made studies of the surfaces and the atmosphere from that vantage point. Instrumented satellites have been orbited around Venus. Two Pioneer and two Voyager spacecrafts have flown past and observed Jupiter, Pioneer 11 and both Voyagers studied Saturn, Voyager 2 studied Uranus, and Voyager 2 will observe Neptune in 1989. All of these space vehicles have had either optical or radio noise detectors, or both, unintentionally or intentionally capable of detecting evidence of lightning.

We consider in the following sections the evidence for lightning on the four planets, Venus, Jupiter, Saturn, and Uranus, for which nearby or in-atmospheric measurements have been interpreted to indicate its presence.

14.3 VENUS

A complete review of our knowledge of all aspects of the planet Venus is found in the book edited by Hunten *et al.* (1983), from which most of the following information is abstracted. Venus is about the size of Earth and has a similar surface structure. A drawing illustrating various surface and atmospheric properties is found in Fig. 14.2. Venus has a surface temperature of about 750 K, and a surface pressure almost 100 times that of Earth. Venus has no surface water. The atmosphere is predominantly carbon dioxide. Because of the high surface pressure, electric fields about 100 times greater than on Earth are required to produce electrical breakdown. The surface winds on Venus are relatively light. While there is considerable evidence in the surface features of previous volcanic activity, no ejected material has been observed as yet in the atmosphere which appears to be clear below about

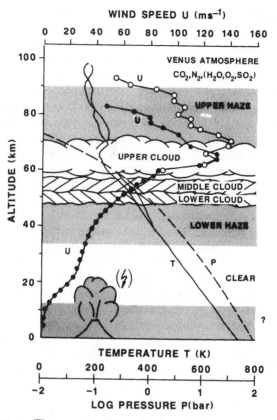

Fig. 14.2 Pressure (P), temperature (T), and horizontal wind speed (U) height profiles of the Venus atmosphere retrieved from Venera and Pioneer Venus probe measurements, together with a sketch of the cloud system. Adapted from Rinnert (1985).

30 km. As shown in Fig. 14.2, there is a dense global cloud deck between about 45 and 70 km above the surface. At that altitude the pressure is near that on the Earth's surface. Above and below the cloud deck are haze layers. The cloud deck has three layers. Drops of sulfuric acid have been identified in the clouds, but no appreciable turbulence has been observed.

There have been a total of 16 probes sent through the Venuvian atmosphere and two instrumented weather balloons floated in it, 3 orbiters around Venus, and a number of spacecraft flybys, making Venus the most explored of the planets. Colin (1983) lists all of these space vehicles except the balloons and gives launch and encounter dates and mission characteristics for each. The Russian probes Venera 11, 12, 13, and 14 carried instrumentation specificially to detect electrical activity. The first reported evidence for the possible existence of extraterrestrial lightning was the impulses in the radio frequency range detected by Venera 11 and 12 in December 1978 (Ksanfomality, 1980). Additionally, measurements of signals that were interpreted as lightning whistlers were obtained by the U.S. Pioneer Venus orbiter (Taylor *et al.*, 1979). Optical data from Venera 9, placed in orbit around Venus in 1975, were subsequently interpreted by Krasnopolsky (1983a,b) as being due to lightning. On the other hand, Borucki *et al.* (1981) evaluated star sensor data from the Pioneer Venus orbiter, placed in orbit in 1978, and observed no light output that could be attributed to lightning; and the two VEGA balloons, instrumented for transient light measurements, detected no lightning optical signals during a total sampling time of about 45 hr (Sagdeev *et al.*, 1986). We now look in more detail at the observations outlined above.

14.3.1 OPTICAL

Borucki *et al.* (1981) used the navigation star sensor in the Pioneer Venus orbiter to search for transient light signals indicative of lightning when the orbiter was on the dark side of Venus. The bandwidth of the optical detector was roughly 0.5–1.0 μm. When the data from 36 orbits were compared to the false alarm rate obtained from the gamma ray burst detector on the orbiter, no statistical difference could be found between the two signals.

Williams *et al.* (1982) and Williams and Thomason (1983) used a Monte Carlo program to model the effect of the Venus clouds on an optical lightning signal. Williams *et al.* (1982) found that the fraction of photons of visible light that escapes into space ranges from 0.1 to 0.4 for cloud discharges and is about 0.05 for red photons produced by discharges near the ground. Williams and Thomason (1983), using an improved model, found about twice as many photons escaping for a cloud discharge and four times as much for red light near the ground. About 3% of blue photons emitted by lightning at the surface escaped (Williams and Thomason, 1983). Since the star sensor used by Borucki *et al.* (1981) was sensitive to red light, lightning flashes that occurred within the clouds, or even at the surface, should have been detectable unless they were much weaker than Earth lightning (Williams *et al.*, 1982, 1983). Thus the false alarm rate on the orbiter gamma ray burst detector could be used to set an upper limit of 30 flashes $km^{-2} yr^{-1}$ for the average planetary lightning flash density, that is, a greater flash density should have been detected by the star sensor experiment. Williams *et al.* (1983) note, however, that the average flash density on Earth is only about 2 $km^{-2} yr^{-1}$, for which they reference the satellite measurements of Turman (1978) (Section 2.4), and that since this value is less than the lowest detectable limit set by the false alarm rate, the same experiment presumably could not have detected Earth lightning. It should be noted, however, that some overland regions on Earth have a yearly flash density an order of magnitude or more greater than the Earth average and that the yearly lightning for such locations generally occurs in a period of months, although it is averaged over a year in the published statistics (Sections 2.2 and 2.3).

Using the Venera 9 scanning spectrometer, Krasnopolsky (1980, 1983a,b) reported observing one 70-sec period of irregular optical pulses averaging 100 bursts per second with a characteristic burst duration of about 0.25 sec. No impulsive events were recorded during a time period three times as long as the surface observations when the spectrometer was pointing into space away from the planet. Krasnopolsky argues that the pulses were real and not attributable to noise or other instrumental problems in the detector because the characteristics of the observed spectrum were distorted in the way expected for an external light source. The total area viewed by the spectrometer was 3.5×10^7 km^2. The area with optical signals was 5×10^4 km^2. The average burst occurrence rate was 2×10^{-3} km^{-2} sec^{-1} or 0.12 km^{-2} min^{-1} in the active area, which can be compared to the flash density for localized Earth storms of about 0.01–0.02 km^{-2} min^{-1} for storm areas one to two orders of magnitude smaller (see Section 2.5 and Table 2.1). The spectrum was relatively flat with a weak peak near 6000 Å. Spectral power was 2.6×10^4 W Å^{-1} in the visible. Thus a typical received burst is indicative, if there were no losses in propagation, of about 3×10^7 J of optical energy. If the source was in or below the Venus clouds, the radiated energy would be greater because of losses in power from scattering and absorption. For an efficiency factor of 3×10^{-3} for the conversion of total energy to optical radiation (see Section 7.2.4), the total energy of the source would be 10^{10} J or greater, which can be compared to the 10^8–10^9 J thought to be dissipated in a typical 5-km channel on Earth (Section 7.2.4). The 0.25 sec length of the individual pulses reported by Krasnopolsky has a similar duration to Earth lightning but lacks the detailed structure such as that due to K-changes (see Chapter 13) that should have been detectable given the characteristics of the instrument. Williams *et al.* (1983) point out that it is difficult to understand how the high burst rate could be due to lightning, since it is not obvious how the Venus meteorology could produce such an active storm, which, as we have noted above, would have to have a flash density about an order of magnitude greater than a localized storm on Earth.

The two VEGA Venus balloons were instrumented to measure lightning optical signals as well as a variety of meterological data (Sagdeev *et al.*, 1986). Their equilibrium float altitudes were in the middle cloud layer. No lightning events were recorded during the 22.5 hr of data taken on June 11, 1985 by VEGA-1. One possible lightning optical signal was recorded during an equal time period 4 days later by VEGA-2, but the validity of the data is questionable (Sagdeev et al., 1986). All of the optical data for Venus are summarized in Table 14.1.

14.3.2 RADIO FREQUENCY NOISE

The Venera 11, 12, 13, and 14 landers carried instruments designed to detect and analyze radio frequency noise that occurred in the lower atmosphere of Venus (Ksanfomality, 1979, 1980; Ksanfomality et al., 1983). We consider only the results from Venera 11 and 12, although preliminary results from Venera 13 and 14 indicate similar observations. The detector consisted of a loop antenna with filters to measure the magnetic field at 10, 18, 36, and 80 kHz. The measurement was made between 60 km and the surface. The Venera 11 detector transmitted data for 76 min after landing; Venera 12 transmitted for 110 min. Both probes followed almost the same trajectory during descent, landing 4 days apart. Throughout their descents, both spacecraft detected impulsive RF signals. Venera 11, however, recorded higher pulse rates and higher pulse amplitudes. Further, Venera 11 registered periodic bursts, each burst containing several hundred pulses, while the spacecraft was between 13 and 9 km above the surface. The bursts were separated by a time of the order of 1 min. Ksanfomality et al. (1983) attribute the observed periodic signals to the presence of a localized source detected by the rotating probe. With this hypothesis, they estimated the angular diameter of the localized source to be about 5°. Pulses from that source occurred at a typical rate of about 20 sec^{-1} with a maximum of about 55 sec^{-1}.

The distance to the sources of the periodic bursts was estimated by Ksanfomality et al. (1983) using several techniques. From measured field strengths, up to 100 μV m^{-1} Hz$^{-1/2}$ in the 10-kHz channel and smaller at the higher frequencies, they conclude that if a source were at a minimum distance of several tens of kilometers, that is, on the surface under the descending spacecraft, the energy in the source discharge would be three orders of magnitude less than for Earth lightning. Such a source might be due to volcanic eruptions. Ksanfomality et al. (1983) prefer to view their measured signal strengths as due to distant lightning, and they use an emperical propagation formula and the magnitude of the 10-kHz signal from a particular burst to localize the source at a range of 1250–1350 km. At this range the 5° source angle obtained from the probe rotation data yields a horizontal source extent of 120–150 km. Similar analysis of field strength at other observed frequencies indicates ranges as large as 2000 km. A different burst is identified as being closer, 700–1000 km, and other bursts are assumed to be from other spatially isolated sources.

Table 14.1

Summary of Indications of Lightning on Venus, Jupiter, Saturn, and Uranus from Spacecraft Observations[a]

Vehicle/instrument	Observations	Inference
	Venus	
Pioneer Venus orbiter, star sensor	No positive identification[b]	Large area average of less than $30 \text{ km}^{-2} \text{yr}^{-1}$ or $10^{-6} \text{ km}^{-2} \text{sec}^{-1 b}$
VEGA Balloons, optical detector Venera 9, spectrometer	No positive identification[c] For 70 sec, 0.25-sec optical bursts, 100 sec^{-1}, $5 \times 10^4 \text{ km}^2$ storm area from $3.5 \times 10^7 \text{ km}^2$ total observed $\pm 32°$ latitude, $2.6 \times 10^4 \text{ W Å}^{-1}$ (4500–7500 Å), relatively flat spectrum[d]	$2 \times 10^{-3} \text{ km}^{-1} \text{sec}^{-1}$ (storm area), 3×10^7 J per burst in the visible, 10^{10} J total energy[d]
Venera 11, 12 (13, 14), "Groza" instrument, magnetic loop, 10, 18, 36, 80 kHz	Bursts of RF pulses, average about 20 impulses/sec, some received from 5° sector, maximum intensity in 10-kHz channel $100 \mu\text{V m}^{-1} \text{ Hz}^{-1/2}$, spectral index -1 to $-2^{e,f}$	Localized storms, one 120–150 km across at 1200- to 1400-km range, $1.5 \times 10^{-3} \text{ km}^{-2} \text{sec}^{-1}$ (storm area)[e,f]
Pioneer Venus orbiter, plasma wave experiment, electric field detector, 0.1, 0.73, 5.4, 30 kHz	RF impulses,[f,g,h] 567 events within 1185 orbits, nightside[i,j]	Whistlers originating from $\pm 30°$ latitude, clustered in region of volcanic origin,[i,j] for the most active 5° × 5° square: about $10^{-7} \text{ km}^{-2} \text{sec}^{-1}$,[j] but possibility of local generation in ionosphere[k,l]
	Jupiter	$10^{-10} \text{ km}^{-2} \text{sec}^{-1} \text{ }^m$ or $3 \times 10^{-3} \text{ km}^{-2} \text{yr}^{-1}$; 2.5×10^9 J optical per flash,[n] 1.7×10^{12} J total per flash,[n] $4 \text{ km}^{-2} \text{yr}^{-1} \text{ }^n$, 3–$30 \text{ km}^{-2} \text{yr}^{-1o}$
Voyager 1, imaging	20 luminous events in 192 sec over 10^9 km^2, 30° to 55° N, most at 45° N[m]	
Voyager 1, 2, plasma wave experiment	167 whistler signals from 2 to 7 kHz, $0.12 \text{ sec}^{-1 q,r}$	Whistlers originating from 66° latitude,[p] $40 \text{ km}^{-2} \text{yr}^{-1 r}$, $4 \times 10^{-2} \text{ km}^{-2} \text{yr}^{-1 s}$

262

Saturn

Voyager 1, 2, planetary radio astronomy instrument, 20 kHz to 40 MHz

Saturn electrostatic discharges (SED), about 3 hr duration and 10-hr period, individual bursts are wideband (0.02–40 MHz) with duration, 30–450 msec and rate 0.2 sec^{-1} [u,v,w,x]

Average burst power, 10^9–10^{10} W,[t] 10^8–10^9 W[u] Two possible sources:
(1) Electrical activity in superrotating clouds, equatorial region, 60° longitude $\times$ 3° latitude, 3×10^{-2} km^{-2} yr^{-1} [x,y]
(2) Electrical activity in B-ring,[v,w] 60° longitude, other two dimensions small[w]

Uranus

Voyager 2, planetary radio astronomy instrument, 1 kHz to 40 MHz

Uranus electrostatic discharges (UED), 140 events in 30 hr, individual bursts are wideband (0.9–40 MHz), duration 100–300 msec, mean 120 msec[z]

Typical burst power, 10^8 W, typical energy, 1–2 $\times$ 10^7 J[z]

[a] Adapted from Rinnert (1985).
[b] Borucki et al. (1981).
[c] Sagdeev et al. (1986).
[d] Kranopolsky (1983a,b).
[e] Ksanfomality et al. (1979).
[f] Ksanfomality et al. (1983).
[g] Taylor et al. (1979).
[h] Scarf et al. (1980).
[i] Scarf et al. (1982).

[j] Scarf and Russell (1983).
[k] Taylor et al. (1985).
[l] Taylor et al. (1986).
[m] Cook et al. (1979).
[n] Borucki et al. (1982a).
[o] Williams et al. (1983).
[p] Menietti and Gurnett (1980).
[q] Gurnett et al. (1979).
[r] Scarf et al. (1981).

[s] Lewis (1980a).
[t] Evans et al. (1983).
[u] Zarka and Pedersen (1983).
[v] Evans et al. (1981, 1982).
[w] Warwick et al. (1981, 1982).
[x] Kaiser et al. (1983, 1984).
[y] Burns et al. (1983).
[z] Zarka and Pedersen (1986).

For an RF burst with an estimated characteristic source size of 120–150 km the active source area would be about $2 \times 10^4 \, km^2$. For an average burst impulse rate of $30 \, sec^{-1}$, Ksanfomality *et al.* (1983) find a flash density of $1.5 \times 10^{-3} \, km^{-2} \, sec^{-1}$, very similar to that observed by the spectrometer on Venera 9, and, as noted in Section 14.3.1, significantly larger than for storms on Earth. The reported observed RF pulse rate is, however, not necessarily a lightning flash rate but could be a measure of the intermittent discharges (e.g., K-change in cloud discharges, see Chapter 13) that comprise flashes.

Ksanfomality *et al.* (1983) fit their measured RF spectra to the functional form f^α. The spectral index α was about -2 for Venera 11 and about -1 for Venera 12. For close return strokes on Earth spectra index values near -1 are generally obtained to a frequency of about 1 MHz (see Section 7.2.1 and Figs. 7.5 and 7.6).

It is possible that some or all of the radio signals observed by the Venera probes were due to electrostatic discharging of the probes. Ksanfomality *et al.* (1983) argue that the different height dependencies of the RF signal intensities measured by the two probes, the periodic nature of the signal on Venera 11, and the observation of bursts after landing (the Venera 12 instrument detected a burst, at least 150 pulses within a single 8-sec interval, 30 min after landing) all combine to exclude the possibility that the electrical signals are related to electrostatic discharge of the spacecraft. The RF data are summarized in Table 14.1.

14.3.3   WHISTLERS

The plasma wave experiment in the Pioneer Venus orbiter detected signals that were propagating in the whistler mode along magnetic field lines on the nightside of Venus (Taylor *et al.*, 1979; Scarf *et al.*, 1980; Ksanfomality *et al.*, 1983). The detector responded to radio waves at frequencies of 100 Hz, 730 Hz, 3.4 kHz, and 30 kHz. The data from early orbits show no impulsive radio signals. The reason was apparently because during this period the orbiter's periapsis was within the day ionosphere which exhibits a turbulent magnetic field that would only support propagation of signals with frequencies less than about 100 Hz. Additionally, noise associated with sunlight on the orbiter and its solar panels complicated the measurements. When the periapsis changed to the nightside, strong impulses were detected in the 100- and 730-Hz channels when the electron gyrofrequency, roughly 28 B where B is the magnetic flux density in gammas (1 gamma is $10^{-9} \, Wb/m^2$), was 400–800 Hz. Pulses were observed when the magnetic field was relatively stable and pointing down into the atmosphere. Although these signals could well have been propagating in the whistler mode, it is neither possible to establish with certainty that the source was lightning nor is it possible to rule out other mechanisms for the signal generation. For example, Taylor *et al.* (1985, 1986) present data to show that the observed signals are locally generated by changes in the ambient plasma density and hence are not caused by lightning. On the other hand, Scarf (1986) argues that there is no cause and effect relationship between the 100-Hz whistlerlike noise bursts and the observed occurrence of so-called ion troughs. Additional discussion of this controversy is found in Luhmann and Nagy (1986). Scarf and Russell (1983) reported that the data from 185 orbits strongly indicate the clustering of whistler sources: near the Beta and Phoebe Regios and near the eastern edge of Aphrodite Terra. Maps with source locations are given by Scarf and Russell (1983) and by Ksanfomality *et al.* (1983). Most of the whistler sources were within $\pm 30°$ latitude. The clustering of source locations implies that the sources may be in the lower atmosphere or even at the surface. However, Borucki (1982) has pointed out that the Pioneer Venus probes found no evidence of dust or ash as would appear to be necessary if the whistler sources were volcanic activity, Levine *et al.* (1983) argue that the determination of whistler source location is extremely inaccurate, and Taylor *et al.* (1985, 1986) claim that ion troughs should occur naturally above or near mountainous topography. Scarf and Russell (1983) surveyed about 14% of Venus as an origin for whistlers. They calculate the occurrence rate for the observed pulses. The most active $5° \times 5°$ square had only 24 events in 1000 sec corresponding to a whistler detection rate per unit area of about $10^{-7} \text{ km}^{-2} \text{ sec}^{-1}$. Although only a fraction of any lightning occurring might produce detectable whistlers, the calculated rate for an active storm area is four orders of magnitude smaller than the estimates in Table 14.1 for spectroscopic and RF detection but consistent with the upper limit set by the star sensor experiment. These data are summarized in Table 14.1.

14.3.4   ASHEN LIGHT

Occasional brightenings of the night hemisphere of Venus, known as ashen light, have been reported for some 300 years. Many theories, including earthshine (Napier, 1971), auroral activity (Levine, 1969), and lightning (Ksanfomality, 1979) have been proposed to explain the ashen light. The following argument against the ashen light being due to lightning is given by Borucki *et al.* (1981), Williams *et al.* (1982), and Ksanfomality *et al.* (1983): about 1300 W m^{-2} of visible solar radiation is incident on Venus, so that for an albedo of 0.76 the scattered power is about 1000 W m^{-2}. On Earth, lightning return strokes dissipate $10^8 - 10^9$ J of energy, but only about 0.1-1% of this energy is converted into light (Section 7.2.4). If there are Venus lightnings, they are probably cloud discharges (which on Earth are not as bright as return strokes), so 10^6 J is a generous estimate of the optical energy radiated by a lightning discharge on the planet. The upper limit to the lightning flash rate on Venus is 30 km^{-2} yr^{-1} (Borucki et al., 1981). This corresponds to 3×10^7 J km^{-2} yr^{-1} of source energy if the lightning is below the clouds. Since only about 20% of the optical lightning signal is expected to escape from the clouds to space (Williams et al., 1982), lightning is capable of producing only about 10^{-7} W m^{-2} of optical power. The daylit side of Venus is 10^{10} times brighter. Therefore lightning cannot be a reasonable explanation for the ashen light since the ashen light would not be visible if it were 10^{10} times less bright than the sunlit part of Venus.

14.4 JUPITER

Jupiter is the largest planet in the solar system with a radius of about 71,000 km, roughly 11 times that of Earth. Four spacecraft have observed Jupiter as they flew by: Pioneer 10 and 11 in December 1973 and 1974, respectively, and Voyager 1 and 2 in March and May 1979, respectively. The book edited by Gehrels (1976) gives a thorough review of knowledge of Jupiter after Pioneer 10 and 11. Figure 14.3 contains a drawing showing a variety of atmospheric properties. All of the present evidence for lightning on Jupiter has come from optical and whistler measurements made from Voyager 1.

Stone (1976), Ingersoll (1976), and Smith et al. (1979) provide recent reviews of what is known about the Jovian atmosphere. That atmosphere is composed of about 90% hydrogen and about 10% helium with fractions of a percent of CH_4, and NH_3, and traces of CH_3, CO, C_2H_6, and C_2H_2. The atmosphere of Jupiter has a banded structure consisting of about 10 alternating white zones and dark belts. It is believed that the white zones are upward-moving portions of the atmosphere and the dark belts are downward moving. Eastward and westward winds alternate with latitude. The absolute wind velocities range from about 20 to 150 m sec^{-1}. High-resolution imaging of white plumes found near the equator shows that they contain small (100 km) puffy elements resembling cumulus clouds. Jupiter has an internal energy source that provides a heat flux larger than the planet receives from the sun. Surprisingly, there is little variation in temperature from equator to pole.

All knowledge of the atmosphere below the upper clouds necessarily comes from atmospheric models. The chemical equilibrium model of Weidenschilling and Lewis (1973) predicts uppermost clouds of NH_3 ice at 0.5 bar and 150 K, lower clouds of NH_4SH ice at 3 bar and 200 K, and a bottom cloud

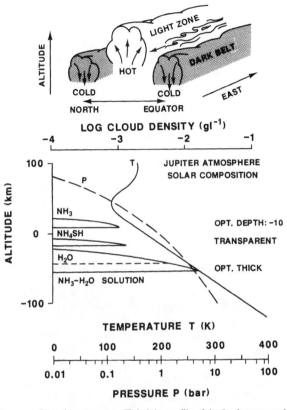

Fig. 14.3 Pressure (P) and temperature (T) height profile of the Jupiter atmosphere, together with the cloud model of Weidenschilling and Lewis (1973). The top panel sketches the global convective motions. Adapted from Rinnert (1985).

of H_2O ice at 5 bar and 270 K. If, in the model calculations, the NH_3, H_2O, and H_2S abundances are allowed to increase by a factor of 5, the H_2O cloud base becomes a liquid water and NH_3 solution at the 6-bar level. Spectroscopic studies have led Sato and Hansen (1979) to conclude that the upper NH_3 cloud has an optical depth of about 10 and that the NH_4SH cloud is almost transparent. Apparently, the H_2O cloud is optically thick. The Weidenschilling and Lewis (1973) model gives an upper limit to the H_2O cloud liquid water content of 10 g m^{-3}, a value substantially higher than the liquid water content of terrestrial thunderstorms. This high water content suggests the presence of precipitation which in turn could be an element in cloud charge generation and separation.

14.4.1   OPTICAL

With a TV vidicon imaging instrument on the Voyager 1 spacecraft, Cook *et al.* (1979) have detected groups of bright transient optical signals in the 380- to 580-nm wavelength range in the night hemisphere of Jupiter. A photograph is reproduced in Fig. 14.4. These transient signals are believed to be lightning flashes. Twenty transient luminous events were recorded between latitudes 30° and 55° N, most at about 45° N, during a time exposure of 192 sec. The images were in an area of about 10^9 km^2, and therefore the flash density was about 3×10^{-10} km^{-2} sec^{-1} or about 3×10^{-3} km^{-2} yr^{-1} during the brief sampling period. The short-term flash density is about 10^8 times lower than that for an active storm region on Earth, although the area covered by the Jovian luminous events is about 10^6 times greater than the area

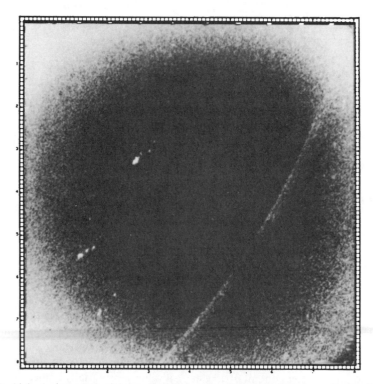

Fig. 14.4   A photograph of the transient light sources likely to be lightning on Jupiter. The photograph is a multiple exposure and, since the spacecraft altitude changed slightly between exposures, the limb of Jupiter appears multiple times. The north pole of Jupiter is near the middle of the observed limb. Adapted from Cook *et al.* (1979). Courtesy, NASA.

of a storm system on Earth (Section 2.5 and Table 2.1). When viewed as a yearly average flash density, the value is about 10^3 times smaller than the average overall of Earth (Section 2.4).

Boruchi and Williams (1986) used the measured optical spot sizes and scattering theory to infer that the lightning activity occurred in a lower cloud composed of water or a mixture of water and ammonium hydrosulfide at a pressure of 5 bar. The average optical energy in the 380- to 580-nm passband radiated by each flash was originally estimated to be 10^{10} J by Smith *et al.* (1979), but that value was revised to about 10^9 J with a range from 4.3×10^8 to 6.6×10^9 J by Borucki *et al.* (1982a). Borucki *et al.* (1982a) also estimate the total flash energy assuming a conversion coefficient of 3.8×10^{-3} (from data in Section 7.2.4 with an adjustment for the passband of the experiment) of 1.7×10^{12} J per flash. Williams *et al.* (1983) point out that about 1 in 10^3 lightning flashes on Earth radiates more than 10^8 J of optical energy and that 5 in 10^7 radiate more than 10^{10} J (Turman, 1978) so that if the Voyager spacecraft only imaged the very bright flashes from a population of lightning whose optical output is similar to that of terrestrial lightning discharges, then perhaps only 1 in 10^3 or 10^4 flashes was observed. If this is true, then, according to Williams *et al.* (1983), the actual lightning rate would be about $3–30 \, km^{-2} \, yr^{-1}$. A similar analysis by Borucki *et al.* (1982a) yielded a flash density of $4 \, km^{-2} \, yr^{-1}$. Of course, Jovian lightning may well be more energetic and brighter than terrestrial lightning, and even if this were not the case, any estimates of flash density could easily be off by orders of magnitude because of the nonrepresentative sampling time and location of observation. The optical data are summarized on Table 14.1.

14.4.2 WHISTLERS

Scarf and Gurnett (1977) describe the plasma wave experiment aboard Voyager 1 which was used to search for whistlers in Jupiter's magnetosphere. The experiment consisted of a dipole antenna and a 16-channel spectrum analyzer covering the frequency range from 10 Hz to 56.2 kHz. Gurnett *et al.* (1979) reported the observation of whistlerlike signals in the frequency range from about 2 to 7 kHz in two separate regions at 5.5 and 6.0 Jovian radii. Examples of these data and, for comparison, whistler signals from Earth are shown in Fig. 14.5. The event rate in both regions of the Jovian magnetosphere was about one whistler every 8 sec. Menietti and Gurnett (1980) have examined whistler propagation in the Jovian magnetosphere for 90 of 167 observed events and have concluded that these whistlers originated near 66° N and propagated without attenuation along a field line to the equatorial plane.

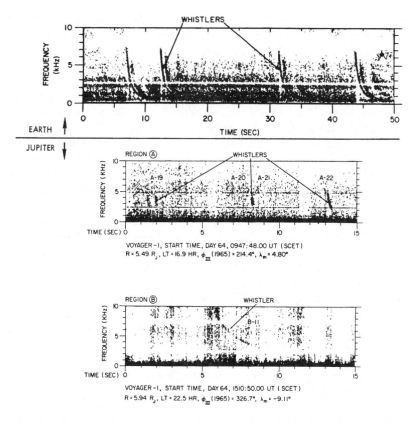

Fig. 14.5 Frequency time spectrograms of representative whistlers observed in two regions of the Jupiter magnetosphere and similar data from Earth. Adapted from Gurnett *et al.* (1979). Courtesy, F. Scarf.

Lewis (1980a) used the observed whistler event rate to estimate an average lightning rate of 4×10^{-3} km^{-2} yr^{-1}. Scarf *et al.* (1981) calculate a rate four orders of magnitude larger, primarily because they localize the area from which the whistlers eminate whereas Lewis (1980a) uses a much larger overall area. Basing their estimate on an analysis of RF propagation in Jupiter's atmosphere by Rinnert *et al.* (1979), Scarf *et al.* (1981) argue that the area over which lightning can be coupled to the magnetic field lines is about 10^6 km^2 and that only flashes 10 times more energetic than the mean on Earth could be detected. They assume that one-tenth of the lightning has been detected and obtain a flash density of 40 km^{-2} yr^{-1}, which considering the uncertainties in our understanding of Jovian lightning, is in remarkable, though probably fortuitous, agreement with optical estimates of Borucki *et al.* (1982a) and Williams *et al.* (1983). The whistler data are summarized in Table 14.1.

14.5 SATURN

Saturn is the second largest planet in the solar system with a radius of about 60,000 km, roughly 10 times that of Earth. The book edited by Gehrels (1984) presents an excellent review of our knowledge of that planet. Some properties of the Saturn atmosphere are shown in Fig. 14.6. Saturn has a weaker

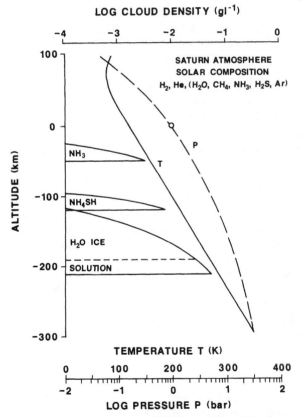

Fig. 14.6 Pressure (P) and temperature (T) height profiles of the Saturn atmosphere, together with the cloud model of Weidenschilling and Lewis (1973). Adapted from Rinnert (1985).

magnetic field than Jupiter and exhibits a ring structure visible from Earth with a small telescope. In September 1979, Pioneer 11 passed Saturn; Voyager 1 and Voyager 2 flew by in November 1980 and August 1981, respectively. The heat flow from the interior of the planet is more than that received from the sun and would be expected to produce atmospheric motions similar to those on Jupiter. Only the top cloud layer has been observed to date; the vertical cloud structure has been estimated from computer modeling. The Saturnian cloud system exhibits a substantially less distinct band structure than Jupiter, with wind directions alternating with latitude in the bands. The wind speed reaches a maximum of the order of 500 msec^{-1} in the equatorial region, considerably larger than on Jupiter, and decreases with increasing latitude. The Saturnian atmosphere is apparently less turbulent than the Jovian atmosphere (Smith *et al.*, 1981, 1982).

Although Saturn would appear to be as good a candidate for lightning as Jupiter, the Voyager imaging and plasma wave instruments, which provided strong evidence for lightning on Jupiter, did not find similar evidence for lightning on Saturn. No impulsive optical signals were recorded by the imaging cameras, and no whistlers were recorded by the plasma wave instrument. RF noise was observed in the magnetosphere near Saturn, but these signals did not have the dispersive characteristics of lightning-generated whistlers (Gurnett *et al.*, 1981; Scarf *et al.*, 1982; Kurth *et al.*, 1983a,b). Among the various RF noises recorded, intense impulsive RF signals were observed by Voyager 2 during the crossing of a ring of Saturn (Scarf *et al.*, 1982, 1983). Similar signals had been observed as rare events by Voyager 1 far from the rings, but no such noise has been found at Jupiter. This impulsive RF noise has been attributed to dust impacts on the antenna of the radio receiver (Scarf *et al.*, 1982, 1983; Gurnett *et al.*, 1983).

The first evidence of electrical activity on Saturn was made by the Voyager 1 planetary radio astronomy (PRA) experiment which consists of a pair of 10-m orthogonal monopole antennas mounted on the conductive structure of the spacecraft and connected to a broadband receiver (Warwick *et al.*, 1981; Evans *et al.*, 1981). Additional discussion and analysis of the PRA data are given by Evans *et al.* (1982), Kaiser *et al.* (1983, 1984), Warwick *et al.* (1982), Zarka and Pedersen (1983), and Zarka (1985). The PRA experiment recorded strong, discrete, wideband bursts of radio emission that were termed Saturn electrostatic discharges (SED). The pulses spanned the entire PRA bandwidth (20 kHz to 40 MHz). No similar signals were observed from Jupiter. Perhaps the most striking feature of these bursts, besides their broad bandwidth, was their periodic occurrence with a repeat time of about 10 hr. The duration of one episode of SEDs was about 3 hr during which time the number of detected events rose to a peak value and fell back to the instrument noise level. The SED intensity depended on the distance of the spacecraft from Saturn and was a maximum at closest approach. The SEDs were observed to have similar characteristics during both Voyager missions, but were less intense and less frequent during Voyager 2, 9 months later than Voyager 1. The distribution of the time between the RF bursts during the hour or so of maximum occurrence closely fits a Poisson distribution with a mean period of 5 sec. SED bursts lasted 30–450 msec with a mean value of about 55 msec. The number of bursts decreased exponentially with increasing burst duration with an e-folding time of about 40 msec (Zarka and Pedersen, 1983). Assuming an isotropic source with 100-MHz bandwidth, Evans *et al.* (1983) determined a radiated power of 10^9–10^{10} W for a single burst. For an assumed 40-MHz bandwidth, Zarka and Pedersen (1983) found an average power of 2×10^9 W for the Voyager 1 observations and an order of magnitude less for the Voyager 2. Peak powers were 5 times higher. Warwick *et al.* (1981) found burst power one to two orders of magnitude lower than the above, which Zarka and Pedersen (1983) attribute to a calibration error.

It is reasonable to associate the periodic nature of the SEDs with a localized source that would appear only when the rotation of Saturn placed the source near the spacecraft. However, the repetition period of the SEDs is significantly shorter than the Saturnian rotational period of 10 hr, 39.4 min, and, in fact, was about 10 min different for Voyager 1 and 2 (Zarka and Pedersen, 1983). Two regions with a rotational period near that of the observed SED repetition period have been identified, one within the B-ring at 1.8 Saturn radii where the revolution period equals 10 hr, 10 min, and the other in the atmosphere at equatorial latitudes where winds with high wind speed at the cloud level produced the necessary superrotation (Smith *et al.*, 1981). Not surprising, there are two opinions as to the location of the source of the SEDs: within the atmosphere and in the B-ring. There does not appear to be any strong reason to prefer one source location over the other at the present time.

Kaiser *et al.* (1983) and Burns *et al.* (1983) argue that the SEDs originate from extended thunderstorms in the equatorial atmosphere of Saturn. To fit the time duration and periodicity of the SED episodes, an active storm region of relatively large longitudinal extent ($\approx 60°$) but narrow latitudinal extent had to be assumed. Further, such a region would have had to remain active for at least 9 months to account for the fact both Voyager spacecraft observed similar SEDs. The interpulse time of about 5 sec and an assumed area of $60° \times 3°$ yield a mean event rate of 3×10^{-2} km^{-2} yr^{-1}.

One serious problem with the interpretation of the source of the SEDs as atmospheric in origin is that of how the lower end of the SED frequency spectrum can propagate through an ionosphere that should reflect radio waves downward below a certain critical frequency. Observations indicate that the night ionosphere should not transmit signals below about 200 kHz.

Burns *et al.* (1983) suggest that, because of the shadow of the rings, there exist deep depressions in the ionospheric electron density in the equatorial region where the electrical storms are assumed to occur. Such depressions could lower the electron density by a factor of about 100 so as to lower the cutoff frequency by a factor of about 10 from 200 to 20 kHz (the lower limit of the frequencies at which PRA signals were observed).

Because of the expected ionospheric shielding of waves of the observed frequency from sources occurring within the atmosphere and the periodicity of the received radio noise, Evans *et al.* (1981, 1982) and Warwick *et al.* (1981, 1982) initially proposed that the SEDs must originate in a localized source in the B-ring. Zarka and Pedersen (1983) discuss the requirements for a source in the B-ring to be able to produce the observed SED characteristics. They conclude that, similar to an in-atmospheric source, the SED radiation must eminate from a region greater than 60° in longitude and very much smaller in height and radial extent. They also argue that a nonisotropic source, perhaps due to propagation, is necessary, whether the source is in the B-ring or the atmosphere, to account for the shape of the SED emission vs time curve.

If the SEDs originate from lightning, that lightning must be quite different from lightning on Earth. The PRA measurements indicate a flat spectrum up to 40 MHz, which necessitates that extremely short electromagnetic pulses, less than about 0.01 μsec in width, comprise the SED RF bursts whose duration is tens to hundreds of milliseconds. In view of the considerable differences between the SEDs and Earth lightning, it is probably not appropriate to compare the SED RF power with that of Earth lightning, but it is interesting to note that the return-stroke peak RF power is of the same order of magnitude as the peak SED power (Section 7.2.1), although the SED has a considerably longer duration. Zarka (1985) argues that Saturn lightning with a radio efficiency of 3×10^{-5} is 1–10 times more energetic than Jovian lightning with an optical efficiency of 3×10^{-3} and that both are more energetic than Earth lightning. The Voyager 2 PRA experiment observed 140 SED-like events from Uranus (23,000 SEDs were observed in the two Saturn encounters) (Zarka and Pederson, 1986). Zarka and Pederson (1986) argue that these so-called UEDs are due to Uranian lightning comparable in energy to Jovian.

14.6 SUMMARY OF INFORMATION ON PLANETARY LIGHTNING

Table 14.1, adapted from Rinnert (1985), summarizes the observations and resultant calculations for optical and electromagnetic events interpreted as lightning on the planets Venus, Jupiter, Saturn, and Uranus.

The evidence for lightning on Jupiter is reasonably strong. Both optical and whistler signals have been observed. With some hand-waving, the flash densities derived from the independent observational techniques can be made to agree. The Voyager 1 and 2 PRA experiment did not observe RF pulses at Jupiter, but, according to Zarka (1985), this is expected to be the case due to ionospheric absorption at Jupiter.

The evidence for lightning on Venus is less certain. There is no convincing optical data. The Venera 11 and 12 probes detected RF noise pulses in Venus's lower atmosphere. The whistler propagation of radio waves was observed by Pioneer Venus. Even though it is possible that the whistler signals were initiated by lightning, it is also possible that some process not related to lightning was the source. Borucki et al. (1981) did succeed, however, in showing that the average flash rate cannot greatly exceed Earth's unless the light output is very low. According to Williams et al. (1983), application of terrestrial cloud electrification mechanisms to the Venus clouds does not suggest strong electrification. Electrification mechanisms near the surface could exist, but Venus's dense atmosphere requires a relatively large electric field for breakdown to occur. Additionally, there is no evidence of ongoing vulcanism capable of generating intense lightning. In spite of the above, calculated flash densities from the spectrometer on Venera 9 and the RF measurements from Venera 11 and 12 are an order of magnitude higher than for storm systems on Earth. If lightning on Venus is eventually confirmed, its discovery would require a major reconsideration of our understanding of cloud electrification.

The evidence for lightning on Saturn is also inconclusive. A wideband periodic noise source has been observed by the Voyager 1 and 2 PRA experiment. The source appears to cover 60° or more of longitude near the equator and to be either in the atmosphere or in the B-ring. No optical signals were observed at Saturn, perhaps because of the high threshold level of optical detection caused by ring light (Zarka, 1985), nor were whistlers detected, perhaps because the spacecraft was in the wrong position relative to the source for whistlers to propagate to it (Burns et al., 1983; Zarka, 1985). The evidence for lightning on Uranus is only the similarity of SEDs and UEDs.

If lightning is indeed present on planets other than Earth, that fact implies that a large-scale separation of electric charge is occurring in an atmosphere of radically different composition and structure from that of the Earth. Additionally, lightning could play a role in the chemistry of nonequilibrium trace gases in a planetary atmosphere (see Chapter 1). Laboratory experiments show that copious amounts of N_2O and CO are produced by electrical discharges (Levine et al., 1979), and investigators have hypothesized that lightning discharges on Venus could significantly affect the production of atmospheric CO and NO_x (Chameides et al., 1979; Bar-Nun, 1980; Levine et al., 1982). Furthermore, Bar-Nun (1975) has argued that the presence of acetylene in Jupiter's atmosphere is due to lightning synthesis.

REFERENCES

Anderson, R., S. Bjornsson, D. C. Blanchard, S. Gathman, J. Hughes, S. Jonasson, C. B. Moore, H. J. Survilas, and B. Vonnegut: Electricity in Volcanic Clouds. *Science*, **148**:1179–1189 (1965).

Bar-Nun, A.: Thunderstorms on Jupiter. *Icarus*, **24**:86–94 (1975).

Bar-Nun, A.: Acetylene Formation on Jupiter: Photolysis or Thunderstorms. *Icarus*, **38**:180–191 (1979).

Bar-Nun, A.: Production of Nitrogen and Carbon Species by Thunderstorms on Venus. *Icarus*, **42**:338–342 (1980).

Bar-Nun, A., N. Noy, and M. Podolak: An Upper Limit to the Abundance of Lightning-Produced Amino Acid in the Jovian Water Clouds. *Icarus*, **59**:162–168 (1984).

Borucki, W. J.: Comparison of Venusian Lightning Observations. *Icarus*, **52**:354–364 (1982).

Borucki, W. J.: Estimate of the Probability of a Lightning Strike to the Galileo Probe. *J. Spacecraft Rockets*, **22**:220–221 (1985).

Borucki, W. J., J. W. Dyer, G. Z. Thomas, J. C. Jordan, and D. A. Comstock: Optical Search for Lightning on Venus. *Geophys. Res. Lett.*, **8**:233–236 (1981).

Borucki, W. J., A. Bar-Nun, F. L. Scarf, A. F. Cook II, and G. E. Hunt: Lightning Activity on Jupiter. *Icarus*, **52**:492–502 (1982a).

Borucki, W. J., Z. Levin, R. C. Whitten, R. G. Keesee, L. A. Capone, O. B. Toon, and J. Dubach: Predicted Electrical Conductivity. between 0 and 80 km in the Venusian Atmosphere. *Icarus*, **51**:302–321 (1982b).

Borucki, W. J., R. E. Orville, J. S. Levine, G. A. Harvey, and W. E. Howell: Laboratory Simulation of Venusian Lightning. *Geophys. Res. Lett.*, **10**:961–964 (1983).

Borucki, W. J., C. P. McKay, and R. C. Whitten: Possible Production by Lightning of Aerosols and Trace Gases in Titan's Atmosphere. *Icarus*, **60**:260–273 (1984).

Borucki, W. J., R. L. McKenzie, C. P. McKay, N. D. Duong, and D. S. Boac: Spectra of Simulated Lightning on Venus, Jupiter, and Titan. *Icarus*, **64**:221–232 (1985).

Borucki, W. J., and M. A. Williams: Lightning in the Jovian Water Cloud. *J. Geophys. Res.*, **91**:9893–9903 (1986).

Briggs, G., K. Klaasen, T. Thorpe, J. Wellman, and W. Baum: Martian Dynamical Phenomena during June–November 1976: Viking Orbiter Imaging Results. *J. Geophys. Res.*, **82**:4121–4149 (1977).

Brook, M., C. B. Moore, and J. Sigurgeirsson: Lightning in Volcanic Clouds. *J. Geophys. Res.*, **79**:472–475 (1974).

Burns, J. A., M. R. Showalter, J. N. Cuzzi, and R. H. Durisen: Saturn Electrostatic Discharges: Could Lightning Be the Cause? *Icarus*, **54**:280–295 (1983).

Chameides, W. J., J. C. G. Walker, and A. F. Nagy: Possible Chemical Impact of Planetary Lightning in the Atmospheres of Venus and Mars. *Nature (London)*, **280**:820–822 (1979).

Colin, L.: Basic Facts about Venus. *In* "Venus" (D. M. Hunten, L. Colin, T. M. Donahue, and V. I. Moroz, eds.), pp. 10–26. Univ. of Arizona Press, Tucson, Arizona, 1983.

Cook, A. F., T. C. Duxbury, and G. E. Hunt: First Results on Jovian Lightning. *Nature (London)*, **280**:794 (1979).

Croft, T. A., and G. H. Price: Evidence for a Low-Attitude Origin of Lightning on Venus. *Icarus*, **53**:548–551 (1983).

Eden, H. F., and B. Vonnegut: Electrical Breakdown Caused by Dust Motion in Low Pressure Atmospheres. *Science*, **180**:962–963 (1973).

Evans, D. R., J. W. Warwick, J. B. Pearce, T. D. Carr, and J. J. Schauble: Impulsive Radio Discharges near Saturn. *Nature (London)*, **292**:716–718 (1981).

Evans, D. R., J. H. Romig, C. W. Hord, K. E. Simmons, J. W. Warwick, and A. L. Lane: The Source of Saturn Electrostatic Discharges. *Nature (London)*, 299:236-237 (1982).

Evans, D. R., J. H. Romig, and J. W. Warwick: Saturn Electrostatic Discharges: Properties and Theoretical Considerations. *Icarus*, 54:267-279 (1983).

Finkelstein, D., and J. Powell: Earthquake Lightning. *Nature (London)*, 228:759-760 (1970).

Foster, H.: An Unusual Observation of Lightning. *Bull. Am. Meteorol. Soc.*, 31:40 (1950).

Gehrels, T. (ed.): "Jupiter." Univ. of Arizona Press, Tucson, Arizona, 1976.

Gehrels, T. (ed.): "Saturn." Univ. of Arizona Press, Tucson, Arizona, 1984.

Gurnett, D. A., R. R. Shaw, R. R. Anderson, W. S. Kurth, and F. L. Scarf: Whistlers Observed by Voyager 1: Detection of Lightning on Jupiter. *Geophys. Res. Lett.*, 6:511-514 (1979).

Gurnett, D. A., W. S. Kurth, and F. L. Scarf: Plasma Waves near Saturn: Initial Results from Voyager 1. *Science*, 212:235-239 (1981).

Gurnett, D. A., E. Grün, D. Gallagher, W. S. Kurth, and F. L. Scarf: Micron-Size Particles Detected near Saturn by the Voyager Plasma Wave Instrument. *Icarus*, 53:236-256 (1983).

Helliwell, R. A.: "Whistlers and Related Ionospheric Phenomena." Stanford Univ. Press, Stanford, California, 1965.

Hunten, D. M., L. Colin, T. M. Donahue, and V. I. Moroz (eds.): "Venus." Univ. of Arizona Press, Tucson, Arizona, 1983.

Ingersoll, A. P.: The Atmosphere of Jupiter. *Space Sci. Rev.*, 18:603-639 (1976).

Kaiser, M. L., J. E. P. Connerney, and M. D. Desch: The Source of Saturn Electrostatic Discharges: Atmospheric Storms. NASA Technical Memorandum 84966, January, 1983a.

Kaiser, M. L., J. E. P. Connerney, and M. D. Desch: Atmospheric Storm Explanation of Saturnian Electrostatic Discharges. *Nature (London)*, 303:50-53 (1983b).

Kaiser, M. L., M. D. Desch, W. S. Kurth, A. Lecacheux, F. Genova, and B. M. Pedersen: Saturn as a Radio Source. *In* "Saturn" (T. Gehrels, ed.). Univ. of Arizona Press, Tucson, Arizona, 1984.

Kamra, A. K.: Measurements of the Electrical Properties of Dust Storms. *J. Geophys. Res.*, 77:5856-5869 (1972).

Knollenberg, R., and D. M. Hunten: Clouds of Venus: A Preliminary Assessment of Microstructure. *Science*, 205:70-74 (1979).

Knollenberg, R. G., and D. M. Hunten: The Microphysics of the Clouds of Venus: Results of the Pioneer Venus Particle Size Spectrometer Experiment. *J. Geophys. Res.*, 85:8039-8058 (1980).

Knollenberg, R., L. Travis, M. Tomasko, P. Smith, B. Ragent, L. Esposito, D. McCleese, J. Martewchick, and R. Beer: The Clouds of Venus: A Synthesis Report. *J. Geophys. Res.*, 85:8059-8081 (1980).

Krasnopolsky, V. A.: On Lightning in the Venus Atmosphere According to the Venera 9 and 10 Data, Report. *Space Res. Inst., Acad. Sci. USSR, Moscow* (1980).

Krasnopolsky, V. A.: Venus Spectroscopy in the 3000-8000 Å Region by Veneras 9 and 10. *In* "Venus" (D. M. Hunten, L. Collin, T. M. Donahue, and V. I. Moroz, eds.), pp. 459-483. Univ. of Arizona Press, Tucson, Arizona, 1983a.

Krasnopolsky, V. A.: Lightnings and Nitric Oxide on Venus. *Planet. Space Sci.*, 31:1363-1370 (1983b).

Ksanfomality, L. V.: Lightning in the Cloud Layer of Venus. *Kosm. Issled.*, 17:747-762 (1979).

Ksanfomality, L. V.: Discovery of Frequent Lightning Discharges in Clouds on Venus. *Nature (London)*, 284:244-246 (1980).

Ksanfomality, L. V., N. M. Vasil'chilkov, O. F. Ganpantserova, E. V. Petrova, A. O. Souvorov, G. F. Fillipov, O. V. Vablonskava, and L. V. Yabrova: Electrical Discharges in the Atmosphere of Venus. *Pisma Astron. Zh.*, 5:229-236 (1979).

Ksanfomality, L. V., F. L. Scarf, and W. W. L. Taylor: The Electrical Activity of the Atmosphere of Venus. *In* "Venus" (D. M. Hunten, L. Colin, T. M. Donahue, and V. I. Moroz, eds.), pp. 565-603. Univ. of Arizona Press, Tucson, Arizona, 1983.

Kurth, W. S., F. L. Scarf, D. A. Gurnett, and D. D. Barbosa: A Survey of Electrostatic Waves in Saturn's Magnetosphere. *J. Geophys. Res.*, **88**:8959–8970 (1983a).

Kurth, W. S., D. A. Gurnett, and F. L. Scarf: A Search for Saturn Electrostatic Discharge in the Voyager Plasma Data. *Icarus*, **53**:255–261 (1983b).

Lanzerotti, L. J., K. Rinnert, E. P. Krider, M. A. Uman, G. Dehmel, F. O. Gliem, and W. I. Axford: Planetary Lightning and Lightning Measurements on the Galileo Probe to Jupiter's Atmosphere. *In* "Proceedings in Atmospheric Electricity" (L. H. Ruhnke and J. Latham, eds.). Deepak, Hampton, Virginia, 1983.

Levin, Z., W. J. Borucki, and O. B. Toon: Lightning Generation in Planetary Atmospheres. *Icarus*, **56**:80–115 (1983).

Levine, J. S.: The Ashen Light: An Auroral Phenomenon on Venus. *Planet. Space Sci.*, **17**:1081–1087 (1969).

Levine, J. S., R. E. Hughes, W. L. Chameides, and W. E. Howell: N_2O and CO Production by Electric Discharge: Atmospheric Implications. *Geophys. Res. Lett.*, **6**:557–559 (1979).

Levine, J. S., G. L. Gregory, G. A. Harvey, W. E. Howell, W. J. Borucki, and R. E. Orville: Production of Nitric Oxide by Lightning on Venus. *Geophys. Res. Lett.*, **9**:893–896 (1982).

Lewis, J. S., Lightning on Jupiter: Rate, Energetics, and Effects. *Science*, **210**:1351–1352 (1980a).

Lewis, J. S.: Lightning Synthesis of Organic Compounds on Jupiter. *Icarus*, **43**:89–95 (1980b).

Luhmann, J. G., and A. F. Nagy: Is There Lightning on Venus? *Nature (London)*, **319**:266 (1986).

Meinel, A. B., and D. T. Hoxie: On the Spectrum of Lightning in the Atmosphere of Venus. *Commun. Lunar Planet. Lab.*, **1**:35–38 (1962).

Menietti, J. D., and D. A. Gurnett: Whistler Propagation in the Jovian Magnetosphere. *Geophys. Res. Lett.*, **7**:49–52 (1980).

Michnowski, S.: On the Observation of Lightning in Warm Clouds. *Indian J. Meteorol. Geophys.*, **14**:320–322 (1963).

Moore, C. B., B. Vonnegut, B. A. Stein, and H. J. Survilas: Observations of Electrification and Lightning in Warm Clouds. *J. Geophys. Res.*, **65**:1907–1910 (1960).

Napier, W. M.: The Ashen Light on Venus. *Planet. Space Sci.*, **19**:1049–1051 (1971).

Park, C. G.: Whistlers. *In* "Handbook of Atmospherics" (H. Volland, ed.), pp. 21–79. CRC Press, Boca Raton, Florida, 1982.

Pierce, E. T.: Atmospheric Electricity—Some Themes. *Bull. Am. Meteorol. Soc.*, **55**:1186–1194 (1974).

Pietrowski, E. L.: An Observation of Lightning in Warm Clouds. *J. Meteorol.*, **17**:562–563 (1960).

Pounder, C.: Volcanic Lightning. *Weather*, **35**:357–360 (1980).

Rinnert, K.: Lightning within Planetary Atmospheres. *In* "Handbook of Atmospherics" (H. Volland, ed.), pp. 99–132. CRC Press, Boca Raton, Florida, 1982.

Rinnert, K.: Lightning on Other Planets. *J. Geophys. Res.*, **90**:6225–6237 (1985).

Rinnert, K., L. J. Lanzerotti, E. P. Krider, M. A. Uman, G. Dehmel, F. O. Gliem, and W. I. Axford: Electromagnetic Noise and Radio Wave Propagation below 100 kHz in the Jovian Atmosphere, 1, The Equatorial Region. *J. Geophys. Res.*, **84**:5181–5188 (1979).

Rinnert, K., L. J. Lanzerotti, G. Dehmel, F. O. Gliem, E. P. Krider, and M. A. Uman: Measurements of the RF Characteristics of Earth Lightning with the Galileo Probe Lightning Experiment. *J. Geophys. Res.*, **90**:6239–6244 (1985).

Sagdeev, R. Z., V. M. Linkin, V. V. Kerzhanovich, A. N. Lipatov, A. A. Shurupov, J. E. Blamont, D. Crisp, A. P. Ingersoll, L. S. Elson, R. A. Preston, C. E. Hilderrand, B. Ragent, A. Seiff, R. E. Young, G. Petit, L. Boloh, Yu. N. Alexandrov, N. A. Armand, R. V. Bakitko, and A. S. Selvanov: Overview of VEGA Venus Balloon *in Situ* Meteorological Measurements. *Science*, **231**:1411–1414 (1986).

Sato, M., and J. E. Hansen: Jupiter's Atmospheric Composition and Cloud Structure Deduced from Absorption Bands in Reflected Sunlight. *J. Atmos. Sci.*, **36**:1133-1167 (1979).

Scarf, F. L.: Comments on "Venus Nightside Ionospheric Troughs: Implications for Evidence of Lightning and Volcanism" by H. A. Taylor, Jr., J. M. Grebowsky, and P. A. Cloutier. *J. Geophys. Res.*, **91**:4594-4598 (1986).

Scarf, F. L., and D. A. Gurnett: A Plasma Wave Investigation for the Voyager Mission. *Space Sci. Rev.*, **21**:289-308 (1977).

Scarf, F. L., and C. T. Russell: Lightning Measurements from the Pioneer Venus Orbiter. *Geophys. Res. Lett.*, **10**:1192-1195 (1983).

Scarf, F. L., W. L. Taylor, C. T. Russell, and L. H. Brace: Lightning on Venus: Orbiter Detection of Whistler Signals. *J. Geophys. Res.*, **85**:8158-8166 (1980).

Scarf, F. L., D. A. Gurnett, W. S. Kurth, R. R. Anderson, and R. R. Shaw: An Upper Bound to the Lightning Flash Rate in Jupiter's Atmosphere. *Science*, **213**:683-685 (1981).

Scarf, F. L., D. A. Gurnett, W. S. Kurth, and R. L. Poynter: Voyager 2 Plasma Wave Observations at Saturn. *Science*, **215**:587-594 (1982).

Scarf, F. L., D. A. Gurnett, W. S. Kurth, and R. L. Poynter: Voyager Plasma Wave Measurements at Saturn. *J. Geophys. Res.*, **88**:8971-8984 (1983).

Singh, R. N., and C. T. Russell: Further Evidence for Lightning on Venus. *Geophys. Res. Lett.*, **13**:1051-1054 (1986).

Smith, B. A., L. A. Soderblom, T. V. Johnson., A. P. Ingersoll, S. A. Collins, E. M. Shoemaker, G. E. Hunt, H. Masursky, M. H. Carr, M. E. Davies, A. F. Cook II, J. Boyce, G. E. Danielson, T. Owen, C. Sagan, R. F. Beebe, J. Veverka, R. G. Strom, J. F. McCauley, D. Morrison, G. A. Briggs, and V. E. Suomi: The Jupiter System Through the Eyes of Voyager 1. *Science*, **204**:951-972 (1979).

Smith, B. A., L. Soderblom, R. Beebe, J. Boyce, G. Briggs, A. Bunker, S. A. Collins, C. J. Hansen, T. V. Johnson, J. L. Mitchell, R. J. Terrile, M. Carr, A. F. Cook II, J. Cuzzi, J. B. Pollack, G. E. Danielson, A. Ingersoll, M. E. Davies, G. E. Hunt, H. Masusky, E. Shoemaker, D. Morrison, T. Owen, C. Sagan, J. Veverka, R. Strom, and V. E. Suomi: Encounter with Saturn: Voyager 1 Imaging Science Results. *Science*, **212**:163-191 (1981).

Smith, B. A., L. Soderblom, R. Batson, P. Bridges, J. Inge, H. Masursky, E. Shoemaker, R. Beebe, J. Boyce, G. Briggs, A. Bunker, S. A. Collins, C. J. Hansen, T. V. Johnson, J. L. Mitchell, R. Terrile, A. F. Cook II, J. Cuzzi, J. B. Pollack, G. E. Danielson, A. P. Ingersoll, M. A. Davies, G. A. Hunt, D. Morrison, T. Owen, C. Sagan, J. Veverka, R. Strom, and V. E. Suomi: A New Look at the Saturn System: The Voyager 2 Images. *Science*, **215**:504-537 (1982).

Stone, P. H.: The Dynamics of the Atmosphere of the Major Planets. *Space Sci. Rev.*, **14**:444-459 (1973).

Stone, P. H.: The Meterology of the Jovian Atmosphere. *In* "Jupiter" (T. Gehrels, ed.), Univ. of Arizona Press, Tucson, Arizona, 1976.

Taylor, H. A., J. M. Grebowsky, and P. A. Cloutier: Venus Nightside Ionospheric Troughs: Implications for Evidence of Lightning and Volcanism. *J. Geophys. Res.*, **90**:7415-7426 (1985).

Taylor, Jr., H. A., J. M. Grebowsky, and P. A. Cloutier: Reply. *J. Geophys. Res.*, **91**:4599-4605 (1986).

Taylor, W. L., F. L. Scarf, C. T. Russell, and L. H. Brace: Evidence for Lightning on Venus. *Nature (London)*, **279**:614-616 (1979).

Toon, O. B., B. Ragent, D. Colburn, J. Blamont, and C. Cot: Large, Solid Particles in the Clouds of Venus: Do They Exist? *Icarus*, **57**:143-160 (1984).

Turman, B. N.: Analysis of Lightning Data from the DMSP Satellite. *J. Geophys. Res.*, **83**:5019-5024 (1978).

Warwick, J. W., J. B. Pearce, D. R. Evans, T. D. Carr, J. J. Schaubli, J. K. Alexander, M. L. Kaiser, M. D. Desch, M. Pederson, A. Lecacheux, G. Daigne, A. Boischot, and C. H. Barrow: Planetary Radio Astronomy Observations from Voyager 2 near Saturn. *Science*, **212**:239–243 (1981).

Warwick, J. W., D. R. Evans, J. H. Ronig, J. K. Alexander, M. D. Desch, M. L. Kaiser, M. Aubier, Y. Leblanc, A. Lecacheux, and B. M. Pedersen: Planetary Radio Astronomy Observations from Voyager 2 near Saturn. *Science*, **215**:582–587 (1982).

Weidenschilling, S. J., and J. S. Lewis: Atmospheric and Cloud Structure of the Jovian Planets. *Icarus*, **20**:465–476 (1973).

Weinheimer, A. J., and A. A. Few, Jr.: The Spokes in Saturn's Rings: A Critical Evaluation of Possible Electrical Processes. *Geophys. Res. Lett.*, **9**:1139–1142 (1982).

Williams, M. A.: On the Energy of Possible Saturian Lightning. *Icarus*, **56**:611–612 (1983).

Williams, M. A., E. P. Krider, and D. M. Hunten: Planetary Lightning: Earth, Jupiter, and Venus. *Rev. Geophys. Space Phys.*, **21**:892–902 (1983).

Williams, M. A., and L. W. Thomason: Optical Signature of Venus Lightning as Seen from Space. *Icarus*, **55**:185–186 (1983).

Williams, M. A., L. W. Thomason, and D. M. Hunten: The Transmission to Space of the Light Produced by Lightning in the Clouds of Venus. *Icarus*, **52**:166–170 (1982).

Zarka, P.: On Detection of Radio Bursts Associated with Jovian and Saturnian Lightning. *Astron. Astrophys.*, **146**:L15–L18 (1985).

Zarka, P., and B. M. Pedersen: Statistical Study of Saturn Electrostatic Discharges. *J. Geophys. Res.*, **88**:9007–9018 (1983).

Zarka, P., and B. M. Pedersen: Radio Detection of Uranian Lightning by Voyager 2. *Nature (London)*, **323**:605–608 (1986).

Chapter 15 | Thunder

15.1 INTRODUCTION

We define thunder as the acoustic radiation associated with lightning. We exclude from our definition those sources of thunderstorm-produced acoustic noise that are not electrical in origin (e.g., Georges, 1973).

Thunder can be divided into two categories: (1) audible: acoustic energy that we can hear, and (2) infrasonic: acoustic energy that is below the frequency that the human ear can detect, generally a few tens of hertz. The reason for making such a division is that the physical mechanisms responsible for audible and for infrasonic thunder are thought to be different. The origin of audible thunder is the expansion of the rapidly heated lightning channel, while the origin of infrasonic thunder is thought to be associated with the conversion to sound of the energy stored in the electrostatic field of the thundercloud when lightning rapidly reduces that cloud field.

A review of the early history of thunder research has been given by Remillard (1960) and by Uman (1984). Review papers concerning the latest experiments and theory have been published by Bhartendu (1969a), Few (1974, 1975, 1981), and Hill (1977c, 1979). Since these authors often express divergent views regarding both the interpretation of the experimental data and the physical mechanisms responsible for the generation of thunder, the review papers make particularly interesting reading.

15.2 OBSERVATIONS AND MEASUREMENTS

Given the fact that thunder is the most common of loud natural noises, it is perhaps surprising that there have been relatively few measurements of its characteristics. In this section we review the experimental data.

15.2.1 SOUNDS HEARD

The terms *clap*, *peal*, *roll*, and *rumble* are commonly used to describe audible thunder. Unfortunately, these terms are used inconsistently, both in the scientific literature and in everyday speech. Often clap and peal are use synonymously, as are roll and rumble. Claps or peals are sudden loud sounds that occur against a background of prolonged roll or rumble. In this chapter we shall refer to all loud impulsive components of thunder as claps. The term roll is sometimes used to describe irregular sound variations whereas rumble is used to describe a relatively weak sound of long duration.

Recordings of the pressure variations due to thunder from a lightning ground flash at a distance of about 8 km are given in Fig. 15.1.

Latham (1964) working in New Mexico reported that a low-intensity sound with duration between 0.1 and 2.2 sec existed in almost every case prior to the main thunder and that thunder in general consisted of 3 or 4 claps, the pressure oscillations within these claps being about 100 Hz. A histogram of clap duration, generally 0.2–2.0 sec, for ground and cloud flashes is given in Fig. 15.2, and a histogram of the time interval from the beginning of one clap to the beginning of the next, generally 1–3 sec, is given in Fig. 15.3. Latham (1964) found that the relative amplitude of the various claps did not change much with clap order, as illustrated in Fig. 15.4. Latham (1964) reported that the initial thunder signal was a compression, as were the initial portion of the claps, and that thunder from cloud and ground flashes had similar

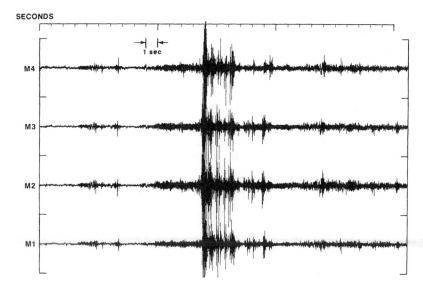

Fig. 15.1   Thunder at a four microphone array from a multiple-stroke ground flash. The four microphones were located at the corners of a 50-m square at the Kennedy Space Center, Florida. The time delay from the flash to the arrival of the first clap was about 25 sec. The acoustic signal was high-pass filtered with a cutoff frequency of 15 Hz. Courtesy, A. A. Few.

characteristics, although the magnitude of the pressure from ground flashes was generally larger.

Uman and Evans (1977), in Florida, found that there were generally 2 or 3 claps per ground flash, as illustrated in Fig. 15.5. They also showed that the first clap in ground flashes was generally the largest, the second clap the second largest, and the third clap the third largest, as illustrated in Fig. 15.6, a result not consistent with the data of Latham (1964) shown in Fig. 15.4 for a combination of ground and cloud flashes. Additionally, from Fig. 15.6, we see that the first clap was the largest in about 55% of the thunder records, the second clap was the largest in about 25% of the records, while the third clap was the largest in about 20% of the records.

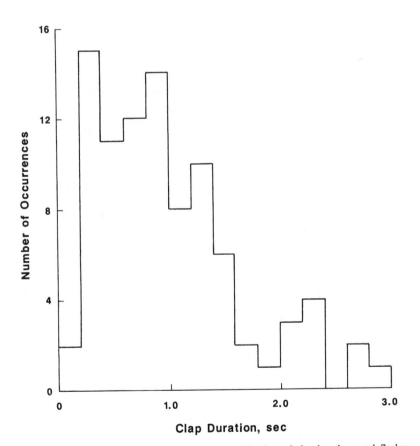

Fig. 15.2 Histogram of clap duration for a combination of cloud and ground flashes. Adapted from Latham (1964).

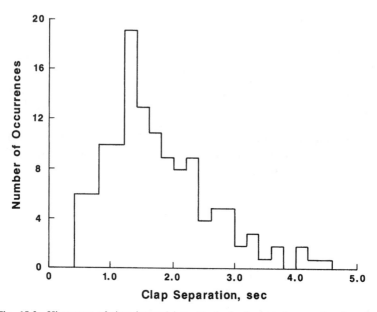

Fig. 15.3   Histogram of time interval between the beginning of successive claps for a combination of cloud and ground flashes. Adapted from Latham (1964).

Although there have been no detailed measurements of lightning at very close range, many observers near lightning have reported common sounds prior to the first loud clap. When lightning strikes a few hundred meters away, the first sounds are like the tearing of cloth. This tearing sound lasts an appreciable fraction of a second and merges into the louder first clap. The origin of the tearing sound has been attributed (1) to a straight channel section of length of the same order as the distance to the observer (Brook, quoted in Hill, 1977c; also see discussion of sound from straight sources in Section 15.3.1) and (2) to a multitude of upward-going connecting discharges from earth (Malan, 1963). When lightning is within about 100 m, according to Malan (1963), one first hears a click, then a whiplike crack, and finally a continuous rumbling thunder. Malan (1963) views the click as due to the major upward leader, the crack as due to the shock wave from the closest part of the return stroke, and the rumble as sound from the higher regions of the tortuous channel.

The sounds of the thunder associated with upward-initiated lightning from tall structures and with lightning that is artificially initiated via the firing of small rockets trailing grounded wires are discussed in Section 12.3.3.1. Thunder from the lightning-like electrical discharges of about 0.5 km length

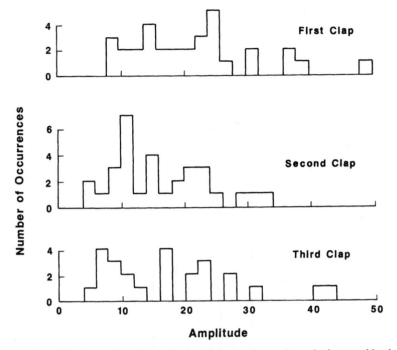

Fig. 15.4   Histograms of relative amplitudes of thunder claps vs clap order for a combination of cloud and ground flashes. Adapted from Latham (1964).

which occur in the material ejected from some volcanoes (Section 14.1) is reported to take the form of "a sharp noise like the firing of artillery" (Anderson *et al.*, 1965).

15.2.2   TIME TO AND DURATION OF THUNDER

The time interval between the arrival of the optical signal from the lightning channel, which travels at about 300 m/μsec and hence arrives in a time of the order of 10 μsec for an observer some kilometers away, and the corresponding thunder, which travels at about 330 m/sec and hence arrives in a time of the order of 10 sec, is essentially determined by the distance to the closest channel point divided by the speed of sound. This time is approximately 3 sec/km of distance to the lightning. The technique, widely used by both the layman and the serious researcher, of determining lightning

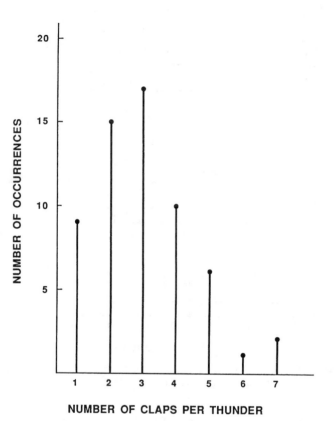

NUMBER OF CLAPS PER THUNDER

Fig. 15.5 Histogram of the number of claps per flash for 60 ground flashes (Uman and Evans, 1977).

distance from the time to the first sound of thunder is termed *thunder ranging*. When thunder is simultaneously ranged from three or more spatially separated stations, acoustic source locations can be determined, as discussed in Section 15.5.

The duration of thunder is a measure of the difference in the distances between the closest point of the lightning channel and the farthest. It therefore represents the minimum possible length for the channel. Thus, for example, for a straight vertical channel to ground, the thunder duration observed at ground should decrease as the distance to the lightning increases, as illustrated for a 7-km channel by the curve in Fig. 15.7. That thunder does not behave in this way is evident, for example, from the experimental data of (1) De L'Isle (1738) in France, apparently a mixture of ground and cloud flashes, (2) Latham (1964) in New Mexico, identified ground and cloud

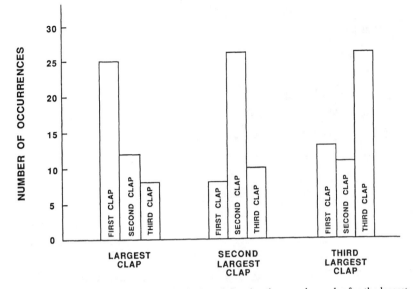

Fig. 15.6 Histograms of relative amplitudes of thunder claps vs clap order for the largest clap in ground flashes having 3 or more claps, for the second largest clap, and for the third largest clap (Uman and Evans, 1977).

discharges, both (1) and (2) being reproduced by Uman (1984), and (3) Uman and Evans (1977) in Florida, ground flashes only, shown in Fig. 15.7. It follows that lightning channels are not simply vertical, a conclusion verified by the acoustic source reconstructions discussed in Section 15.5 and by, for example, the stroke charge locations of Krehbiel *et al.* (1979) discussed in Section 3.2 and illustrated in Figs. 3.3 and 3.4. Teer and Few (1974) give statistics on thunder duration without regard to range or type of flash at a number of different geographic locations. The median thunder duration they report is about 15 sec in Socorro, New Mexico, 29 sec in Roswell, New Mexico, 18 sec in Tucson, Arizona, and 41 sec in Houston, Texas. From acoustic channel reconstructions in one Tucson storm with 17 ground and 20 cloud flashes, Teer and Few (1974) found an average horizontal channel length in the cloud for those 37 flashes of about 10 km.

15.2.3 DISTANCE THUNDER CAN BE DETECTED

The first observation of the fact that thunder can seldom be heard from lightning flashes more than about 25 km distant is apparently due to

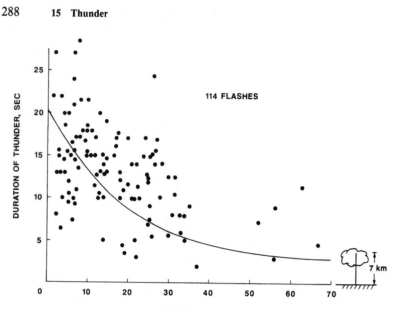

Fig. 15.7   Thunder duration vs time interval between lightning and thunder for 114 ground flashes. The data were obtained during summer convective storms in periods of relatively low activity so that thunder from individual events could be unambiguously identified. Solid curve shows the theoretical value for a straight vertical channel to ground of 7 km length (Uman and Evans, 1977).

De L'Isle (1783). A hundred years later, Veenema (1917, 1918, 1920) studied nearly every thunderstorm occurring near him during the years 1895 to 1916 with a view toward determining how far thunder could be heard. He confirmed De L'Isle's conclusion but documented occasional examples of thunder from lightning up to and over 100 km distant. Isolated reports of thunder heard at distances greater than about 25 km have been given by Brooks (1920), Cave (1919), Page (1944), Taljaard (1952), and Thomson (1980). On the other hand, Ault (1916), captain of the research ship *Carnegie*, reported that thunder could not be heard from one storm when it was beyond about 8 km.

Fleagle (1949) has shown that the inaudibility of thunder beyond 25 km can be ascribed to the upward curving of sound rays resulting from the usual atmospheric temperature gradient. Since the velocity of sound is proportional to the square root of the temperature and since the temperature in general decreases as a function of height, Snell's law indicates that sound waves will normally be refracted upward. Fleagle (1949) calculated that, in the presence of a linear lapse rate (temperature decreasing linearly with height), those sound rays that leave the channel and at some point become tangent to the ground plane exhibit a trajectory that is very nearly parabolic. For a typical lapse rate of 7.5 K km^{-1}, sound that originates at a height of 4 km has a maximum range of audibility of 25 km if wind shear is ignored. That is, the sound rays from a 4 km height are tangent to the ground 25 km distant from the discharge channel. All sounds originating from heights below 4 km will not be heard at 25 km; sounds originating from above 4 km will be heard. Only the very close observer can hear the sound from the base of the discharge channel. Fleagle has also shown that wind shear, a change in wind velocity with height, can provide a refraction of the thunder that is of the same order of magnitude as that due to temperature gradients. Sound rays may be refracted upward or downward depending on the relation between the wind shear and the sound ray direction. A wind shear of 4 m sec^{-1} km^{-1} can yield a sound ray trajectory almost equivalent to a lapse rate of 7.5 K km^{-1}. Similar analyses are found in Fleagle and Businger (1963) and in Few (1981).

Fleagle (1949) cautions that factors other than the lapse rate and wind shear may affect the audibility of thunder. For example, a region of temperature inversion will tend to increase the range of audibility; and features of the terrain that hinder the essentially horizontal propagation of the critical sound ray in its final several kilometers will decrease the range of audibility. Additional effects of the atmosphere on acoustic wave propagation are considered in Section 15.4.

15.2.4 OVERPRESSURE AND ACOUSTIC ENERGY

The most quantitative and detailed measurements of thunder overpressure and energy have been made by Holmes *et al.* (1971a) on a mountain top (about 3 km above sea level) in central New Mexico. Thunder from a total of 40 cloud and ground flashes, most apparently at a range of a few kilometers, was studied. Intracloud thunder spectrums showed a mean peak value of acoustic power at 28 Hz with a mean total acoustic energy of 1.9×10^6 J and a range 1.8–3.1×10^6 J. Ground-flash thunder spectra showed a mean peak value of acoustic power at 50 Hz with a mean total acoustic energy of 6.3×10^6 J and a range 1.1–17×10^6 J. Holmes *et al.* (1971a) state that there appear to be significant differences in total acoustic energy and frequency spectrum between the thunder from cloud and from ground flashes. Total acoustic energy W in joules was determined from the expression

$$W = \int P(t)4\pi R(t)^2 \, dt \qquad (15.1)$$

where $P(t)$ is the recorded total power flux at a given time t in J m^{-2} sec^{-1}, $R(t)$ is the distance to the acoustic source, $v(t - t_0)$, where v is the speed of the acoustic wave, and t_0 is the time the lightning occurs as indicated from electric field records. Among the assumptions made in using Eq. (15.1) is that atmospheric attenuation and refraction are small and that a spherical (isotropic) acoustic wave is radiated from each point on the channel. For all the thunder data, the average total power flux ranged from 0.17 to 19.3×10^{-3} J m^{-2} sec^{-1} and the average rms pressure from 0.22 to 2.4 N m^{-2}. Note that a pressure of 1 standard atmosphere or 1 bar is about 10^5 N m^{-2}.

Holmes *et al.* (1971a) calculated the efficiency for the conversion of electrical energy to acoustic for a ground flash as follows. They assumed that the energy dissipated per unit length of channel by a first stroke was 2.3×10^5 J m^{-1} (Krider *et al.*, 1968; Section 7.2.4). The length of an average lightning channel to the mountain was assumed to be 4 km yielding a first stroke energy of 9.2×10^8 J. All subsequent strokes together were assumed to have the energy of the first stroke for a total flash energy of 1.8×10^9 J. This total electrical energy was divided into the average acoustic energy of 11 ground flashes (presumably the best data), 3.26×10^6 J, to yield an acoustic efficiency of 0.18%. Additional discussion of acoustic efficiency is found in Section 15.3.1.

Bhartendu (1968, 1969b, 1971b) has measured sound pressure for thunder from nearby lightning. For discharges in the range 2–9 km, Bhartendu (1971b) found that the maximum rms acoustic pressure for each individual flash varied from about 0.2 to about 6.0 N m^{-2}. Few *et al.* (1967) show one plot of thunder pressure vs time in which the maximum pressure is about 10 N m^{-2}.

Bohannon *et al.* (1977) found single infrasonic pulses of amplitude about 0.1 N m^{-2} and period 0.5 sec superimposed on the audible thunder signals. The pulse pressures were initially compressional, and their origins were reported to be from the cloud. Balachandran (1983) measured sound pressure from infrasonic thunder pulses originating in overhead thunderstorms and found amplitudes of a few tenths of a Newton per square meter, with the pulses being primarily compressional.

Hill and Robb (1968) measured the shock wave overpressure 0.35 m from the breakdown channel of a 0.1-m spark gap placed in series with a rocket-initiated lightning. They reported maximum overpressures of the order of 2 bar or 2×10^5 N m^{-2}. Some comments on the interpretation of these measurements are given by Dawson *et al.* (1968b) and are noted in Section 15.3.1. Measured overpressures near long laboratory sparks are discussed in Section 15.3.1 in relation to the attempts to validate proposed thunder generation mechanisms.

15.2.5 FREQUENCY SPECTRUM

The definitive work on the measured thunder frequency spectrum is by Holmes *et al.* (1971a) and will be discussed in the following paragraph. One isolated thunder spectrum has been presented by Few (1969a) and another by Nakano and Takeuti (1970). Some additional spectral data have been given by Bhartendu (1969b). Prior to these studies considerable literature was published which, at best, is of questionable validity. According to Few (1969a) and to Hill (1977c), the experimental thunder frequency spectrum published by Few *et al.* (1967), an average of 23 spectra showing a broad maximum near 200 Hz, is in error, the peak frequency being about a factor of two too high, because of errors in the data analysis. The one published spectrum of Few (1969a) has a peak at about 40 Hz. According to McCrory and Holmes (1968) and Bhartendu (1969b), spectra published by Bhartendu (1968) are of doubtful validity because of problems associated with analyzing data taken with a hot-wire microphone. Earlier data from Schmidt (1914) and from Arabadzi (1952, 1957) have questionable aspects and are reviewed by Uman (1984). Significantly, both Schmidt (1914) and Arabadzi (1952, 1957) identified the existence of infrasonic thunder. They also argued that the maximum of the thunder frequency spectrum was in the infrasonic region. The argument regarding whether thunder is dominated by an audible or by an infrasonic component (Uman, 1984) has been settled by Holmes *et al.* (1971a) who found that for individual events either can be the case and that, in fact, the thunder spectrum is not stationary. It changes as a function of time during the thunder event.

Holmes *et al.* (1971a) found, for 40 thunder records, that the thunder power spectrum peaked at frequencies from less than 4 to 125 Hz and that the ratio of the peak frequency to the width of the power spectrum at half amplitude varied between 0.5 and 2, this ratio being termed the Q of the spectrum. A histogram of the peak frequency in the power spectrum of thunder for 24 ground flashes is given in Fig. 15.8. A typical power spectrum with peak frequency in the audible range near 100 Hz is shown in Fig. 15.9, along with the spectrum of the ambient wind noise measured during a 2-sec interval prior to the thunder. A spectrum that peaked in the infrasonic is shown in Fig. 15.10A. The peak power flux in the recorded spectra for total events of the type shown in Figs. 15.9 and 15.10A ranged from 4.0 to 0.03×10^{-4} J m^{-2} sec^{-1} Hz^{-1}. The variation of thunder frequency content as a function of time is illustrated (Fig. 15.10B). There the spectrum was calculated for successive 1-sec time windows. Early dominant frequencies are in the 100-Hz range whereas 9 sec after the first thunder the peak signal is infrasonic. The possibility that in some cases nonstationary wind noise contributes to the apparent infrasonic signal cannot be completely ruled out

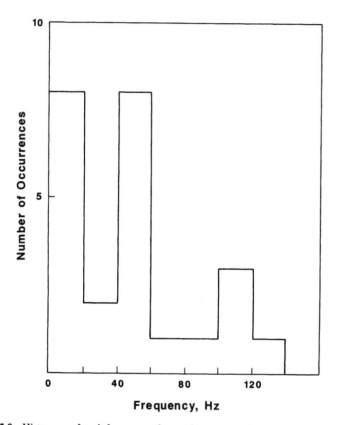

Fig. 15.8 Histogram of peak frequency of acoustic power spectrum for 24 ground flashes. Adapted from Holmes *et al.* (1971a).

since proof that an infrasonic signal is acoustic requires the measurement of its propagation velocity across an array of microphones.

15.3 GENERATION MECHANISMS

As indicated in Section 15.1, different mechanisms are thought to be responsible for (1) audible and for (2) infrasonic thunder: (1) the initial expansion of the lightning channel is thought to produce shock waves that ultimately become the time sequence of sound pressure waves we hear as audible thunder, and (2) the relaxation of the electrostatic stress in the cloud after a lightning changes the cloud charge distribution is thought to produce the infrasonic thunder component. In this section we examine the physical

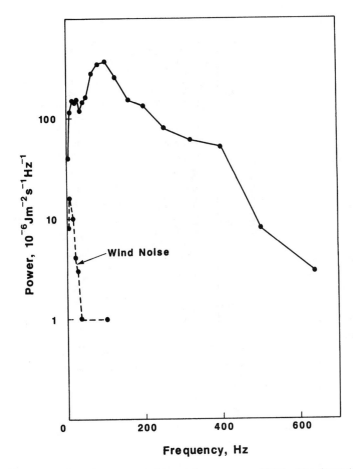

Fig. 15.9   Typical thunder spectrum with peak frequency near 100 Hz. Also shown (dashed line) is the spectrum of the wind noise. Adapted from Holmes *et al.* (1971a).

models proposed to describe these two processes. As we shall see, there is still considerable room for the improvement of our understanding of the thunder generation mechanisms.

15.3.1   AUDIBLE THUNDER: THE ACOUSTIC RADIATION FROM HOT CHANNELS

As we have discussed in Section 7.3.4, it has been determined from the analysis of optical spectroscopy that a given section of the return stroke

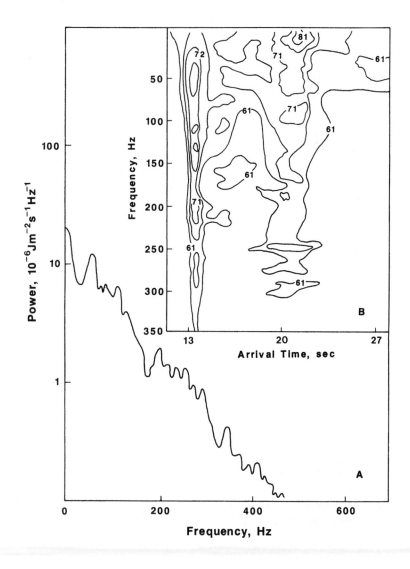

Fig. 15.10 (A) Typical thunder spectrum with peak frequency in the infrasonic. Wind noise was insignificant. (B) Power spectrum as a function of thunder arrival time calculated for windows of 1 sec duration and displayed as power contours in db above a standard level of $10^{-12}\,\mathrm{J\,m^{-2}\,sec^{-1}\,Hz^{-1}}$. Unlabeled contours are 5 db above or below the nearest labeled contour. Adapted from Holmes *et al.* (1971a).

channel attains a temperature of about 30,000 K in a time less than 10 μsec.
It follows that the channel pressure must rise in response to the temperature
rise since there is insufficient time for the channel particle density to change
appreciably. The spectroscopic data indicate an average channel pressure of
about 10 bar during the first 5 μsec. Such a channel overpressure results in
an expansion of the channel behind a shock wave. As the channel expands,
its pressure decreases toward atmospheric on a time scale of tens of
microseconds. While the discussion in this chapter is primarily directed to
return strokes that are expected to be the strongest generators of audible
thunder, thunder can be generated by the presence of any impulsive current,
for example, as part of a breakdown into virgin air such as in a stepped-leader
step or via the traversal of an existing channel by the current associated with
a K-process.

Return stroke wavefronts traverse 1 m of leader channel in a time of the
order of 0.01 μsec. For a channel expanding at 10 times the speed of sound,
a reasonable order of magnitude for the speed of expansion as indicated, for
example, in Fig. 15.11, the channel radius increases at about 3 mm/μsec. A
fully formed channel has a radius of the order of centimeters (Section 7.2.4).
Thus the bulk of the energy input processes to an appreciable length of
return stroke channel, those processes being associated with the current rise

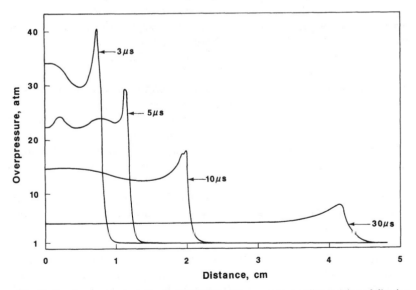

Fig. 15.11   Computer calculation by Hill (1971) of the pressure vs radius at 4 times following
the injection of a current $I = I_0[(\exp - \alpha t) - (\exp - \beta t)]$, where $I_0 = 30,000$ A, $\alpha = 3 \times 10^4$ sec^{-1}, $\beta = 3 \times 10^5$ sec^{-1}, into a cylinder of atmospheric air of radius 0.6 mm.

to peak and the corresponding collapse of the axial electric field as the channel resistance decreases, takes place on a more-or-less instantaneous scale compared to the scale on which the hydrodynamic processes associated with the channel expansion occur. Energy input to the hot channel after the expansion stage is probably relatively small because the channel resistance at that time is relatively low. Examples of computer-calculated channel pressure profiles at different times are given in Fig. 15.11 for the case of a cylindrical channel in air of initial radius 0.6 mm into which is injected a current rising to a peak of near 30 kA in a few microseconds and having a total energy input of 1.5×10^4 J/m. Calculations of the type shown in Fig. 5.11 done specifically to model return strokes have been published by Jones *et al.* (1968), Troutman (1969), Colgate and McKee (1969), Hill (1971), and Plooster (1971b). Similar calculations, which may be applicable to lightning, are given by Drabkina (1951), Sakurai (1953, 1954, 1955a,b, 1959), Lin (1954), Braginskii (1958), Rouse (1959), Swigart (1960), and Plooster (1970a,b, 1971a) for cylindrical geometry and by Brode (1955) for spherical geometry.

In modeling thunder, one must not only take account of the physics of the expansion of short sections of channel as calculated in the above references but also of the length and tortuosity of the lightning channel. Hill (1968), in measurements of cloud-to-ground channel segments between 5 and 70 m in length, found that direction changes for successive segments were randomly distributed, essentially independent of segment length, with a mean absolute value of channel direction change of about $16°$. Tortuosity on a much smaller scale is evident in the photographs of Evans and Walker (1963). Because of this observed tortuosity, Few *et al.* (1967, 1970) and Few (1969a,b, 1981) proposed that for the purpose of thunder generation, the lightning channel could best be modeled as a connected series of short cylindrical segments. At a radius shorter than the length of a given segment, cylindrical shock wave theory should apply; at a radius larger than the length of the channel segment, there should occur a roughly spherical divergence of the shock wave. We will consider the quantitative details, predictions, and problems of this approach below. A contrary view is held by the cylindrical model advocates (Jones *et al.*, 1968; Troutman, 1969, 1970; Remillard, 1969; Plooster, 1971a,b) who suggest that thunder is basically a cylindrical shock and acoustic wave. Few *et al.* (1970) have criticized this point of view both on the grounds that it produces a thunder pressure larger than measured and that the observed long duration of thunder is not consistent with the acoustic wave from a single cylindrical source. Later we will compare both the Few approach and the cylindrical model approach to the only adequate experiment available for evaluating the theories, pressure measurements made near a 4-m laboratory spark.

We now present the Few theory as given in Few (1969a, 1981). Relaxation radii R_s for spherical and R_c for cylindrical energy inputs are defined as the radii of the volumes generated if all the input electrical energy is used to perform PV work on the surrounding atmosphere which is at constant pressure P_0. Thus

$$P_0 \tfrac{4}{3}\pi R_s^3 = E_t \tag{15.2}$$

where E_t is the total input energy in the spherical case, and

$$P_0 \pi R_c^2 = E_L \tag{15.3}$$

where E_L is the energy per unit length in the cylindrical case. It is assumed that essentially all of the lightning input energy is transferred to the hot gas driving the shock wave since the radiated radio frequency and the optical radiation had been measured to be relatively small (Sections 7.2.1 and 7.2.4). A nondimensional coordinate $X_{s,c}$ for each geometry is defined as

$$X_s = r_s/R_s, \qquad X_c = r_c/R_c \tag{15.4}$$

where r_s and r_c are the distances from the spherical and cylindrical sources, respectively, and R_s and R_c are as defined in Eqs. (15.2) and (15.3). The solution to the hydrodynamic equations for the shock wave in either geometry in terms of $X_{s,c}$ is independent of the input energy because $X_{s,c}$ is normalized in terms of that energy. When $X_c = X_s = 1$, the pressure pulses from cylindrical and spherical sources have similar magnitudes and waveshapes. A parameter χ is defined in terms of the length of the straight line segment L assumed to be contributing to thunder

$$\chi = L/R_c \tag{15.5}$$

Channel segments for which $\chi \ll 1$ are termed microtortuous and will be engulfed by the expanding shock waves. Channel segments for which $\chi \gg 1$ are termed macrotortuous and will be composed of long segments that produce claps when oriented perpendicular to the observer. These large segments must themselves exhibit a smaller scale mesotortuosity with values of χ of the order of unity. It is values of L associated with χ near unity that are assumed responsible for the observed frequency spectrum of thunder. It is assumed that for a value of $X > \chi$ the shock wave diverges as a spherical wave with source energy $E_L L$, consistent with the calculation of the similar properties of spherical and cylindrical waves near $X = 1$. On the basis of one measured thunder spectrum, the sound waveshapes near a 4-m laboratory spark, and convenience [Eq. (15.2)], a value of 4/3 was chosen for χ. Spherical pressure waves from Brode (1956) at $X = 10$ were Fourier analyzed to obtain a thunder power spectrum. One result of this analysis is the prediction that the acoustic spectrum has a maximum at frequency f_m

$$f_m = 0.63C_0(P_0/E_L)^{1/2} \tag{15.6}$$

where C_0 is the speed of sound. For one spectrum Few (1969a) finds a frequency peak of 40 Hz which requires $E_L = 2 \times 10^6$ J/m (for which R_c is about 2 m) an order of magnitude greater than believed typical (Section 7.2.4). Since observed peak frequencies will decrease with distance from the channel (Otterman, 1959; Few, 1969a, 1981; Wright and Medendorp, 1967; Section 15.4), for measurements made in the kilometer or greater range the constant in Eq. (15.6) should be up to a factor of two smaller than 0.63 (Section 15.4), which was derived for $X = 10$, 20 m for $E_L = 2 \times 10^6$ J/m. Thus for the spectrum of Few (1969a), the resultant calculated energy per unit length should be up to a factor of four smaller than given above. Additionally, it is clear from Eq. (15.6) with any reasonable constant that the Few theory cannot predict infrasonic frequency peaks without requiring unrealistically large channel energies per unit length and hence can only be used to describe the sonic peaks.

Hill (1977c) has questioned the arbitrary assumption of 4/3 for the factor χ, but Few (1969a) points out that the rough value is reasonable and that the predictions of the theory are a weak function of χ while Few (1981) considers the effects on the theory of the presence of channel segments larger than $\chi = 4/3$.

The thunder theory of Few has been tested by Holmes et al. (1971a) as part of measurements discussed in Section 15.2.5 and by Dawson et al. (1968a) and Uman et al. (1970) on long laboratory sparks. While the test results generally indicate support for the model, other interpretations are possible, as we shall see. Holmes et al. (1971a) used their measured average acoustic efficiency of 0.18% based on an assumed 2.3×10^5 J/m energy input and a 4-km channel and the measured total acoustic energy to determine the energy per unit length to use in Eq. (15.6). For one set of data with measured acoustic power spectrum peaks between 40 and 100 Hz the agreement between theory and measurement was reasonably good with the calculated values being up to a factor of about 2 larger than the measured. For measured peaks in the 10-Hz range, the calculated peak frequency was an order of magnitude greater. The infrasonic peaks, however, may well be due to a source other than the hot channel, as we shall discuss in the next section. Holmes et al. (1971a) also pointed out that while the acoustic energy received from cloud flashes was a factor of 3 less than from ground flashes, the average peak frequency for cloud flashes was a factor of 2 lower than for ground flashes, opposite to the dependence required by Eq. (15.6), although, again, this result could be explained in terms of the simultaneous presence of two different modes of thunder production.

Dawson *et al.* (1968a) related the energy in a 4-m laboratory spark, 5×10^3 J/m, to the observed dominant frequency in the acoustic signal, between 1350 and 1650 Hz, and found the data were fit by Eq. (15.6) with the constant equal unity. Uman *et al.* (1970) Fourier transformed a typical shock waveform from the spark and found a peak frequency of 1400 Hz. The fact that Eq. (15.6) gives correct, at least to a factor of 2, peak frequencies for discharge input energies that are two orders of magnitude different would tend to lend confidence to the general validity of the theory. Additional support is supplied by the detailed overpressure and acoustic waveshape measurements of Uman *et al.* (1970) at distances from 0.34 to 16.5 m from the tortuous laboratory 4-m spark. The measured overpressure data are reproduced in Fig. 15.12 as are theoretical curves for cylindrical and spherical sources. For distances less than 2 m, both the shock overpressures and the duration of the overpressure are between a factor of 1.5 and 5 less than predicted by cylindrical shock wave theory. Close to the spark a single shock wave was observed, as illustrated in Fig. 15.13, whereas farther away multiple, generally 3 or 4, shock waves overlapped, as shown in Fig. 15.14, as might be expected if individual channel segments were generating separate shock waves. At 16.5 m, the number of apparent shock waves was less than at intermediate distances, apparently since various points on the spark channel become more nearly equidistant to the point of measurement and so previously separated shock waves merged together. As can be seen in Fig. 15.12, the data are fairly well fit by assuming the individual shock overpressure pulses were due to spherical waves with input energy 2.5×10^3 J derived from 0.5-m segments of spark channel. Similarly, the duration of the initial shock compression is better fit by the Few thunder model than by cylindrical shock wave theory (Uman *et al.*, 1970). The duration of the shock wave rarefaction differs from the predictions of either theory.

Plooster (1971a) has attempted to computer model the laboratory spark data and finds that when the current used by Uman *et al.* (1970) is put into his model, only about 0.1 of the energy input observed by Uman *et al.* (1970) is dissipated in raising the channel temperature to values consistent with those measured by Orville *et al.* (1967). Similarly, in the modeling of natural lightning, Plooster (1971b) found that only 10^3 to 6×10^3 J/m of input energy, corresponding to peak currents of 10–40 kA, respectively, was needed to simulate adequately channel properties, values significantly less than the 10^5 J/m deduced from other techniques (see Section 7.2.4). Plooster (1971a) argues that a cylindrical energy input to the spark of 4.2×10^2 J/m, an order of magnitude below that measured but consistent with his calculations and the measured input current, lowers the cylindrical curve in Fig. 15.12 so that it passes through the data points. He therefore concludes that Uman *et al.* (1970) must have made an order of magnitude error in

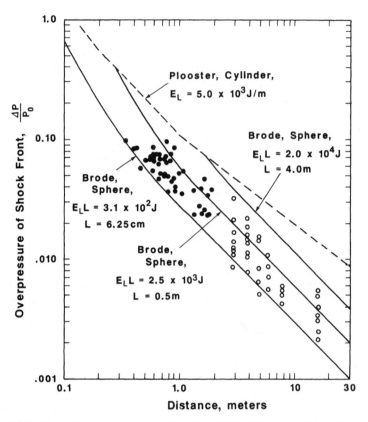

Fig. 15.12 Shock-front overpressure as a function of distance from a 4-m laboratory spark. The solid circles represent data obtained with a piezoelectric microphone and the open circles with a capacitor microphone. Also shown are calculations for cylindrical and for spherical shock waves. Adapted from Uman *et al.* (1970).

determining the spark input energy. Hill (1977b), on the other hand, suggests that Uman *et al.* (1970) have measured the input energy correctly but that only about 10% of that energy subsequently appeared in the hot spark channel producing a cylindrical shock wave, the remainder being diffusely deposited around the spark channel by processes preceding the spark's return stroke and producing no appreciable acoustic emission.

Dawson *et al.* (1968b) have analyzed the close overpressure measurements of Hill and Robb (1968) on artificially initiated lightning discussed in Section 15.2.4 and find the measured overpressure of about 2 bar a distance of 0.55 m from a 0.1 m gap consistent with a shock wave in transition between cylindrical and spherical expansion for an energy input of 10^5 J/m.

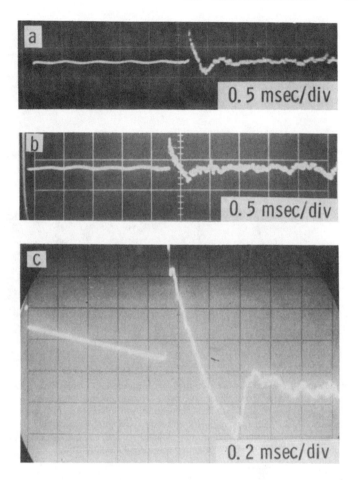

Fig. 15.13 Typical shock waves observed close to spark: (a) piezoelectric microphone record, 88 cm from spark at midgap height, $\Delta P/P_0 = 0.085$; (b) piezoelectric microphone record, 84 cm from spark at midgap height, $\Delta P/P_0 = 0.075$; (c) capacitor microphone record, 3 m from spark at a height of 1 m above the bottom electrode, time delay of 8 msec between spark initiation and oscilloscope triggering, $\Delta P/P_0 = 0.020$; the acoustic signal is superimposed on the electrical response of a microphone to ambient electric fields. Adapted from Uman *et al.* (1970).

A few additional comments on tortuosity are in order. Calculations by Uman *et al.* (1968) indicate that the acoustic signal at ground from an exploding straight line oriented vertically above ground takes the form of a compression followed by a period of relatively little sound followed by a rarefaction. From a physical point of view, a compressional pulse is first

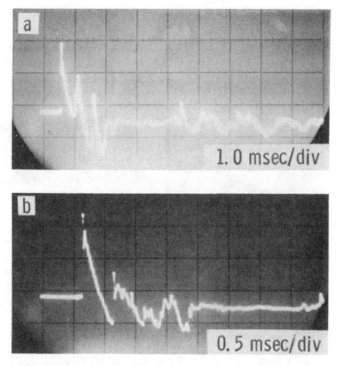

Fig. 15.14 Typical shock waves recorded 8 m from spark at midgap height by a capacitor microphone. Reflections from the ground plane arrive at the microphone about 4 msec after the direct signal. There is a 23-msec delay between spark initiation and oscilloscope triggering. (a) $\Delta P/P_0 = 0.0049$, from the same spark whose shock wave at a close distance is shown in Fig. 15.13b, with the capacitor microphone on the opposite side of the spark from the piezoelectric microphone; (b) $\Delta P/P_0 = 0.0057$. Adapted from Uman *et al.* (1970).

heard from the closest end of the line due to the explosion at that end. The sound arriving from the main body of the line is weak because compressions from one section are canceled by rarefactions from the adjacent one due to the differing distances of adjacent sections. A final rarefactional pulse is heard from the farthest end of the line. Experiments have shown that the sounds from one long section of the linear explosive primacord suspended vertically in the air are not thunder-like, but the primacord segments arranged in a tortuous pattern do produce sounds that are more suggestive of thunder (Brook, 1969). The sound signal from the explosion of a long straight line has an analog in the electromagnetic radiation field generated as a current pulse propagates along a long straight wire (Uman *et al.*, 1975; Section 7.3). Here, again, a positive radiation field is received from the close end of the wire, cancellation of signals occurs from the main length of the wire, and a negative field is received from the far end.

Laboratory measurements on 1-cm sparks and accompanying theory by Wright (1964) and Wright and Medendorp (1967) have illustrated a result similar to those expected from long straight sources as noted above and have further experimentally characterized the sound wave overpressure and waveform as a function of angle from a 1-cm spark. Observations made perpendicular to the spark yielded a single acoustic N-wave similar in shape to that shown in Fig. 15.13 since all of the spark channel sections were essentially equidistant from the observer. The N-wave is so named for its N-like shape. The effect discussed above for long line explosions, an acoustic signal composed of two pulses with a quiet space between, was observed for angles off the perpendicular to the spark.

Ribner and Roy (1982) have used the acoustic waveshapes observed by Wright and Mendendorp (1967) as the starting point for a computer generation of acoustic signals from model tortuous channels. Few (1974, 1981) describes a similar study. The calculated acoustic signals have much in common with observed thunder and illustrate the crucial roll played by tortuosity and channel orientation in producing claps and rumble.

Few (1981) discusses in detail the generation of thunder claps that are to be associated with sound emitted by sections of the main channel and channel branches that are approximately perpendicular to the line of sight from the observer, a fact verified experimentally by, for example, Few (1970). Thunder generally consists of several discrete claps (Section 15.2.1) super-imposed on a rumbling noise. It is reasonable to expect branches of first strokes to be powerful sources of sound since, according to the data of Malan and Collens (1937), branches may be instantaneously brighter than the channels above those branches. Further, it is interesting to note that there are roughly the same number of claps per thunder as there are branches in a first stroke (Schonland *et al.*, 1935), so that the branches may well account for a significant fraction of the claps.

15.3.2 INFRASONIC THUNDER: CONVERSION OF
ELECTROSTATIC ENERGY TO ACOUSTIC

As we have seen in the preceding section, it is difficult on physical grounds to ascribe the origin of the infrasonic component of thunder to a hot expanding channel. Holmes *et al.* (1971a) established that the peak acoustic power could be in the audible at one time during a thunder record and in the infrasonic at another time, as illustrated in Fig. 5.10. Holmes *et al.* (1971a) did not observe an infrasonic component in many of their records. Measurements indicate that infrasonic thunder apparently arrives in discrete pulses, is characterized by an initial compression, and is preferentially observed beneath thunderstorms (Bohannon *et al.*, 1977; Balachandron, 1983). The conversion of stored electrostatic energy to acoustic energy, as an explanation for the infrasonic component of thunder, has been examined from a theoretical point of view by Wilson (1920), McGehee (1964), Colgate and McKee (1969), Dessler (1973), and Few (1985). We examine some of that theory now.

Wilson (1920) first suggested that the electrical stress in a cloud, when relieved by lightning, would provide "a by no means negligible contribution to thunder." Analyses adopting this physical model have been provided for a variety of geometries in the papers referenced above. McGehee (1964) considered spherical charge volumes in the cloud. Dessler (1973) considered spherical, cylindrical, and disk geometries for the cloud charge. Colgate and McKee (1969) examined the acoustic effect of the charge distributed around the stepped leader. The electrostatic sound predicted by their analysis had a peak frequency of 130 Hz and was two orders of magnitude smaller in pressure than the acoustic wave from the hot channel. The analysis of Dessler (1973) is of particular interest in that he found that the infrasound emission from the collapse of the cloud electric field was highly directional, primarily propagating upward and downward. He suggested that this fact might explain the variability in the observations of infrasound in that one needs to be beneath a thundercloud for efficient detection, a prediction apparently experimentally verified by both Bohannon *et al.* (1977) and Balachandron (1983). On the other hand, Dessler (1973), as well as all others who have modeled the cloud charge collapse as an origin for infrasound, predicts an initial infrasonic rarefaction. All measurements indicate a compression. Few (1985) has attempted to remedy this serious inconsistency by proposing a model similar to Dessler's but with the addition of an extensive network of small electrical discharges in the cloud, the heating of whose channels is responsible for the initial compression.

15.4 PROPAGATION

Few (1981) has reviewed propagation effects on sound waves in air as they pertain to thunder. Some of these effects we have previously discussed: the refraction of sound waves in the atmosphere (Section 15.2.3) and the lengthening of sound waves with propagation distance (Section 15.3.1). We now consider these and other effects of propagation important to an understanding of the characteristics of thunder.

The three primary propagation effects of interest are waveshape change associated with finite-amplitude sound waves, sound wave attenuation, and thermal refraction. All of these effects can in principle be satisfactorily considered in a general theory of thunder propagation. Refraction due to wind shear which does not vary appreciably with time can also be modeled if the horizontal wind velocity is somehow known as a function of height. Other factors influencing propagation paths such as transient winds, aerosols, turbulence, and reflections from irregular terrain such as mountains are more difficult to handle analytically.

Otterman (1959) has published a theoretical treatment of the propagation of large-amplitude acoustic signals that can be applied to thunder. A single pulse evolves into the shape of a N-wave of the type observed in the laboratory by Wright and Medendorp (1967) and Uman et al. (1970) and illustrated in Fig. 15.13. The length of the N-wave increases with propagation distance. Few (1981) has reproduced an expression due to Otterman (1959) for pulse length and used it to evaluate the lengthening of the Brode (1956) waveshape from $X = 10$, the most distant waveshape calculated by Brode (1956), outward. Between Brode's most distant calculation and about 1 km, the N-wave lengthens about a factor of 2. Beyond that range its length remains roughly constant as expected in small-amplitude linear propagation. Since attenuation is not included in the Otterman (1959) theory, Few (1981) views the pulse length increase of a factor of 2 as a maximum value. If the pulse length does increase by a factor of 2 then the constant in Eq. (15.6) should be halved and the resultant energy per unit length found by Few (1969a), 2×10^6 J/m for a 40-Hz maximum frequency in the thunder power spectrum, should be reduced by a factor of 4 to a value of 5×10^5 J/m, in better agreement with the generally accepted values of the order of 10^5 J/m for that parameter (Section 7.2.4).

Attenuation of acoustic signals in air is primarily due to the interaction of the sound waves with air molecules and is a function of the water vapor content of the air. Harris (1967) has produced tables of attenuation constants for the acoustic signal amplitude decrease with propagation distance as a function of signal frequency and atmospheric temperature and humidity. Few (1981) has used this and other literature to calculate the attenuation for typical atmospheric conditions and finds that there is little attenuation for frequencies below about 100 Hz for distances as large as 10 km. For a frequency of 1 kHz and a range of 10 km, the attenuation is a factor of 2. On the other hand, Bass and Losely (1975) calculate that for 50% humidity at 20°C and a range of 5 km, the attenuation at 400 Hz is about a factor of 3 but is negligible at 50–100 Hz. They also show that changing the relative humidity from 20 to 100% for a discharge at 2 km increases the attenuation at 400 Hz by about a factor of 3. It follows that observed thunder frequency spectra will be affected by attenuation on the high-frequency end, that attenuation could possibly play a role in determining the peak frequency observed, and that differences in atmospheric conditions at different times or locations could result in different thunder spectra for similar sources.

Another type of attenuation, that due to the scattering of acoustic waves from cloud particles, has been discussed by Few (1981). This type of scattering also preferentially attenuates the higher frequencies. A variety of other processes, such as turbulence, which can result in attenuation, are also reviewed by Few (1981).

Refractive effects as they relate to thunder have been discussed by Few (1981), as well as in Section 15.2.3.

15.5   ACOUSTIC RECONSTRUCTION OF LIGHTNING CHANNELS

If significant features of a given thunder record can be recognized at three or more noncolinear microphones, these data and a knowledge of the time interval between the electromagnetic signal associated with the discharge and the arrival time of each thunder feature at the microphones allow a determination of the location of the source of the feature. Two different techniques have been used. The more accurate technique and that capable of giving many locations per thunder event is called *ray tracing*. The time *difference* between the arrival of significant features at each of the network of microphones located relatively close together, typically tens of meters apart, is used to determine the direction of the incoming sound wave at the network and that directional ray is mathematically traced back to the source given the atmospheric conditions and the time between the lightning and the arrival of the particular feature at the network. A discussion of the accuracy of ray tracing has been given by Few and Teer (1974), and channels reconstructed using this technique have been published, for example, by Few (1970), Nakano (1973, 1976), Few and Teer (1974), Teer and Few (1974), Winn et al. (1978), Christian et al. (1980), MacGorman et al. (1981), and Weber et al. (1982). In ray tracing the signals on separate microphones are very similar because the microphones are relatively close together. A second, but less accurate, source location technique is called *thunder ranging*, mentioned previously in Section 15.2.2. In thunder ranging the three noncolinear microphones are separated by a relatively large distance, of the order of a kilometer. According to Few (1981), thunder signals become spatially incoherent at microphone separations greater than about 100 m due to differences in perspective and propagation path. However, gross features such as claps remain coherent for microphone separations of the order of kilometers. To thunder range on a clap, the time difference between the electromagnetic field and clap arrival at each station is used to define a spherical surface of possible source locations. The three spherical surfaces from each of the three microphones intersect at a single point, the clap location. The implementation of this technique is discussed by Bohannon (1978). The technique of thunder ranging was used to reconstruct the lightning channel studied by Uman *et al.* (1978).

Perhaps the most interesting feature of the lightning channels that have been reconstructed from thunder records is the fact that they are generally oriented more horizontally in the cloud than vertically (e.g., MacGorman *et al.*, 1981) although vertical distribution of thunder sources does occur (e.g., Christian *et al.*, 1980). This horizontal orientation is probably associated primarily, as noted in Section 15.2.2, with the negative charge generally found between -10 and $-25°C$ and spread horizontally (Section 3.2; Figs. 3.3 and 3.4). Few (1970) observed a horizontal channel about 20 km in horizontal extent at about 5 km altitude. Nakano (1973, 1976) found that horizontal channels at 7 or 8 km height were oriented mainly along the wind direction. Few and Teer (1974) have favorably compared lightning photographs with thunder channel reconstructions. Teer and Few (1974) for 17 cloud-to-ground and 20 intracloud discharges occurring during 30 min at the end of a storm found the typical ratio of long horizontal axis to short horizontal axis to vertical axis for intracloud flashes and the in-cloud part of ground flashes was 3 : 2 : 1. The intracloud events and the in-cloud portion of the cloud-to-ground flashes were generally aligned in the same direction. Winn *et al.* (1978) show thunder source locations as part of an overall study of one thunderstorm. MacGorman *et al.* (1981) reconstructed all the lightning channels in three storms, one each in Arizona, Colorado, and Florida. They found that the in-cloud lightning activity was in layers 2–3 km thick and that in the Arizona and Florida storms there appeared to be two separate layers of activity. They interpret the two layers as being associated with the upper positive and lower negative charges of the cloud dipole structure (see Section 3.2).

REFERENCES

Ajayi, N. O.: Acoustic Observation of Thunder from Cloud–Ground Flashes, *J. Geophys. Res.*, 77:4586–4587 (1972).

Anderson, R., S. Bjornsson, D. C. Blanchard, S. Gathman, J. Hughes, S. Jonasson, C. B. Moore, H. J. Survilas, and B. Vonnegut: Electricity in Volcanic Clouds. *Science*, **148**:1179–1189 (1965).

Arabadzhi, V.: Certain Characteristics of Thunder. *Dokl. Akad. Nauk SSSR*, **82**:377–378 (1952). Translation available as RJ-1058 from Associated Technical Services, Inc., Glen Ridge, New Jersey.

Arabadzhi, V.: Some Characteristics of the Electrical State of Thunderclouds and Thunderstorm Activity. *Uch. Zap. Minsk. Gos. Ped. Inst., im A.M. Gorkogo Yubil. Vypusk, Ser. Fiz-Mat.*, (7) (1957). Translation available as RJ-1315 from Associated Technical Services, Inc., Glen Ridge, New Jersey.

Arabadzhi, V.: The Spectrum of Thunder. *Priroda (Moscow)*, 54:74–75 (1965).

Arabadzhi, V. I.: Acoustical Spectra of Electrical Discharge. *Sov. Phys. Acoust.*, 14:92–93 (1968).

Arnold, R. T., H. E. Bass, and A. A. Atchley: Underwater Sound from Lightning Strikes to Water in the Gulf of Mexico. *J. Acoust. Soc. Am.*, 76 (1):320–322 (1984).

Ault, C.: Thunder at Sea. *Sci. Am.*, 218:525 (1916).

Balachandran, N. K.: Infrasonic Signals from Thunder. *J. Geophys. Res.*, 84:1735–1745 (1979).

Balachandran, N. K.: Acoustic and Electrical Signals from Lightning. *J. Geophys. Res.*, 88:3879–3884 (1983).

Bass, H. E.: The Propagation of Thunder through the Atmosphere. *J. Acoust. Soc. Am.*, 67:1959–1966 (1980).

Bass, H. E., and R. E. Losely: Effect of Atmospheric Absorption on the Acoustic Power Spectrum of Thunder. *J. Acoust. Soc. Am.*, 57:882–823 (1975).

Beasley, W. H., T. M. Georges, and M. W. Evans: Infrasound from Convective Storms: An Experimental Test of Electrical Source Mechanisms. *J. Geophys. Res.*, 81:3133–3140 (1976).

Bhartendu: A Study of Atmospheric Pressure Variations from Lightning Discharges. *Can. J. Phys.*, 46:269–231 (1968).

Bhartendu: Thunder—A Survey. *Nat. Can.*, 96:671–681 (1969a).

Bhartendu: Audio Frequency Pressure Variations from Lightning Discharges. *J. Atmos. Terr. Phys.*, 31:743–747 (1969b).

Bhartendu: Comments on Paper "On the Power Spectrum and Mechanisms of Thunder" by C. R. Holmes, M. Brook, P. Krehbiel, and R. McCrory. *J. Geophys. Res.*, 76:7441–7442 (1971a).

Bhartendu: Sound Pressure of Thunder. *J. Geophys. Res.*, 76:3515–3516 (1971b).

Bhartendu, and B. W. Currie: Atmospheric Pressure Variations from Lightning Discharges. *Can. J. Phys.*, 41:1929–1933 (1963).

Bohannon, J. L.: Infrasonic Pulses from Thunderstorms. M.S. thesis, Rice University, Houston, Texas, 1978.

Bohannon, J. L.: Infrasonic Thunder: Explained. Ph.D. thesis, Department of Space Physics and Astronomy, Rice University, Houston, 1980.

Bohannon, J. L., A. A. Few, and A. J. Dessler: Detection of Infrasonic Pulses from Thunderclouds. *Geophys. Res. Lett.*, 4:49–52 (1977).

Braginskii, S. I.: Theory of the Development of a Spark Channel. *Sov. Phys. JETP (Engl. Transl.)*, 34:1068–1074 (1958).

Brode, H. L.: Numerical Solutions of Spherical Blast Waves. *J. Appl. Phys.*, 26:766–775 (1955).

Brode, H. L.: The Blast Wave in Air Resulting from a High Temperature, High Pressure Sphere of Air. RAND Corp. Res. Memorandum RM-1825-AEC (1956).

Brook, M.: Discussion on the Few-Dessler Paper. *In* "Planetary Electromagnetics" (S. C. Coroniti and J. Hughes, eds.), Vol. 1, p. 579. Gordon & Breach, New York, 1969.

Brooks, C. F.: Another Case. *Mon. Weather Rev.*, 48:162 (1920).

Brown, E. H., and S. F. Clifford: On the Attenuation of Sound by Turbulence. *J. Acoust. Soc. Am.*, 60:788–794 (1976).

Cave, C. J. P.: The Audibility of Thunder. *Nature (London)*, 104:132 (1919).

Christian, H., C. R. Holmes, J. W. Bullock, H. Gaskell, A. J. Illingworth, and J. Latham: Airborne and Ground-Based Studies of Thunderstorms in the Vicinity of Langmuir Laboratory. *Q. J. R. Meteorol. Soc.*, 106:159–174 (1980).

Colgate, S. A., and C. McKee: Electrostatic Sound in Clouds and Lightning. *J. Geophys. Res.*, 74:5379–5389 (1969).

Dawson, G. A., C. N. Richards, E. P. Krider, and M. A. Uman: The Acoustic Output of a Long Spark. *J. Geophys. Res.*, 73:815–816 (1968a).

Dawson, G. A., M. A. Uman, and R. E. Orville: Discussion of Paper by E. L. Hill and J. D. Robb, "Pressure Pulse from a Lightning Stroke." *J. Geophys. Res.*, 73:6595–6597 (1968b).

De L'Isle, J. N.: "Memoires pour Servir à l'Histoire et au Progres de l'Astronomie de la Géographie et de la Physique." L'Imprimerie de l'Académie des Sciences, St. Petersbourg, 1738.

Dessler, J.: Infrasonic Thunder. *J. Geophys. Res.*, 78:1889–1896 (1973).

Drabkina, D. I.: The Theory of the Development of the Spark Channel. *J. Exp. Theor. Phys. (USSR)*, 21:473–483 (1951). English translation, AERE Lib/Trans. 621, Harwell, Berkshire, England.

Evans, W. H., and R. L. Walker: High Speed Photographs of Lightning at Close Range. *J. Geophys. Res.*, 68:4455 (1963).

Few, A. A.: Thunder. Ph.D. thesis, Rice University, Houston, Texas (1968).

Few, A. A.: Power Spectrum of Thunder. *J. Geophys. Res.*, 74:6926–6934 (1969a).

Few, A. A.: Reply to Letter by W. J. Remillard. *J. Geophys. Res.*, 74:5556 (1969b).

Few, A. A.: Lightning Channel Reconstruction from Thunder Measurements. *J. Geophys. Res.*, 75:7517–7523 (1970).

Few, A. A.: Thunder Signatures. *Trans. Am. Geophys. Union*, 55:508–513 (1974).

Few, A. A.: Thunder. *Sci. Am.*, 233:80–90 (1975).

Few, A. A.: Acoustic Radiations from Lightning. *In* "Handbook of Atmospherics" (H. Volland, ed.), Vol. II. CRC Press, Boca Raton, Florida, 1981.

Few, A. A.: The Production of Lightning-Associated Infrasonic Acoustic Sources in Thunderclouds. *J. Geophys. Res.*, 90:6175–6180 (1985).

Few, A. A., and T. L. Teer: The Accuracy of Acoustic Reconstructions of Lightning Channels. *J. Geophys. Res.*, 79:5007–5011 (1974).

Few, A. A., A. J. Dessler, D. J. Latham, and M. Brook: A Dominant 200-Hertz Peak in the Acoustic Spectrum of Thunder. *J. Geophys. Res.*, 72:6149–6154 (1967).

Few, A. A., H. B. Garrett, M. A. Uman, and L. E. Salanave: Comments on Letter by W. W. Troutman, "Numerical Calculation of the Pressure Pulse from a Lightning Stroke." *J. Geophys. Res.*, 75:4192–4195 (1970).

Fleagle, R. G.: The Audibility of Thunder. *J. Acoust. Soc. Am.*, 21:411–412 (1949).

Fleagle, R. G., and J. A. Businger: "An Introduction to Atmospheric Physics." Academic Press, New York, 1963.

Georges, T. M.: Infrasound from Convective Storms: Examining the Evidence. *Rev. Geophys. Space Phys.*, 11:571–594 (1973).

Georges, T. M., and W. H. Beasley: Refraction of Infrasound by Upper-Atmospheric Winds, *J. Acoust. Soc. Am.*, 61:28–34 (1977).

Goyer, G. G., and M. N. Plooster: On the Role of Shock Waves and Adiabatic Cooling in the Nucleation of Ice Crystals by Lightning Discharge. *J. Atmos. Terr. Phys.*, 25:857–862 (1968).

Harris, C. M.: Absorption of Sound in Air Versus Humidity and Temperature. NASA-CR-647, Columbia University, New York, 1967.

Hill, E. L., and J. D. Robb: Pressure Pulse from a Lightning Stroke. *J. Geophys. Res.*, 73:1883–1888 (1968).

Hill, R. D.: Analysis of Irregular Paths of Lightning Channels. *J. Geophys. Res.*, 73:1897–1906 (1968).

Hill, R. D.: Channel Heating in Return Stroke Lightning. *J. Geophys. Res.*, 76:637–645 (1971).

Hill, R. D.: Comments on "Quantitative Analysis of a Lightning Return Stroke Diameter and Luminosity Changes as a Function of Space and Time" by Richard E. Orville, John H. Helsdon, Jr., and Walter H. Evans. *J. Geophys. Res.*, **80**:1138 (1975).

Hill, R. D.: Energy Dissipation in Lightning. *J. Geophys. Res.*, **82**:4967–4968 (1977a).

Hill, R. D.: Comments on "Numerical Simulation of Spark Discharges in Air" by M. N. Plooster. *Phys. Fluids*, **20**:1584–1586 (1977b).

Hill, R. D.: Thunder. *In* "Lightning, Vol. 1, Physics of Lightning" (R. H. Golde, ed.), pp. 385–408. Academic Press, New York, 1977c.

Hill, R. D.: A Survey of Lightning Energy Estimates. *Rev. Geophys. Space Phys.*, **17**:155–164 (1979).

Hill, R. D.: Investigation of Lightning Strikes to Water Surfaces. *J. Acoust. Soc. Am.*, **78**:2096–2099 (1985).

Holmes, C. R., M. Brook, P. Krehbiel, and R. McCrory: On the Power Spectrum and Mechanism of Thunder. *J. Geophys. Res.*, **76**:2106–2115 1971a).

Holmes, C. R., M. Brook, P. Krehbiel, and R. McCrory: Reply to Comment by Bhartendu on "On the Power Spectrum and Mechanisms of Thunder." *J. Geophys. Res.*, **76**:7443 (1971b).

Jones, D. L.: Comments on Paper by A. A. Few, A. J. Dessler, D. J. Latham, and M. Brook, "A Dominant 200-Hertz Peak in the Acoustic Spectrum of Thunder." *J. Geophys. Res.*, **73**:4776–4777 (1968a).

Jones, D. L.: Intermediate Strength Blast Wave. *Phys. Fluids*, **11**:1664–1667 (1968b).

Jones, D. L., G. G. Goyer, and M. N. Plooster: Shock Wave from a Lightning Discharge. *J. Geophys. Res.*, **73**:3121–3127 (1968).

Kitagawa, N.: Discussion. *In* "Problems of Atmospheric and Space Electricity" (S. C. Coroniti, ed.), pp. 350–351. American Elsevier, New York, 1965.

Krehbiel, P. R., M. Brook, and R. A. McCrory: An Analysis of the Charge Structure of Lightning Discharges to the Ground. *J. Geophys. Res.*, **84**:2432–2456 (1979).

Krider, E. P., G. A. Dawson, and M. A. Uman: Peak Power and Energy Dissipation in a Single-Stroke Lightning Flash. *J. Geophys. Res.*, **73**:3335–3339 (1968).

Latham, D. J.: A Study of Thunder from Close Lightning Discharges. M.S. thesis, Physics Department, New Mexico Institute of Mines and Technology, Socorro, New Mexico, 1964.

Lin, S. C.: Cylindrical Shock Waves Produced by an Instantaneous Energy Release. *J. Appl. Phys.*, **25**:54–57 (1954).

MacGorman, D. R.: Lightning Location in a Colorado Thunderstorm. M.S. thesis, Rice University, Houston, Texas, 1977.

McCrory, R. A., and C. R. Holmes: Comment on Paper by Bhartendu, "A Study of Atmospheric Pressure Variations from Lightning Discharges." *Can. J. Phys.*, **46**:2333 (1968).

McGehee, R. M.: The Influence of Thunderstorm Space Charges on Pressure. *J. Geophys. Res.*, **69**:1033–1035 (1964).

MacGorman, D. R.: Lightning Location in a Storm with Strong Wind Shear. Ph.D. thesis, Department of Space Physics and Astronomy, Rice University, Houston, Texas, 1978.

MacGorman, D. R., A. A. Few, and T. L. Teer: Layered Lightning Activity. *J. Geophys. Res.*, **86**:9900–9910 (1981).

Malan, D. J.: "Physics of Lightning," pp. 162–163. English Univ. Press, London, 1963.

Malan, D. J., and H. Collens: Progressive Lightning, Pt. 3, The Fine Structure of Return Lightning Strokes. *Proc. R. Soc. London Ser. A*, **162**:175–203 (1937).

Nakano, M.: Lightning Channel Determined by Thunder. *Proc. Res. Inst. Atmos., Nagoya Univ., Japan*, **20**:1–9 (1973).

Nakano, M.: Characteristics of Lightning Channel in Thunderclouds Determined by Thunder. *J. Meteorol. Soc. Japan*, **54**:441–447 (1976).

Nakano, M., and T. Takeuti: On the Spectrum of Thunder. *Proc. Res. Inst. Atmos., Nagoya Univ., Japan*, **17**:111–113 (1970).

Newman, M. M., J. R. Stahmann, and J. D. Robb: Experimental Study of Triggered Natural Lightning Discharges. Rep. DS-67-3, Project 520-002-03X, Federal Aviation Agency, Washington, D.C., March, 1967a.

Newman, M. M., J. R. Stahmann, J. D. Robb, E. A. Lewis, S. G. Martin, and S. V. Zinn: Triggered Lightning Strokes at Very Close Range. *J. Geophys. Res.*, **72**:4761–4764 (1967b).

Orville, R. E., M. A. Uman, and A. M. Sletten: Temperature and Electron Density in Long Air Sparks. *J. Appl. Phys.*, **38**:895–896 (1967).

Otterman, J.: Finite-Amplitude Propagation Effect on Shock-Wave Travel Times from Explosions at High Altitudes. *J. Acoust. Soc. Am.*, **31**:470–474 (1959).

Page, D. E.: Distance to Which Thunder Can Be Heard. *Bull. Am. Meteorol. Soc.*, **25**:366 (1944).

Pain, H. J., and E. W. E. Rogers: Shock Waves in Gases. *Rep. Prog. Phys.*, **25**:287–336 (1962).

Plooster, M. N.: Shock Waves from Line Sources. Numerical Solutions and Experimental Measurements. *Phys. Fluids*, **13**:2665–2675 (1970a).

Plooster, M. N.: Erratum: Shock Waves from Line Sources. Numerical Solutions and Experimental Measurements. *Phys. Fluids*, **13**:2248 (1970b).

Plooster, M. N.: Numerical Simulation of Spark Discharges in Air. *Phys. Fluids*, **14**:2111–2123 (1971a).

Plooster, M. N.: Numerical Model of the Return Stroke of the Lightning Discharge. *Phys. Fluids*, **14**:2124–2133 (1971b).

Plooster, M. N.: On Freezing of Supercooled Droplets Shattered by Shock Waves. *J. Appl. Meteorol.*, **11**:161–165 (1972).

Reed, J. W.: Airblast Overpressure Decay at Long Ranges. *J. Geophys. Res.*, **77**:1623–1629 (1972).

Remillard, W. J.: The Acoustics of Thunder. Tech. Mem. 44, Acoustic Research Laboratory, Division of Engineering and Applied Physics, Harvard University, Cambridge, Massachusetts, September 1960.

Remillard, W. J.: Comments on Paper by A. A. Few, A. J. Dessler, Don J. Latham, and M. Brook, "A Dominant 200-Hertz Peak in the Acoustic Spectrum of Thunder." *J. Geophys. Res.*, **74**:5555 (1969).

Remillard, W. J.: Pressure Disturbances from a Finite Cylindrical Source. *J. Acoust. Soc. Am.*, **59**:744–748 (1976).

Ribner, H. S., and D. Roy: Acoustics of Thunder; A Quasilinear Model for Tortuous Lightning. *J. Acoust. Soc. Am.*, **72**(b):1911–1925 (1982).

Ribner, H. S., F. Lam, K. A. Leung, D. Kurtz, and N. D. Ellis: Computer Model of the Lightning—Thunder Process with Audible Demonstration. *Prog. Astronaut. Aeronaut.*, **46**:77–87 (1976).

Rouse, C. A.: Theoretical Analysis of the Hydrodynamic Flow in Exploding Wire Phenomena. *In* "Exploding Wires" (W. G. Chace and H. K. Moore, eds.). Plenum, New York, 1959.

Sakurai, A.: On the Propagation and Structure of the Blast Wave (1). *J. Phys. Soc. Japan*, **8**:662–669 (1953).

Sakurai, A.: On the Propagation and Structure of the Blast Wave (2). *J. Phys. Soc. Japan*, **9**:256–266 (1954).

Sakurai, A.: On Exact Solution of the Blast Wave Problem. *J. Phys. Soc. Japan*, **10**:827–828 (1955a).

Sakurai, A.: Decrement of Blast Wave. *J. Phys. Soc. Japan*, **10**:1018 (1955b).

Sakurai, A.: On the Propagation of Cylindrical Shock Waves. *In* "Exploding Wires" (W. G. Chace and H. K. Moore, eds.). Plenum, New York, 1959.

Schmidt, W.: Über den Donner. *Meteorol. Z.*, **31**:487–498 (1914).

Schonland, B. F. J., D. J. Malan, and H. Collens: Progressive Lightning, Pt. 2. *Proc. R. Soc. London Ser. A*, **152**:595–625 (1935).

Swigart, R. J.: Third Order Blast Wave Theory and Its Application to Hypersonic Flow Past Blunt-Nosed Cylinders. *J. Fluid Mech.*, **9**:613–620 (1960).

Taljaard, J. J.: How Far Can Thunder Be Heard? *Weather*, **7**:245–246 (1952).

Taylor, G. L.: The Formation of a Blast Wave by a Very Intense Explosion, 2, the Atomic Explosion of 1945. *Proc. R. Soc. London Ser. A*, **201**:175–186 (1950).

Teer, T. L.: Acoustic Profiling: A Technique for Lightning Channel Reconstruction. M.S. thesis, Rice University, Houston, Texas, 1972.

Teer, T. L.: Lightning Channel Structure Inside an Arizona Thunderstorm. Ph.D. dissertation, Department of Space Physics and Astronomy, Rice University, Houston, Texas, 1973.

Teer, T. L., and A. A. Few: Horizontal Lightning. *J. Geophys. Res.*, **79**:3436–3441 (1974).

Temkin, S.: A Model for Thunder Based on Heat Addition. *J. Sound Vibr.*, **52**:401–414 (1977).

Thomson, E. M.: Characteristics of Port Moresby Ground Flashes. *J. Geophys. Res.*, **85**:1027–1036 (1980).

Troutman, W. S.: Numerical Calculation of the Pressure Pulse from a Lightning Stroke. *J. Geophys. Res.*, **74**:4595 (1969).

Troutman, W. S.: Reply to Comments by A. A. Few, H. B. Garrett, M. A. Uman, and L. E. Salanave on "Numerical Calculations of the Pressure Pulse from a Lightning Stroke." *J. Geophys. Res.*, **75**:4196 (1970).

Uman, M. A.: "Lightning." McGraw-Hill, New York, 1969. See also, Dover, New York, 1984.

Uman, M. A., and S. R. Evans: Previously Unpublished Measurements from the University of Florida Research Group, 1977.

Uman, M. A., D. K. McLain, and F. Myers: Sound from Line Sources with Application to Thunder. Westinghouse Research Laboratory Rep. 68-9E4-HIVOL-R1, 1968.

Uman, M. A., A. H. Cookson, and J. B. Moreland: Shock Wave from a Four-Meter Spark. *J. Appl. Phys.*, **41**:3148–3155 (1970).

Uman, M. A., D. K. McLain, and E. P. Krider: The Electromagnetic Radiation from a Finite Antenna. *Am J. Phys.*, **43**:33–38 (1975).

Uman, M. A., W. H. Beasley, J. A. Tiller, Y. T. Lin, E. P. Krider, C. D. Weidman, P. R. Krehbiel, M. Brook, A. A. Few, J. L. Bohannon, C. L. Lennon, H. A. Poehler, W. Jafferis, J. R. Gulick, and J. R. Nicholson: An Unusual Lightning Flash at Kennedy Space Center. *Science*, **201**:9–16 (1978).

Veenema, L. C.: Die Hörweite des Gewitterdonner. *Z. Angew. Meteorol., Wetter*, 127–130, 187–192, 258–262 (1917).

Veenema, L. C.: Die Hörweite des Gewitterdonner. *Z. Angew. Meterol., Wetter*, 56–68 (1918).

Veenema, L. C.: The Audibility of Thunder. *Mon. Weather Rev.*, **48**:162 (1920).

Weber, M. E., H. J. Christian, A. A. Few, and M. F. Stewart: A Thunderstorm Electric Field Sounding: Charge Distribution and Lightning. *J. Geophys. Res.*, **87**:7158–7169 (1982).

Whitham, G. B.: On the Propagation of Weak Shock Waves. *J. Fluid Mech.*, **1**:290–318 (1956).

Wilson, C. T. R.: Investigations on Lightning Discharges and on the Electric Field of Thunderstorms. *Philos. Trans. R. Soc. London Ser. A*, **221**:73–115 (1920).

Winn, W. P., C. B. Moore, C. R. Holmes, and L. G. Byerly: Thunderstorm on July 16, 1975, over Langmuir Laboratory: A Case Study. *J. Geophys. Res.*, **83**:3079–3092 (1978).

Wright, W. M.: Experimental Study of Acoustical N Waves. *J. Acoust. Soc. Am.*, **36**:1032 (1964).

Wright, W. M., and N. W. Medendorp: Acoustic Radiation from a Finite Line Source with N-Wave Excitation. *J. Acoust. Soc. Am.*, **43**:966–971 (1967).

Zhivlyuk, Yu., and S. L. Mandel'shtam: On the Temperature of Lightning and Force of Thunder. *Sov. Phys. JETP (Engl. Transl.)*, **13**:338–340 (1961).

Appendix A | Electromagnetics

A.1 ELECTROSTATICS

A.1.1 SIGN CONVENTION

It has been found by measurement that the fine-weather electric field vector above the Earth is directed downward toward the Earth. That is, the Earth is negatively charged and the atmosphere above the Earth is positively charged (Israel, 1971; Section 1.8). The magnitude of the fine-weather electric field intensity at the ground is of the order of 100 V/m. In most of the atmospheric–electrical literature the fine-weather electric field is termed a positive electric field. We therefore use the following historical sign convention: an electric field at the ground is called positive if it is the same direction as the field due to positive charge above ground level, that is, if the vector field is directed downward toward the Earth; an electric field at the ground is called negative if the vector field is directed upward away from the Earth. An electric field change at the ground is defined as positive if the change is attributable to an increase of positive charge (or decrease of negative charge) overhead, that is, an increase in magnitude of the downward-directed field vector. A negative field change is associated with the increase in magnitude of an upward-directed field vector. The signs of the fields just defined are opposite to those for more standard coordinate systems with either an origin at the center of Earth and radial coordinate outward or an origin on the Earth's surface with z-coordinate upward.

A.1.2 ELECTRIC FIELDS FROM POINT, SPHERICALLY SYMMETRICAL, AND LINEAR CHARGE DISTRIBUTIONS

The electric field intensity $\mathbf{E}$ a distance R from a stationary point charge is (Cheng, 1983)

$$\mathbf{E} = (Q/4\pi\varepsilon_0 R^2)\mathbf{a_R} \qquad (A.1)$$

where $\mathbf{a_R}$ is a unit vector directed along R in the direction away from the charge, Q can be either positive or negative, and ε_0 is the permittivity of vacuum (which is essentially equal to the permittivity of atmospheric air). In SI units $(4\pi\varepsilon_0)^{-1} \cong 9 \times 10^9$. Equation (A.1) also describes the electric field intensity on the surface or outside of a spherically symmetric charge distribution of total charge Q (Cheng, 1983).

To calculate the electric field at the ground due to lightning charges above the Earth, we approximate the Earth as a flat perfectly conducting plane and the charge centers as static point or spherically symmetrical charges. We first calculate the electric field intensity due to a point charge Q located a distance H above a conducting plane. For illustrative purposes in Fig. A.1 we have assumed that the charge is positive and denoted that fact by showing the sign explicitly. By the method of electrical images (Cheng, 1983), the field due to the negative surface charge density induced on the conducting plane by the positive charge above it can be found by replacing the plane and its surface charge with a negative image charge located a distance H below the plane, as shown in Fig. A.1. The magnitude of the electric field intensity on the plane a horizontal distance D from either the real or the image charge is, by Eq. (A.1),

$$E = Q/[4\pi\varepsilon_0(H^2 + D^2)] \quad \text{V/m} \tag{A.2}$$

The field vector direction resulting from each charge is shown in Fig. A.1. The total electric field is obtained by vector addition. The field components

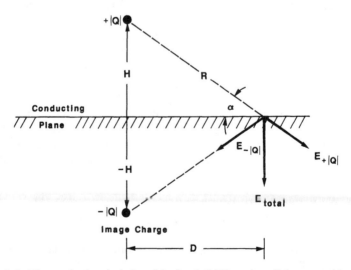

Fig. A.1 Diagram for the calculation of the electric field intensity at D due to a positive point charge $+Q$ at height H above a conducting plane.

parallel to the plane are of equal magnitude and of opposite direction and thus add to zero. (The tangential electric field at any perfectly conducting surface is zero.) The components normal to the plane are both positive, in the sense discussed in Section A.1.1, and thus can be added directly to obtain the total electric field intensity. The normal field component due to either point charge is found by multiplying the total field due to that charge by $\sin \alpha = H/(H^2 + D^2)^{1/2}$. Thus the total electric field magnitude is

$$E_{\text{total}} = \frac{2QH}{4\pi\varepsilon_0(H^2 + D^2)^{3/2}} = \frac{2QH}{4\pi\varepsilon_0 D^3[1 + (H/D)^2]^{3/2}}$$

$$= \frac{2Q}{4\pi\varepsilon_0 D^2}\left\{\frac{H/D}{[1 + (H/D)^2]^{3/2}}\right\} \quad \text{V/m}$$

(A.3)

with a direction perpendicular to the plane and sign positive if the charge above the plane is positive.

If the height H of the charge in Fig. A.1 is varied, the field at D will pass through a maximum as illustrated in Fig. A.2. We can explain this field variation from a physical point of view. If H is relatively small, the electric field at D, which is twice the vertical component of the field from $+Q$, will be small because $\sin \alpha$ is small. As H increases, $\sin \alpha$ increases and the field reaches a maximum. The field will decrease for larger H because the square

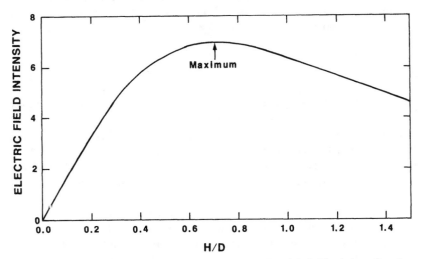

Fig. A.2 Electric field intensity for the charge configuration given in Fig. A.1 as a function of H/D. To obtain electric field in volts per meter for a given H/D multiply the ordinate by $Q \times 10^9/D^2$.

of the distance from the charge to the observation point, $R^2 = H^2 + D^2$, increases faster than $\sin \alpha$ increases. We can determine the value of H/D for which the field is maximum by taking the derivative of E_{total} with respect to H from Eq. (A.3) and setting that derivative equal to zero. The result is $H/D = 1/\sqrt{2}$.

If $(H/D)^2 \ll 1$, Eq. (A.3) can be approximated as

$$E_{\text{total}} \simeq \frac{2QH}{4\pi\varepsilon_0 D^3} = \frac{M}{4\pi\varepsilon_0 D^3} \quad \text{V/m} \tag{A.4}$$

where $M = 2QH$ is defined as the electric dipole moment of the charge Q and its image. If electric field measurements are made about 25 km or farther from a lightning to ground, the measurement of the change in electric field due to the destruction of the charge source in the cloud, a knowledge of D, and the use of Eq. (A.4) allow a calculation of the change in the dipole moment. A similar expression to Eq. (A.4) is derived in Section 13.2 for a vertical or a nonvertical intracloud lightning discharge where, if equal positive and negative charge sources are destroyed, the charge height H appearing in Eq. (A.4) is replaced by the vertical distance between the centers of the two charge sources as indicated in Eqs. (13.5) and (13.6).

The concept of an electric dipole moment of a single charge above a conducting plane can be generalized to include an electric dipole moment for any arbitrary distribution of charge above a conducting plane. Consider, for example, the case in which the charge varies along a vertical line. The generalized dipole moment is defined as

$$M = 2 \int_{H_B}^{H_T} \rho_L(z')z' \, dz' \quad \text{C-m} \tag{A.5}$$

where $\rho_L(z')$ is the charge per unit length along the vertical line whose lower limit is H_B and whose upper limit is H_T. For illustrative purposes in Fig. A.3 we have assumed the line charge is positive and denoted that fact by showing the sign explicitly. The prime on the z' in Eq. (A.5) is used to indicate the location of a source of field (a charge) as opposed to a location at which the field is being calculated, which will be given in unprimed coordinates.

Consider now the electric field at D due to the line charge and its electrical image as shown in Fig. A.3. The electric field due to a small element of charge $\rho_L(z') \, dz'$ within the line charge is identical to the field from a small point charge. Thus the magnitude of the field due to the charge within dz' is

$$dE_{+\rho_L \, dz} = \frac{\rho_L(z') \, dz'}{4\pi\varepsilon_0(z'^2 + D^2)} \quad \text{V/m} \tag{A.6}$$

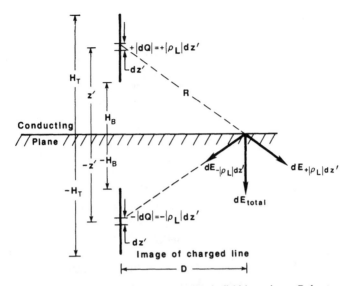

Fig. A.3 Diagram for the calculation of the electric field intensity at D due to a vertical charged line of length $H_T - H_B$ with positive charge per unit length ρ_L located above a conducting plane.

and the total field due to this charge element and its image is

$$dE_{total} = \frac{2\rho_L(z')z'\,dz'}{4\pi\varepsilon_0(z'^2 + D^2)^{3/2}} \quad \text{V/m} \tag{A.7}$$

and is perpendicular to the plane and positive if the line charge is positive. The field due to the total charge distribution in the vertical line is found by integration:

$$E_{total} = \int_{H_B}^{H_T} \frac{2\rho_L(z')z'\,dz'}{4\pi\varepsilon_0(z'^2 + D^2)^{3/2}} \quad \text{V/m} \tag{A.8}$$

If $(z'/D)^2 \ll 1$, Eq. (A.8) becomes identical to Eq. (A.4) with the definition of the dipole moment from Eq. (A.5), and thus moment changes can be derived from field change measurements on distant lightning for any vertically distributed charge sources. Approximately the same electric field and moment change is obtained for a nonvertical charge distribution if the charge density is the same function of height and the horizontal extent of the charge sources is small compared to D. Additional discussion of the relationships between lightning fields and the electric dipole moment for vertical channels is given by McLain and Uman (1971) and for nonvertical channels by Krehbiel et al. (1979).

A.1.3   LEADER MODELS

We now calculate the electric field from a *uniformly* charged vertical line. The uniformly charged line constitutes a part of some simple models of stepped- and dart-leader processes, the return stroke, and some cloud discharge processes. The field from such a charge source is found by integrating Eq. (A.8) with ρ_L assumed constant:

$$E_{\text{total}} = \frac{2\rho_L}{4\pi\varepsilon_0}\left[\frac{1}{(D^2 + H_B^2)^{1/2}} - \frac{1}{(D^2 + H_T^2)^{1/2}}\right] \quad \text{V/m} \qquad (A.9)$$

We consider now a simple leader model. If a charged leader emerges from a volume of charge of the same sign, the charging of the leader as its length is extended results in a decrease of charge in the source volume. If the leader length is l, the height of the source charge is H, and the charge source can be treated as a point charge or as a spherically symmetrical distribution centered at H, the field change at the ground due to the decrease in the source charge can be found from Eq. (A.3)

$$\Delta E_s = -\frac{2\rho_L l H}{4\pi\varepsilon_0(H^2 + D^2)^{3/2}} \quad \text{V/m} \qquad (A.10)$$

where $\rho_L l$ is the amount of charge lost by the source volume. If, for example, ρ_L is positive, the source field has become less positive since positive charge is removed from the source volume, hence the negative sign on the right-hand side of Eq. (A.10). Note that we are now allowing electrostatic fields to change due to the motion of charges and that the electrostatic fields calculated will only be a good approximation to the total fields if the charge motion is sufficiently slow, as discussed in Section A.3. Consider now the case of a negatively charged leader moving downward from a spherically symmetric, negatively charged volume. This situation approximates that existing in the usual dart leader and to some extent in the stepped leader. The upper end of the leader is at H, in the center of the source charge. Hence in Eqs. (A.9) and (A.10), $H_T = H$, and $l = H - H_B$. The total field change due to the extension of the leader and the decrease in the source charge, from Eqs. (A.9) and (A.10) is

$$\begin{aligned}
\Delta E = -\frac{2|\rho_L|}{4\pi\varepsilon_0 D}&\left[\frac{1}{(1 + H_B^2/D^2)^{1/2}} - \frac{1}{(1 + H^2/D^2)^{1/2}}\right.\\
&\left.- \frac{H - H_B}{D}\frac{H}{D}\frac{1}{(1 + H^2/D^2)^{3/2}}\right]
\end{aligned} \qquad (A.11)$$

where $H_B = H$ at $t = 0$, the time of leader initiation. As time increases, H_B decreases until the time the leader touches the ground, $H_B = 0$. The field change for the model leader calculated from Eq. (A.11) is given in Fig. 5.2 for several values of H/D. If the speed v of the leader tip is constant, the variable H_B in Figs. A.3 and 5.2 and in Eq. (A.11) can be replaced by $H - vt$, that is, $l = vt$. Field changes observed close to the leader initially exhibit a negative value; field changes observed at a large distance initially are positive. The field change when the model leader touches the ground, $H_B = 0$, is zero for $H/D = 1.27$, negative for larger values of H/D, and positive for small values of H/D. The field change due to the close model leader exhibits a characteristic hook shape for the same reason that a lowered point charge produces a field maximum (Fig. A.2). Distant leaders for which $H/D \ll 1$ are characterized by field changes that monotonically increase as a function of time, corresponding qualitatively to field changes associated with moving a point charge to lower H in the region to the left of the maximum in Fig. A.2. To see quantitatively that this is so, we expand Eq. (A.11) assuming that $H_B = H - l$ and H are both small compared to D. If all terms of order greater than $(H/D)^2$ are ignored, we find

$$\Delta E \simeq \frac{|\rho_L|l^2}{4\pi\varepsilon_0 D^3} = \frac{|\rho_L|v^2 t^2}{4\pi\varepsilon_0 D^3} \quad \text{V/m} \tag{A.12}$$

Electric field changes from leaders that lower positive charge from positive sources in the cloud can be found by reversing the sign on the ordinate of Fig. 5.2.

Consider now a negatively charged leader moving upward from a negative charge center. A possible example of this type of discharge might be a leader moving upward from the N toward the P region to initiate an intracloud discharge. For this situation $H_B = H$ and $l = H_T - H$ in Eqs. (A.9) and (A.10). The resultant field change is given by

$$\Delta E = -\frac{2|\rho_L|}{4\pi\varepsilon_0 D}\left[\frac{1}{(1 + H^2/D^2)^{1/2}} - \frac{1}{(1 + H_T^2/D^2)^{1/2}}\right.$$
$$\left. - \frac{H_T - H}{D}\frac{H}{D}\frac{1}{(1 + H^2/D^2)^{3/2}}\right] \tag{A.13}$$

where $H_T = H$ at $t = 0$, and H_T increases with time. At close distances the initial field change is positive; far away the initial field change is negative, as shown in Fig. A.4. The field changes shown in Fig. A.4 can be qualitatively predicted from an examination of Fig. A.2. The field change for an upward-moving, positive leader is found by reversing the sign on the ordinate in Fig. A.4.

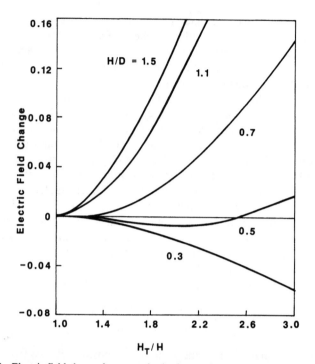

Fig. A.4 Electric field change for a negative leader moving upward from a negative charge center at height H. The length of the leader above the charge center is $H_T - H$. Electric field changes that are negative for small H_T/H become positive for large H_T/H. Multiply field change given in the graph times $\rho_L/2\pi\varepsilon_0 D$ to obtain units of volts per meter.

More complex leader models than the one considered above are discussed, for example, by Huzita and Ogawa (1976) and by Thomson (1985). These studies consider nonvertical channels and nonuniform charge distributions, among other effects.

A.1.4   LEADER TO RETURN-STROKE FIELD CHANGE RATIO

It is instructive, for the simple model discussed in Section A.1.3, to form the ratio of the total field change due to the cloud-to-ground leader process to that due to the return stroke since this ratio has been measured (Section 5.4, Fig. 5.5). We assume there exists a downward-moving, uniformly negatively charged leader whose total field change is given by Eq. (A.11) with $H_B = 0$. We assume that the return stroke removes the negative charge on

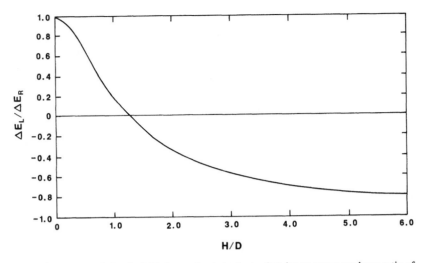

Fig. A.5 Ratio of electric field change due to leader to that due to return stroke vs ratio of height of charge center to horizontal distance from charge center.

the leader and hence causes a positive field change. The field change due to the return stroke is given by the negative of Eq. (A.9) with ρ_L negative, $H_B = 0$, and $H_T = H$. The ratio of these two field changes is plotted in Fig. A.5 as a function of H/D and in Fig. 5.5 as a function of D for $H = 5$ km and $H = 10$ km. Far from the discharge, $(H/D)^2 \ll 1$, the ratio approaches unity, with both the leader and the return-stroke field changes being positive. In this case, the leader field change is approximately the field change caused by moving a point charge equal to the leader charge from height H to the average height of the leader charge, $H/2$, and the return-stroke field change is approximately an equal field change caused by moving the point charge from $H/2$ to ground, both field changes corresponding to the linear portion of the curve near the origin in Fig. A.2. Very close to the discharge the magnitude of the ratio again approaches unity but the sign is negative. In this case, the leader field change is negative due to the proximity of close negative charge on the lower portions of the leader channel, and the return-stroke field change is positive because it removes this close negative charge.

A.1.5   LEADER RADII AND CLOUD CHARGE DIMENSIONS

The charge residing on the dart and stepped leaders and on the charge sources for these leaders must each be contained within a minimum dimension which can be determined by setting the electric field intensity at the surface of a volume containing the known charge to the value that will just cause electrical breakdown. If this charge were to be contained within a smaller dimension, the field at the surface would be in excess of the breakdown value and the breakdown would serve to enlarge the charge volume to the previously determined dimension.

The N region of the thundercloud contains the negative charge associated with both cloud and ground discharges. If we assume the N region charge can be approximated as being spherically symmetrical, the field at the surface of the sphere containing the charge is given by Eq. (A.1). That equation, for the case that $Q = 40$ C, $E = 10^6$ V/m (the breakdown value in dry, atmospheric pressure air, 3×10^6 V/m, reduced to take account crudely of both the reduced pressure at cloud height and the presence of particulate matter), yields a minimum radius R within which the 40 C is contained of 0.6 km. The foregoing is necessarily a rough calculation, but does serve to establish that tens of coulombs of cloud charge cannot be contained in overall dimensions much less than about 1 km. Note that the minimum radius varies, from Eq. (A.1), with the square root of the charge and so is not too sensitive to its exact value. Krehbiel *et al.* (1979) have plotted charge volumes (see Fig. 3.3) assuming a volume charge density of 20 C/km^3. For this assumed charge density, a total charge of 40 C is contained within a sphere of radius about 0.8 km, consistent with the above calculations.

A similar analysis to that given in the previous paragraph, except in cylindrical geometry, can be used to determine minimum leader radii (Uman, 1969). A reasonable value for the charge per unit length along the stepped leader is 10^{-3} C/m (see Section 5.4). For a stepped-leader surface field near ground of 3×10^6 V/m, the minimum radius is 6 m. The minimum radius, in cylindrical coordinates, is linearly related to the charge per unit length, and hence the dart leader, with less charge (Section 8.3) and hence less charge per length, will exhibit a proportionally smaller minimum radius.

A.1.6   CLOUD-TO-GROUND POTENTIAL DIFFERENCE AND ENERGY AVAILABLE FOR LIGHTNING

The potential difference V between two points in an electrostatic field is defined as the line integral of the electric field between the two points and is independent of the path over which that integral is performed (Cheng, 1983). The energy stored when equal and opposite charges, $\pm Q$, reside on two spatially separated conductors is $\frac{1}{2}QV$ J (Cheng, 1983). An order of magnitude approximation to the energy associated with, for example, the negative cloud charge brought to Earth in a flash is QV where Q is the charge involved and V is the potential difference between the Earth and the charge center. Upper and lower limits to the potential difference can be obtained as follows. If the dry-air atmospheric-pressure breakdown electric field of 3×10^6 V/m existed from ground to, say, a height of 5 km, the potential difference would be the product of those two values, 1.5×10^{10} V, an upper limit to the potential difference. Note that from Table 3.1, fields of the magnitude just assumed exceed by an order of magnitude the largest fields measured within a cloud. On the other hand, if a typical value of field observed at ground under a thunderstorm, 10^4 V/m, existed from ground to 5 km, the potential difference would be only 5×10^7 V, a lower limit to the potential difference. The potential difference determined from integration of electric fields associated with cloud charge models such as shown in Fig. 3.2 is in the range 10^8 to 10^9 V. It follows that for a flash charge transfer of the order of tens of coulombs, the total energy dissipated is of the order of 10^9 to 10^{10} J. If the channel is 5 km in length, the flash energy dissipated per meter of channel is 2×10^5 to 2×10^6 J/m. For first return strokes in a flash, values in the lower part of this range are indicated (Section 7.2.4). Additional discussion is found in Uman (1969).

A.2 MAGNETOSTATICS

The motion of charges constitutes an electric current. This current creates a magnetic field that may be measured remotely. We shall consider in this section a simple model to describe the magnetostatic effects due to a spatially uniform but slowly time-varying current. We assume the current flow to be vertical as shown in Fig. A.6.

The magnetic flux density $\mathbf{dB}$ at a distance R from a short element of length dz' carrying current I is (Cheng, 1983)

$$\mathbf{dB} = \frac{\mu_0 I \, dz'}{4\pi R^2} (\mathbf{a_I} \times \mathbf{a_R}) \quad \text{Wb/m}^2 \tag{A.14}$$

where $\mathbf{a_R}$ is a unit vector directed outward along $\mathbf{R}$, $\mathbf{a_I}$ is a unit vector in the direction of the current flow at dz', and μ_0 is the permeability of vacuum (which is essentially equal to the permeability of atmospheric air). In SI units $\mu_0/4\pi \simeq 10^{-7}$ H/m. If, as shown in Fig. A.6, the current is flowing vertically upward above a perfectly conducting plane, then the magnetic flux density is tangent to concentric circles about the wire. On the plane a horizontal distance D from the current the magnetic flux density points into the page, as shown. The magnitude of the magnetic flux density at D due to the

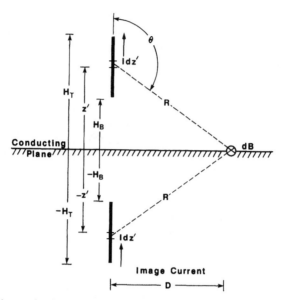

Fig. A.6 Diagram for the calculation of the magnetic flux density at D due to a vertical line of length $H_T - H_B$ carrying current I above a conducting plane.

elemental current $I\,dz'$ is, from Eq. (A.14),

$$dB = \frac{\mu_0 I\,dz'}{4\pi} \frac{D}{(z'^2 + D^2)^{3/2}} \quad \text{Wb/m}^2 \tag{A.15}$$

since

$$|(\mathbf{a}_I \times \mathbf{a}_R)| = \sin\theta = \frac{D}{(z'^2 + D^2)^{1/2}} \tag{A.16}$$

To find the magnitude of the total magnetic flux density on the plane at D we integrate Eq. (A.15) over the length of the current and multiply by 2 to take account of the equal contribution to the flux density from the image current shown in Fig. A.6, with the result that

$$B = \frac{\mu_0 I}{2\pi D}\left[\frac{H_T}{(H_T^2 + D^2)^{1/2}} - \frac{H_B}{(H_B^2 + D^2)^{1/2}}\right] \quad \text{Wb/m}^2 \tag{A.17}$$

If the current flow is between a charge center at height H and the ground ($H_T = H$, $H_B = 0$), Eq. (A.17) becomes

$$B = \frac{\mu_0 I}{2\pi D}\frac{H}{(H^2 + D^2)^{1/2}} \quad \text{Wb/m}^2 \tag{A.18}$$

For the case that an observation is made very close to a discharge, $(H/D)^2 \gg 1$, Eq. (A.18) is approximately

$$B \simeq \mu_0 I / 2\pi D \quad \text{Wb/m}^2 \tag{A.19}$$

the magnetic flux density characteristic of an infinitely long wire. On the other hand, if $(H/D)^2 \ll 1$, that is, far from the discharge, Eq. (A.18) becomes

$$B \simeq \mu_0 I H / 2\pi D^2 \quad \text{Wb/m}^2 \tag{A.20}$$

A.3   TIME-VARYING FIELDS

A.3.1   GENERAL APPROACH

The results obtained in the previous two sections can be applied to lightning charges and currents only for the case that these are relatively slowly varying. Specifically, the significant wavelengths of the electric and magnetic fields generated, roughly found by multiplying the speed of light by the characteristic time at which the sources are changing, must be much larger than the size of the overall system of the lightning and the observer. In this section we derive general expressions that relate the time-varying lightning currents and charges to the electric and magnetic fields they produce, the results of the previous two sections being special cases of these more general results. The following discussion is taken from Uman *et al.* (1975) and Master and Uman (1983).

We begin by writing Maxwell's four equations in free space

$$\nabla \cdot \mathbf{E} = \rho / \varepsilon_0$$

$$\nabla \cdot \mathbf{B} = 0$$

$$\nabla \times \mathbf{E} = -\partial \mathbf{B} / \partial t \tag{A.21}$$

$$\nabla \times \mathbf{B} = \mu_0 \mathbf{J} + \frac{1}{c^2} \frac{\partial \mathbf{E}}{\partial t}$$

where ρ is the volume charge density in C/m^3, J is the volume current density in A/m^2, and c is the speed of light. A general solution to these equations may be obtained for the case of known ρ and J in terms of scalar and vector potentials. For the geometry sketched in Fig. A.7,

$$\mathbf{E}(\mathbf{r}_s, t) = -\nabla \phi - \partial \mathbf{A} / \partial t \tag{A.22}$$

$$\mathbf{B}(\mathbf{r}_s, t) = \nabla \times \mathbf{A} \tag{A.23}$$

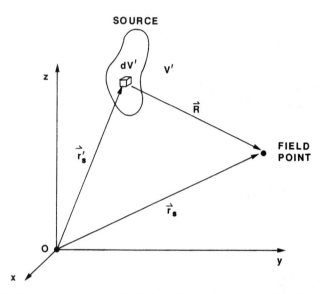

Fig. A.7 The geometry for general solutions of the time-dependent Maxwell equations.

where the scalar potential is

$$\phi(\mathbf{r}_s, t) = \frac{1}{4\pi\varepsilon_0} \int_{V'} \frac{\rho(\mathbf{r}_s', t - R/c)}{R} dV' \qquad (A.24)$$

and the vector potential

$$\mathbf{A}(\mathbf{r}_s, t) = \frac{\mu_0}{4\pi} \int_{V'} \frac{\mathbf{J}(\mathbf{r}_s', t - R/c)}{R} dV' \qquad (A.25)$$

The Lorentz condition

$$\nabla \cdot \mathbf{A} + \frac{1}{c^2} \frac{\partial \phi}{\partial t} = 0 \qquad (A.26)$$

relates the two potential functions. Note again that source locations are identified with a prime, while field locations are unprimed.

Consider an infinitesimal vertical dipole of length dz' having a current $i(z', t)$ located a distance z' above a perfectly conducting plane, as shown in Fig. 7.12. As in the electrostatic case, the plane can be replaced by an image dipole a distance z' below the plane (Stratton, 1941). The electric and magnetic fields at an observation point on or above the plane are the sums of the fields from the real and image dipoles.

A.3.2   CALCULATION OF THE MAGNETIC FLUX DENSITY FROM THE SOURCE DIPOLE

Since the dipole current shown in Fig. 7.12 has only a z-component, the resulting vector potential, from Eq. (A.25), has only a z-component,

$$\mathbf{dA} = \frac{\mu_0}{4\pi} \frac{i(z', t - R/c)}{R} dz' \, \mathbf{a}_z \qquad (A.27)$$

If we adopt a spherical coordinate system with its origin at the dipole and with orthogonal unit vectors $\mathbf{a}_R$, $\mathbf{a}_\theta$, and $\mathbf{a}_\phi$ (see illustration in Fig. 7.12), Eq. (A.27) can be written in that coordinate system as

$$\mathbf{dA}(\mathbf{R}, t) = \frac{\mu_0 \, dz'}{4\pi} \left[i(z', t - R/c) \frac{\cos \theta}{R} \mathbf{a}_R - i(z', t - R/c) \frac{\sin \theta}{R} \mathbf{a}_\theta \right] \quad (A.28)$$

Evaluating the curl of Eq. (A.28), we obtain

$$\nabla \times \mathbf{dA} = \frac{\mu_0 \, dz'}{4\pi} \left[-\frac{\sin \theta}{R} \frac{\partial i(z', t - R/c)}{\partial R} + \frac{\sin \theta}{R^2} i(z', t - R/c) \right] \mathbf{a}_\phi \quad (A.29)$$

By using the identity

$$\frac{\partial i(z', t - R/c)}{\partial R} = -\frac{1}{c} \frac{\partial i(z', t - R/c)}{\partial t} \qquad (A.30)$$

in Eq. (A.29), we obtain the magnetic flux density of the dipole via Eq. (A.23)

$$\mathbf{dB} = \frac{\mu_0 \, dz'}{4\pi} \sin \theta \left[\frac{i(z', t - R/c)}{R^2} + \frac{1}{cR} \frac{\partial i(z', t - R/c)}{\partial t} \right] \mathbf{a}_\phi \quad (A.31)$$

A.3.3   CALCULATION OF THE ELECTRIC FIELD INTENSITY FROM THE SOURCE DIPOLE

The differential electric field due to the source dipole can be determined from Eq. (A.22) using $\mathbf{dA}$ from Eq. (A.28) and the scalar potential $d\phi$ found by substituting Eq. (A.28) in Eq. (A.26)

$$d\phi(\mathbf{R}, t) = \frac{dz' \cos \theta}{4\pi\varepsilon_0} \left[\frac{1}{R^2} \int_0^t i(z', \tau - R/c) \, d\tau + \frac{i(z', t - R/c)}{cR} \right] \quad (A.32)$$

Substituting Eq. (A.28) and (A.32) into Eq. (A. 22) and using Eq. (A.30) to convert the spatial derivatives to time derivatives, we obtain

$$
\begin{aligned}
\mathbf{dE(R,}\ t) = \frac{dz'}{4\pi\varepsilon_0} \Bigg\{ \cos\theta \Bigg[\frac{2}{R^3} \int_0^t i(z', \tau - R/c)\, d\tau \\
+ \frac{2}{cR^2} i(z', t - R/c) \Bigg] \mathbf{a_R} \\
+ \sin\theta \Bigg[\frac{1}{R^3} \int_0^t i(z', \tau - R/c)\, d\tau \\
+ \frac{1}{cR^2} i(z', t - R/c) \\
+ \frac{1}{c^2 R} \frac{\partial i(z', t - R/c)}{\partial t} \Bigg] \mathbf{a_\theta} \Bigg\}
\end{aligned}
\tag{A.33}
$$

Equation (A.33) has been derived in Uman *et al.* (1975). The equivalent expression in cylindrical coordinates has been derived in Master and Uman (1983) and is given in Eq. (7.1).

A.3.4   FIELDS AT THE SURFACE OF A PERFECTLY CONDUCTING EARTH

In order to determine the electric and magnetic fields above a perfectly conducting Earth, the fields from the real dipole above the conducting plane must be added to the fields from the image dipole beneath and this result integrated over all channel segments. Let us consider the special case of the fields at the Earth's surface. Since the distance R from both the source and image dipoles to a point on the plane is the same, the expression for the magnetic field from the image dipole will be the same as Eq. (A.31) except that θ must be replaced by $\pi - \theta$, as is clear from an examination of Fig. 7.12. Since $\sin\theta = \sin(\pi - \theta)$, the effect of the image is to double the magnetic flux density given by Eq. (A.31).

The electric field intensity from the infinitesimal image dipole is given by Eq. (A.33) with θ replaced by $\pi - \theta$, except that the unit vectors $\mathbf{a_R}$ and $\mathbf{a_\theta}$ are in different directions for the real and image dipoles since the coordinate system origins are in different locations. The following expressions relate the unit vectors $\mathbf{a}_R$, $\mathbf{a}_\theta$ and $\mathbf{a}_{Ri}$, $\mathbf{a}_{\theta i}$, the comparable image quantities, to $\mathbf{a}_z$, $\mathbf{a}_H$, unit vectors normal to and parallel to the ground plane, respectively:

$$\mathbf{a}_R = \mathbf{a}_z \cos(\pi - \theta) + \mathbf{a}_H \cos[\theta - (\pi/2)]$$
$$= \mathbf{a}_z \cos\theta + \mathbf{a}_H \sin\theta \tag{A.34}$$

$$\mathbf{a}_\theta = -\mathbf{a}_z \cos[\theta - (\pi/2)] + \mathbf{a}_H \cos\theta$$
$$= -\mathbf{a}_z \sin\theta + \mathbf{a}_H \cos\theta \tag{A.35}$$

$$\mathbf{a}_{Ri} = \mathbf{a}_z \cos(\pi - \theta) + \mathbf{a}_H \cos[\theta - (\pi/2)]$$
$$= -\mathbf{a}_z \cos\theta + \mathbf{a}_H \sin\theta \tag{A.36}$$

$$\mathbf{a}_{\theta i} = -\mathbf{a}_z \cos[\theta - (\pi/2)] + \mathbf{a}_H \cos(\pi - \theta)$$
$$= -\mathbf{a}_z \sin\theta - \mathbf{a}_H \cos\theta \tag{A.37}$$

The result of combining the source and image dipole solutions found in Eq. (A.33) using Eqs. (A.34) through (A.37) for a point on the plane a horizontal distance D from the channel is

$$d\mathbf{E}(D, t) = \frac{dz'}{2\pi\varepsilon_0} \left[\frac{(2 - 3\sin^2\theta)}{R^3} \int_0^t i(z', \tau - R/c)\, d\tau \right.$$
$$+ \frac{(2 - 3\sin^2\theta)}{cR^2} i(z', t - R/c) \tag{A.38}$$
$$\left. - \frac{\sin^2\theta}{c^2 R} \frac{\partial i(z', t - R/c)}{\partial t} \right] \mathbf{a}_z$$

Note that, as expected, the electric field has no component tangential to the plane and that, for small source elevation angles, the total field is positive, that is, it points in the $-\mathbf{a}_z$ direction.

The total magnetic and electric fields at ground level from the complete vertical antenna of height H may be obtained by integrating the infinitesimal dipole solutions. These expressions are given in cylindrical coordinates in Eqs. (7.3) and (7.4).

Note that the first term of Eq. (A.31), the so-called magnetostatic field, is the total magnetic flux density if the current is steady and, in that case, is equivalent to Eq. (A.15); and that the terms in Eq. (A.33) containing the integral over the current (the charge), the so-called electrostatic field, represent the total electric field intensity if the current is zero and are equivalent to the field of a small electric dipole, that is, to the fields of a positive and a negative charge, each given by Eq. (A.1), at either end of a short vertical line. For a general time-varying current, these magnetostatic and electrostatic fields are components of the total magnetic and electric fields, are time varying, and propagate outward at the speed of light, as is evident from the presence of the retarded time in Eqs. (A.31) and (A.33).

A.4   DERIVATION OF CURRENTS FROM FIELDS

A.4.1   GENERAL

The measured values of electric and magnetic fields can be used to determine lightning currents, although assumptions must necessarily be made regarding the spatial variation of the current along the lightning channel. In the following derivation of currents from measured fields, we follow Uman and McLain (1970a), Uman *et al.* (1975), and Lin *et al.* (1980).

Let us assume that each point on the lightning channel is the same distance r (or D) from an observation point on the Earth's surface and that the Earth is a perfectly conducting plane. Thus, we treat the channel as if it were composed of a circular arc above the Earth's surface centered at the observation point and its "image" arc below the Earth's surface. The "arc" approximation to a straight, vertical lightning channel, itself an idealization, is roughly valid for distances from the channel greater than several times the return stroke height. For example, suppose one is solely interested in obtaining values for the maximum rate of rise of current and the peak current associated with the return stroke. These current features occur in the first several microseconds of the return stroke when the maximum return stroke height of interest is less than a few hundred meters (corresponding to an initial return stroke speed of about one-third the speed of light), and hence the theory can reasonably be applied for measurements made at distances beyond about a kilometer. For a complete channel of length H, the "arc" approximation to a straight, vertical channel is valid for $(H/D) \ll 1$.

If the speed of the return stroke wavefront is $v(t)$ and the current behind the wavefront is $i(z'\, t)$, where z' is now the distance measured along the arc, the magnetic flux density **B** at P can be found from Eq. (A.31)

$$B_\phi(r, t) = \frac{\mu_0}{2\pi r^2} \int_0^{l(t')} i(z', t')\, dz' + \frac{\mu_0}{2\pi cr} \int_0^{l(t')} \frac{\partial i(z', t')}{\partial t}\, dz'$$

$$+ \frac{\mu_0}{2\pi cr} i[l(t'), t']v(t') \qquad (A.39)$$

where $l(t')$ is the length of the return stroke channel

$$l(t') = \int_0^{t'} v(\tau)\, d\tau \qquad (A.40)$$

In Eq. (A.39), z' is measured along the channel and has its origin at the Earth's surface, and, for brevity, the retarded time is denoted by t', where $t' = t - r/c$. We have divided the integral of the second term on the right of Eq. (A.31) into the second and third terms on the right of Eq. (A.39) in order to illustrate that radiation is emitted both from along the channel from $z' = 0$ to $l(t')$ and from the wavefront at $l(t)$ if, at the wavefront, the current changes abruptly from a finite value $i[l(t'), t']$ to zero. Thus, the second term on the right of Eq. (A.39) is meant to be integrated only up to the discontinuity in current at the wavefront. The third term on the right side of Eq. (A.39) is the result of integrating across the wavefront current discontinuity and is called the "turn-on" field. The "turn-on" field is nonzero only when the current at the wavefront is nonzero. At this point the time derivative of the current is infinite, but its spatial integral across the wavefront exists (Uman and McLain, 1970b).

We consider in the next two sections simple analytical models that relate return stroke currents to the magnetic fields they produce: (1) the classical model of Bruce and Golde (1941) in which the current is uniform with height at all points below the return stroke wavefront and zero above but may be time varying, and (2) the transmission line model in which a fixed current waveshape propagates up the channel. In the final section of this Appendix we consider a physically realistic return-stroke current model. The simple models discussed in Sections A.4.2 and A.4.3 provide background and perspective for a proper understanding of the more complex model discussed in Section A.4.4 as well as being worthwhile exercises in electromagnetics.

A.4.2   BRUCE–GOLDE MODEL

If we assign $t = 0$ to the time that the current is initiated, Eq. (A.39) can be written

$$B_\phi\left(r, t + \frac{r}{c}\right) = \frac{\mu_0 l(t)}{2\pi c r} \frac{di(t)}{dt} + \frac{\mu_0}{2\pi r}\left[\frac{l(t)}{r} + \frac{v(t)}{c}\right] i(t) \qquad \text{(A.41)}$$

the solution of which is

$$i(t) = \frac{2\pi r c}{\mu_0 l(t)} \int_0^t \exp\left[-\frac{c}{r}(t - \tau)\right] B_\phi\left(\tau + \frac{r}{c}\right) d\tau \qquad \text{(A.42)}$$

where $l(t)$ is given by Eq. (A.40) for $t < t_0$, the time for the return stroke wavefront to reach the end of the channel, and by H, the full channel length, for $t \geq t_0$. For simplicity we have dropped the distance dependence of B in

Eq. (A.42). For the case of a fully formed channel and a slowly varying magnetic field, integration by parts of Eq. (A.42) yields

$$i(t) = \frac{2\pi r^2}{\mu_0 H} B_\phi\left(t + \frac{r}{c}\right)$$ (A.43)

the relation between current and induction field for observation distances several times the channel height, derived previously in Eq. (A.20).

If the observation point P is sufficiently far away from the channel that the induction field, which varies inversely as the square of the distance, is not important, the solution to Eq. (A.41) is

$$i(t) = \frac{2\pi c r}{\mu_0 l(t)} \int_0^t B_\phi\left(\tau + \frac{r}{c}\right) d\tau$$ (A.44)

where $l(t)$ is given by Eq. (A.40) for $t < t_0$ and by H for $t \geq t_0$. Since for radiation fields the magnitude of the electric field intensity is related to the magnitude of the magnetic flux density by $(E_z/B_\phi) = c$, Eq. (A.44) can also be used to compute return stroke currents from measured electric radiation fields. It is interesting to note from Eq. (A.44) that if $l(t)$ increases linearly with time (a constant speed), and if the initial part of the radiation field increases as t^n where n is any integer, the initial current has the same functional form as the initial radiation field.

A.4.3   TRANSMISSION LINE MODEL

We consider now the transmission line model for the return stroke current as described, for example, by Dennis and Pierce (1964) and by Uman and McLain (1969, 1970a). In the transmission line model the current takes the form of a pulse on a lossless transmission line

$$i(z, t) = i[t - (z'/v)]$$ (A.45)

where, to make possible the analytical solution given below, v is assumed constant. In the model of Lin *et al.* (1980) to be discussed in the next section, the current due to electrical breakdown at the return stroke wavefront is described via the transmission line model. This breakdown current pulse is one of three current components comprising the overall model of Lin *et al.* (1980). From Eq. (A.45) it follows that

$$\partial i/\partial t = -v(\partial i/\partial z')$$ (A.46)

When the second term on the right of Eq. (A.39) is integrated using Eq. (A.46), Eq. (A.39), with time referred to the start of the current and $t < t_0$, becomes

$$B_\phi\left(t + \frac{r}{c}\right) = \frac{\mu_0}{2\pi r^2} \int_0^{l(t)} i\left(t - \frac{z'}{v}\right) dz' + \frac{\mu_0 v}{2\pi cr} i(t) \qquad (A.47)$$

Note that the turn-on term and the upper limit of the integral of the second term in Eq. (A.39) have added to zero. If there is no discontinuity in current at the wavefront, each of these terms is individually zero. If the derivative with respect to time of Eq. (A.47) is taken and the resulting integral term evaluated using Eq. (A.46), Eq. (A.47) becomes

$$dB_\phi\left(t + \frac{r}{c}\right)\Big/dt = \frac{\mu_0 v}{2\pi r^2}\left[i(t) + \frac{r}{c}\frac{di(t)}{dt}\right] \qquad (A.48)$$

the solution of which is

$$i(t) = \frac{2\pi cr}{\mu_0 v} \int_0^t \exp\left[-\frac{c}{r}(t - \tau)\right] \cdot \left[dB_\phi\left(\tau + \frac{r}{c}\right)\Big/d\tau\right] d\tau \qquad t < t_0 \quad (A.49)$$

The radiation field or distant solution can be obtained directly from Eq. (A.47) by setting the induction field term, the first term on the right, to zero. The result is

$$i(t) = \frac{2\pi cr}{\mu_0 v} B_\phi\left(t + \frac{r}{c}\right) \qquad t < t_0 \qquad (A.50)$$

As indicated previously Eq. (A.50) can also be used to compute currents from measured electric radiation fields. A different approach to the derivation of Eq. (A.50), which is also Eq. (7.5), is given in Uman *et al.* (1975).

A.4.4 THE MODEL OF LIN *et al.* (1980)

In Section A.4.3 we have examined the theory used to describe one component, the breakdown current pulse, of the return stroke model of Lin *et al.* (1980). This component models the initial current and field rise to peak. In Section 7.3, the total model is discussed. An illustrative drawing of all three current components comprising the model is given in Fig. 7.13. Lin *et al.* (1980) show how to find the peak and other characteristic values of all three current components given electric fields measured simultaneously at close and distant stations. A discussion of the validity of the use of Eq. (A.50)

[Eq. (7.5)] as well as the validity of other aspects of the model of Lin *et al.* (1980) is found in Section 7.3. Currents determined using the model are given in Sections 7.2.2 and 7.3.

REFERENCES

Bruce, C. E. R., and R. N. Golde: The Lightning Discharge. *J. IEEE, London*, **88**(Pt. 2):487–524 (1941).

Cheng, D. K.: "Field and Wave Electromagnetics." Addison-Wesley, Reading, Massachusetts, 1983.

Dennis, A. S., and E. T. Pierce: The Return Stroke of a Lightning Flash to Earth as a Source of VLF Atmospherics. *Radio Sci.*, **68D**:777–794 (1964).

Huzita, A., and T. Ogawa: Electric Field Changes Due to Tilted Streamers in the Cloud Discharge. *J. Meteorol. Soc., Japan*, **54**:289–293 (1976).

Israel, H.: "Atmospheric Electricity," Vols. 1 and 2. U.S. Dept. of Commerce, Clearinghouse for Federal Scientific and Technical Information, TT67-51394/1, 1971. Translation of 2nd Ed. of Atmosphärische Electrizität. Israel Program for Scientific Translation, IPST Catalog No. 1995, Jerusalem, 1971.

Krehbiel, P. R., M. Brook, and R. A. McCrory: An Analysis of the Charge Structure of Lightning Discharges to Ground. *J. Geophys. Res.*, **84**:2432–2456 (1979).

Lin, Y. T., M. A. Uman, and R. B. Standler: Lightning Return Stroke Models. *J. Geophys. Res.*, **85**:1571–1583 (1980).

McLain, D. K., and M. A. Uman: Exact Expression and Moment Approximation for the Electric Field Intensity of the Lightning Return Stroke. *J. Geophys. Res.*, **76**:2101–2105 (1971).

Master, M. J., and M. A. Uman: Transient Electric and Magnetic Fields Associated with Establishing a Finite Electrostatic Dipole. *Am. J. Phys.*, **51**:118–126 (1983).

Stratton, J. A.: "Electromagnetic Theory." McGraw-Hill, New York, 1941.

Thomson, E. M.: A Theoretical Study of Electrostatic Field Waveshapes from Lightning Leaders. *J. Geophys. Res.*, **90**:8125–8131 (1985).

Uman, M. A.: The Earth and Its Atmosphere as a Leaky Spherical Capacitor. *Am. J. Phys.*, **42**:1033–1035 (1974).

Uman, M. A.: "Lightning," pp. 68–73, 211–216. McGraw-Hill, New York, 1969. See also Dover, New York, 1984.

Uman, M. A., and D. K. McLain: Magnetic Field of Lightning Return Stroke. *J. Geophys. Res.*, **74**:6899–6910 (1969).

Uman, M. A., and D. K. McLain: Lightning Return Stroke Current from Magnetic and Radiation Field Measurements. *J. Geophys. Res.*, **75**:5143–5147 (1970a).

Uman, M. A., and D. K. McLain: Radiation Field and Current of the Lightning Stepped Leader. *J. Geophys. Res.*, **75**:1058–1066 (1970b).

Uman, M. A., D. K. McLain, and E. P. Krider: The Electromagnetic Radiation from a Finite Antenna. *Am. J. Phys.*, **43**:33–38 (1975).

Appendix B | Statistics

B.1 PROBABILITY DENSITY FUNCTION, CUMULATIVE PROBABILITY DISTRIBUTION FUNCTION, ARITHMETIC MEAN, STANDARD DEVIATION, MODE, MEDIAN, AND GEOMETRIC MEAN

The *probability density function* $f(x)$ can be defined as follows: $f(x)\,dx$ is the probability that a variable x (e.g., return-stroke peak current, time between strokes) has a value between values x and $x + dx$. It follows that the probability $P(a, b)$ of the parameter occurring between $x = a$ and $x = b$ is

$$P(a, b) = \int_a^b f(x)\,dx \tag{B.1}$$

and that

$$1 = \int_{-\infty}^{+\infty} f(x)\,dx \tag{B.2}$$

An example of the most common density function, the Gaussian probability density function, is plotted in Fig. B.1a.

The *cumulative probability distribution function* $F(x)$ can be defined in terms of the probability density function as follows

$$F(x) = \int_{-\infty}^{x} f(\xi)\,d\xi \tag{B.3}$$

The cumulative distribution is the probability of the variable being smaller than x. The cumulative distribution is shown in Fig. B.1b for the Gaussian density function plotted in Fig. B.1a.

If the probability density function describing a particular variable is known, the *arithmetic mean value* (generally referred to as the mean or

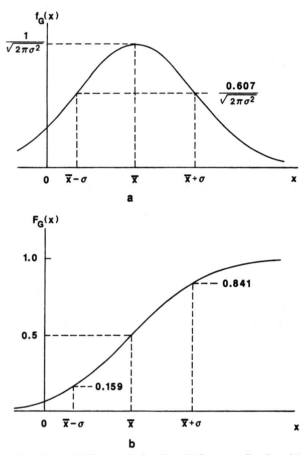

Fig. B.1   (a) Gaussian probability density function. (b) Corresponding Gaussian cumulative probability distribution function.

average value) of the parameter x is

$$\bar{x} = \int_{-\infty}^{+\infty} xf(x)\, dx \tag{B.4}$$

and the *standard deviation* σ, a measure of the spread of values about the mean, is

$$\sigma = \left[\int_{-\infty}^{+\infty} (x - \bar{x})^2 f(x)\, dx \right]^{1/2} \tag{B.5}$$

The square of σ is termed the *variance*.

The *median* of a probability density function $f(x)$ is that value of x for which the cumulative distribution function $F(x)$ is equal to 0.5, that is, half the variable's values are greater than the median and half are less. The *mode* of a probability density function is the value of x that occurs at the peak of the distribution, that is, the value that occurs with the highest probability. A density function can have more than one mode.

If N measurements are made of the parameter x, each termed x_i, the mean value determined from these measurements is

$$\bar{x}_m = \frac{1}{N} \sum_{i=1}^{N} x_i \tag{B.6}$$

and the standard deviation is

$$\sigma_m = \left[\frac{1}{N} \sum_{i=1}^{N} (x_i - \bar{x}_m)^2 \right]^{1/2} \tag{B.7}$$

As the number of measurements becomes very large, if those measurements represent an experimental sample from a theoretical parent distribution $f(x)$, $\bar{x}_m$ from Eq. (B.6) approaches $\bar{x}$ from Eq. (B.4), the mean of the parent distribution, and σ_m from Eq. (B.7) approaches σ from Eq. (B.5), the standard deviation of the parent distribution. If a finite number N measurements of x are made, the best estimate of the standard deviation of the parent distribution, from the sample, is found by replacing N in Eq. (B.7) with $N - 1$. If N measurements are made many separate times, the standard deviation of the $\bar{x}_m$ from the different sets of data is $\sigma/\sqrt{N}$ with σ being the standard deviation of the parent distribution.

The *geometric mean* of a series of N measurements is defined as

$$\bar{x}_g = \sqrt[N]{x_1 \cdot x_2 \cdot x_3 \cdots x_N} = \sqrt[N]{\prod_i x_i} \tag{B.8}$$

A discussion of the relation between the geometric and arithmetic means is given in Section B.3.

The material in this section is considered in more detail in almost all statistics books, good examples being Lipson and Sheth (1973) and Johnson and Leone (1977).

B.2 THE GAUSSIAN OR NORMAL PROBABILITY FUNCTION

The most common probability density function is the Gaussian. The Gaussian distribution is also known as the *normal distribution* and the *normal error function*. When, for example, random experimental errors are described by a Gaussian distribution, as is often the case, those errors are said to be *normally distributed*.

The Gaussian probability density function is plotted in Fig. B.1a and the corresponding cumulative distribution function is plotted in Fig. B.1b. Analytically, the Gaussian probability density function is given by

$$f_G(x) = \frac{1}{\sqrt{2\pi\sigma^2}} e^{-(x-\bar{x})^2/2\sigma^2} \tag{B.9}$$

The function is symmetrical about the mean, $\bar{x}$, with a maximum value $(2\pi\sigma^2)^{-1/2}$ at the mean. Because of this symmetry, the mean value is also the median and the mode. The standard deviation is σ, and at $x = \bar{x} + \sigma$ and $x = \bar{x} - \sigma$ the distribution is at 0.607 of its maximum value. The Gaussian cumulative distribution function is given by

$$F_G(x) = \frac{1}{\sqrt{2\pi\sigma^2}} \int_{-\infty}^{x} e^{-(\xi-\bar{x})^2/2\sigma} \, d\xi \tag{B.10}$$

The integral in Eq. (B.10) must be evaluated numerically, tables being found in almost all books on statistics (e.g., Lipson and Sheth, 1973; Johnson and Leone, 1977). About 68% of the parameters described by the distribution fall within one standard deviation of the mean value, that is, $[F_G(\bar{x} + \sigma) - F_G(\bar{x} - \sigma)] = 0.6826$.

A common method of determining whether a set of measured parameters is normally distributed is to plot those data on special graph paper called cumulative probability paper, derived from Eqs. (B.9) and (B.10), on which the plotted data, if normal, will trace a straight line. An example of this graph paper is shown in Fig. B.2. Note that the horizontal axis represents the percentage of events exceeding a given value of the measured parameter on the vertical axis, and hence the horizontal axis represents $100[1 - F_G]$.

From a theoretical point of view, the Gaussian probability function can be derived by writing mathematically the probability of a given number of successes in a very large number of independent trials if the probability of success in any one trial is a constant. It is the limiting case of the binomial distribution that gives the same probability for a finite number of independent trials. Common examples of the binomial distribution are the probabilities involved in the flipping of a coin many times or the rolling of a die many times. The Gaussian distribution describes well the distribution of random errors in many types of experiments. That this is the case can be explained theoretically if it is assumed that any random error is the result of a large number of elemental errors of equal magnitude but each equally likely to be positive or negative.

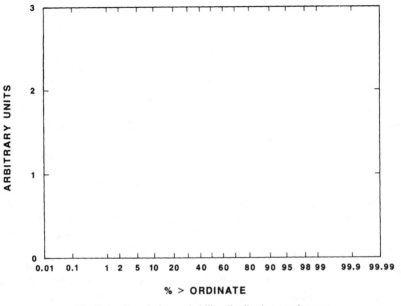

Fig. B.2 Cumulative probability distribution graph paper.

B.3 LOG NORMAL DISTRIBUTION FUNCTION

The log normal probability distribution is of particular importance in lightning research because it provides a good approximation to the distribution of a number of measured lightning properties. Among these are the peak current (see discussion in Section 7.2.2 and Fig. 7.8) and the time interval between strokes (Fig. 1.11). The log normal density function is identical to the normal density function of Eq. (B.9) with x replaced by $\log_{10} x$. The log normal distribution has exactly the shape shown in Fig. B.1a (see also Fig. 1.11b) if $\log_{10} x$ is plotted on the horizontal axis. If, however, x is plotted on the horizontal axis instead of $\log_{10} x$, the shape of the resulting distribution consists of a fast rise to peak at small x followed by an elongated tail at large x (see Fig. 1.11a). There are relatively more very large values of a parameter in a log normal distribution than in a normal distribution.

For any distribution, the arithmetic mean of $\log_{10} x$ [Eq. (B.4)] is related to the geometric mean of x [Eq. (B.8)] by

$$\overline{\log_{10} x} = \log_{10} \overline{x_g} \tag{B.11}$$

or

$$\overline{x_g} = 10^{\overline{\log_{10} x}} \qquad (B.12)$$

For the log normal distribution, the geometric mean, $\overline{x_g}$, is also the median value, as is evident from Eqs. (B.11) or (B.12) and the symmetry of the distribution about the mode and arithmetic mean value $\overline{\log_{10} x}$.

On cumulative probability paper with a logarithmic parameter axis, examples of which are found in Figs. 7.8 and B.3, data distributed log normally trace a straight line. The straight lines in Fig. B.3, in the view of Cianos and Pierce (1972), represent satisfactory approximations to the distribution of many lightning parameters. These approximations do not necessarily agree with measurements discussed elsewhere in this book and are included primarily to illustrate the use of the log normal distribution.

A theoretical example of the occurrence of a log normal distribution in electromagnetics is the distribution of the amplitudes of waves that propagate through many sequential layers of attenuating material where the layer thicknesses and attenuation constants vary randomly and independently from layer to layer (Davenport, 1970). Measured examples of the occurrence of the log normal distribution can be found, for example, in economics (Aitchison and Brown, 1957), in biology (Cramer, 1946), and in meteorology (Biondini, 1976; Lopez, 1976, 1977). In all of these examples, the magnitude of the log normally distributed variable (e.g., wealth, organ size, cloud size) is physically attributed to growth via the so-called law of proportional effects. For example, in the case of the cloud size distribution, one can assume that small convective cloud elements are randomly formed near ground, rise into the atmosphere, and agglomerate. The larger cloud elements, covering a greater area, should intercept a larger number of other elements and hence should grow more rapidly than the smaller ones, with a result that the cloud size distribution is approximately log normal. No such physical argument has yet been advanced to account for the fact that many lightning parameters are log normally distributed.

B.4   OTHER DISTRIBUTIONS

Distributions can be invented having enough parameters so that any set of measured data can be satisfactorily modeled. These distributions do not necessarily have a physical basis. An example of such a distribution is the Weibull probability density function which has three adjustable parameters

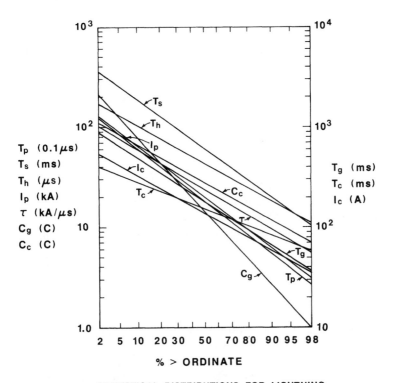

STATISTICAL DISTRIBUTIONS FOR LIGHTNING
PARAMETERS

	OCCURRENCE %				
PARAMETER	2	10	50	90	98
Duration of flash (T_g), msec	850	480	180	68	36
Return stroke interval (T_s), msec	320	170	60	20	11
Return stroke peak current (I_p), kA	140	65	20	6.2	3.1
Charge transfer per flash (C_g), C	200	75	15	2.7	1
Time to r.s. peak current (T_p), μ sec	12	5.8	1.8	0.66	0.25
Rates of r.s. current rise (τ),kA/ μ sec	100	58	22	9.5	5.5
Time to r.s. current half-value (T_h), μ sec	170	100	45	17	10.5
Duration of continuing current (T_c), msec	400	260	160	84	58
Continuing current (I_c), A	520	310	140	60	33
Charge in continuing current (U_c), U	110	64	26	12	7

Fig. B.3 Approximations of the probability distributions of a number of return stroke and ground flash parameters by log normal distributions according to Cianos and Pierce (1972).

(e.g., Weibull, 1951; Lipson and Sheth, 1973; Johnson and Leone, 1977)

$$F(x) = \begin{cases} 0, & x < c \\ \dfrac{b}{a}\left(\dfrac{x-c}{a}\right)^{b-1} \exp\left[-\left(\dfrac{x-c}{a}\right)^b\right], & x > c \end{cases} \tag{B.13}$$

The corresponding Weibull cumulative distribution is

$$F(x) = \begin{cases} 0, & x \leqslant c \\ 1 - \exp\left[-\left(\dfrac{x-c}{a}\right)^b\right], & x > c \end{cases} \tag{B.14}$$

For $b = 1$ the distribution is exponential, for $b = 2$ it is the Raleigh distribution, and for b between 1 and 2, and appropriate a and c, the Weibull distribution provides a reasonable approximation to the log normal distribution.

B.5 χ^2

The χ^2 (chi squared) function provides a quantitative measure of the fit of a given theoretical curve to a set of experimental data. Further, the theoretical curve can be adjusted until the χ^2 is a minimum, indicating a best fit, and the value of χ^2 for that minimum can be used to indicate the adequacy of the fit. For example, to find the magnitude and location of a single cloud charge responsible for the lightning field changes measured at multiple stations on the ground, Jacobson and Krider (1976) and Krehbiel *et al.* (1979) use

$$\chi^2 = \sum_{i=1}^{N} \left[\frac{E_{mi} - E_{ci}(x, y, z, Q)}{\sigma_i} \right]^2 \tag{B.15}$$

where E_{mi} is the field change measured at the ith ground station due to the neutralization by lightning of charge Q at a location in the cloud x, y, z and σ_i^2 is the variance of the measurement at each ground station. The computed or theoretical field change at each station, E_{ci}, is a function of x, y, z, and Q via a model such as that expressed in Eq. (A.3). The χ^2 is minimized (see, for example, Bevington, 1969) to determine the charge magnitude and location. According to Krehbiel *et al.* (1979), the value of χ^2 so obtained should be roughly equal to the number of measurements minus the number of adjustable parameters (here, four adjustable parameters). For electric field changes involving the neutralization of two or more charges at different locations or changes in charge distribution that are not spherically symmetrical, χ^2s with more adjustable parameters (more complex models) must be used (Jacobson and Krider, 1976; Krehbiel *et al.*, 1979).

B.6   LINEAR CORRELATION COEFFICIENT *r*

The linear correlation coefficient *r* is a measure of the goodness of fit of a straight line to two sets of possibly related data (e.g., logarithm of the number of lightning flashes to ground vs storm duration, as shown in Fig. 2.8). The straight line $y = mx + b$ that fits the measured data points (x_i, y_i) is determined by the least-squares method in which the chosen line minimizes the sum of the squares of the *vertical* distances between the line and the data points. If there is no correlation, the sum of the squares will be minimized by a horizontal line with slope $m = 0$. If the data are correlated, the slope will have some value *m*. Now if we use the least-squares method to minimize the sum of the squares of the *horizontal* distances between the line and the data points for the line $x = m'y + b'$, where x_i and y_i are the same two measured parameters being tested for correlation, the slope of the line will be $1/m'$. If, in the ideal case, all of the data points fall on a straight line, that is, there is perfect correlation, then $m = 1/m'$, and $mm' = 1$. If there is no correlation between x_i and y_i, $mm' = 0$. Since the product mm' always falls between 0 and 1, that product is clearly related to the extent to which the variables x_i and y_i are correlated, and thus it is reasonable to define a linear correlation coefficient as $r = \sqrt{mm'}$.

Once *r* is calculated, it is possible to interpret that value of *r* in terms of the likelihood that there is or is not a correlation between x_i and y_i, since, in the presence of measurement error, a given value of *r* could occur by chance. Tables have been published that give the probability of obtaining a given value of *r* for various numbers of pairs of observations in the presence of measurement error (e.g., Johnson and Leone, 1977). For example, a commonly used rule of thumb is to regard a correlation as significant if there is less than 1 chance in 20 that the value of *r* will occur by chance. According to this rule, if there are 10 sets of observations, any value of *r* greater than 0.632 should be regarded as showing a significant correlation. For 5 sets of observations, *r* must be greater than 0.878 to be significant; while for 20 observations, *r* must only be greater than 0.444. If one adopts the more stringent criterion that there must be less than 1 chance in 1000 that the value of *r* will occur by chance, for 5 sets of observation *r* must exceed 0.992 for a significant correlation, for 10 sets, 0.873, and for 20 sets, 0.679.

REFERENCES

Aitchison, J., and J. A. C. Brown: "The Lognormal Distribution." Cambridge Univ. Press, London and New York, 1957.

Bevington, P. R.: "Data Reduction and Error Analysis for the Physical Sciences." McGraw-Hill, New York, 1969.

Biondini, R.: Cloud Motion and Rainfall Statistics. *J. Appl. Meteorol.*, **15**:205-224 (1976).

Cianos, N., and E. T. Pierce: A Ground-Lightning Environment for Engineering Usage. Stanford Research Institute Project 1834, Technical Report 1, Stanford Research Institute, Menlo Park, California 94025, August, 1972.

Cramer, H.: "Mathematical Methods of Statistics." Princeton Univ. Press, Princeton, New Jersey, 1946.

Davenport, W. B., Jr.: "Probability and Random Processes." McGraw-Hill, New York, 1970.

Jacobson, E. A., and E. P. Krider: Electrostatic Field Changes Produced by Florida Lightning. *J. Atmos. Sci.*, **33**:103-117 (1976).

Johnson, N. L., and F. C. Leone: "Statistics and Experimental Design in Engineering and the Physical Sciences," 2nd Ed. Wiley, New York, 1977.

Krehbiel, P. R., M. Brook, and R. A. McCrory: An Analysis of the Charge Structure of Lightning Discharges to Ground. *J. Geophys. Res.*, **84**:2432-2456 (1979).

Lipson, C., and N. J. Sheth: "Statistical Design and Analysis of Engineering Experiments." McGraw-Hill, New York, 1973.

Lopez, R. E.: Radar Characteristics of the Cloud Populations of Tropical Disturbances in the Northwest Atlantic. *Mon. Weather Rev.*, **104**:269-283 (1976).

Lopez, R. E.: The Lognormal Distribution and Cumulus Cloud Population. *Mon. Weather Rev.*, **105**:865-872 (1977).

Weibull, W.: A Statistical Distribution Function of Wide Applicability. *J. Appl. Mech.*, **18**:293-297 (1951).

Appendix C | Experimental Techniques

C.1 ELECTRIC FIELD MEASUREMENTS

Both cloud-to-ground and cloud discharges have a total time duration that can extend to a second or more, while the individual physical processes comprising these discharges can vary on a submicrosecond time scale. It follows that variations of current and charge associated with a lightning flash produce "wideband" electric and magnetic fields with significant frequency content in a frequency range from below 0.1 Hz to well over 10 MHz. With "narrow-band" systems, those systems that accept inputs from a narrow frequency range about a central frequency, relatively small signals associated with various lightning processes can be detected into the thousands of megahertz range (e.g., Pierce, 1977). The overall wideband electric field change, the "electrostatic" field change, due to a lightning flash at 100 km is of the order 1 V/m and for a flash at 5 km is of the order of 10^4 V/m, the electrostatic field varying roughly as D^{-3}, where D is distance to the flash, over this range of distances (Sections 7.3 and A.3). Initial radiation field peaks from return strokes in ground discharges have zero-to-peak risetimes in the microsecond range and fast transitions in the tenth of microsecond range with typical peak values of 5 V/m at 100 km (Tables 7.1 and 11.2). Radiation field peaks due to stepped leaders (Section 5.2) and K-changes (Sections 10.4 and 13.5) are typically an order of magnitude smaller at the same distance. Radiation fields vary as D^{-1} if propagation effects are ignored (Sections 7.3 and A.3). At a given distance, the lightning electric fields usually can be expected to fall in a range about five times greater or less than the typical values, due to a similar natural variation in the lightning current and charge (Tables 7.2 and 7.3).

In order to measure the fields discussed above, two different types of field sensors have been developed.

The first is an electrostatic fluxmeter or field mill (Malan and Schonland, 1950) in which a grounded and segmented top plate rotates so as to cover and uncover a fixed, similarly segmented field-detecting plate beneath it, as illustrated in Fig. C.1 where metal studs comprise the segmented field-detecting plate. The field mill operates by sensing the charge induced on the fixed plate by the ambient electric field. The boundary condition on the component of the electric flux density $\mathbf{D} = \varepsilon_0 \mathbf{E}$ normal to the plate requires that $\varepsilon_0 E_n = Q/A$ where E_n is the normal electric field intensity, Q is the induced charge, and A is the plate area, assuming that the charge is uniformly distributed over the area. That charge flows to ground through a resistor when the fixed plate is covered by the rotating grounded top plate and flows back to the fixed plate when the plate is uncovered. Thus the rotating top plate converts a relatively slowly time-varying electric field into an ac signal, a voltage across the resistor, whose amplitude is proportional to the ambient electric field. A field mill can sense a dc or relatively steady field such as exists in fine weather or beneath clouds and can sense lightning field changes with an upper frequency response in the 1- to 10-kHz range. The response time is determined by the product of the rotation speed of the top plate and the

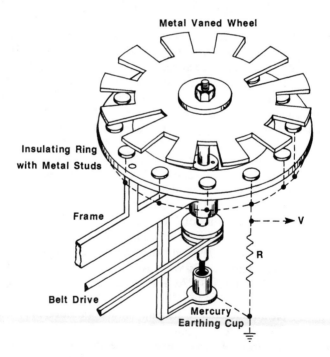

Metal Vaned Wheel

Insulating Ring with Metal Studs

Frame

Belt Drive

Mercury Earthing Cup

V

R

Fig. C.1 Drawing of a field mill. Charge induced by the ambient electric field on the metal studs is periodically shielded by the grounded, segmented plate producing a current flow through and hence a voltage across the resistor, that voltage being proportional to the ambient field. Adapted from Malan (1963).

number of segments into which the two plates are divided, i.e., the frequency at which an individual sensing element is shielded and unshielded.

An extensive network of field mills is operated by the NASA Kennedy Space Center in Florida. The individual mills as well as the network are described by Jacobson and Krider (1976). The stator and rotor of the mills are composed of 8 sectors, and the grounded rotor turns at 1800 rpm. The associated electronics is designed to limit the system response to about 0.25 sec, sufficient only to measure the overall flash field change.

The second type of sensor is a flat plate or other metallic surface such as a sphere or vertical wire (whip) on which the electric field can terminate. Whip antennas are generally not used to measure the fields of close lightning because the antenna can enhance the already large thunderstorm and lightning fields to a degree that corona discharge occurs from the antenna. An elevated flat plate antenna is illustrated in Fig. C.2 and a flat plate antenna with surrounding grounded structure suitable for flush mounting in the earth is illustrated in Fig. C.3. The charge $Q(t)$ induced on these antennas by the electric field is sensed by an electronic circuit. In the circuits shown in Figs. C.2, C.3, and C.4 either a capacitor to ground or an electronic

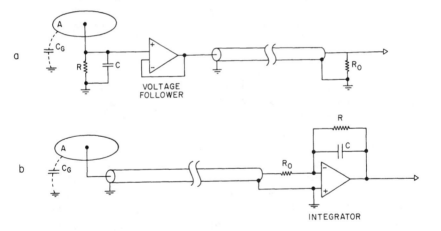

Fig. C.1   Diagram of two electric field antenna systems. The electronics integrates the current to the plate antenna providing an output voltage proportional to the charge on the plate and hence the ambient electric field. In a the integration is performed by the capacitor to ground at the base of the antenna; in b by the electronic integrator. C_G is the capacitance between the antenna plate and the ground. R_0 serves to terminate the coaxial cable in its characteristic impedance so there are no reflections of the signal back to the antenna. The relatively large resistor R serves to discharge the integrating capacitor C so that the output voltage decays toward zero with a time constant RC. Adapted from Krider et al. (1975).

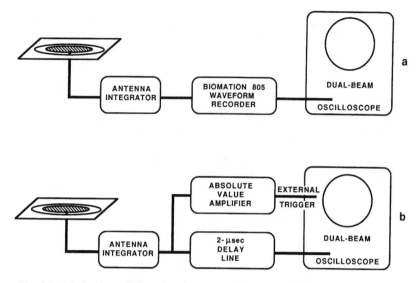

Fig. C.3 Block schematic diagrams of (a) a waveform recording system using a device that continuously digitizes and stores voltages proportional to the incoming field so that a programmable time before and after a given trigger signal can be displayed on the oscilloscope and (b) a direct waveform photography system employing a delay line so that 2 μsec of the signal prior to the amplitude of signal that provided the trigger can be displayed on the oscilloscope. Adapted from Krider *et al.* (1975).

integrator is used to integrate the current, dQ/dt, flowing to the antenna plate. Since $E_n = Q/\varepsilon_0 A$ from the boundary condition on the component of the electric flux density normal to the plate and since the circuit voltage after integration is $V = Q/C$, it follows that $V = (\varepsilon_0 A/C)E_n$, where E_n is the actual field normal to the plate and not the field that would be present in the absence of the plate unless the plate is flush mounted in the ground, the only antenna configuration that does not increase the value of the ambient field. The relatively large resistor R in parallel with the integrating capacitor C determines the decay time constant, RC, of the output signal. It should be made larger than any time variation of interest. For example, RC should be milliseconds if return stroke fields are being measured and seconds if total flash fields are being studied. A drawing of a resistive feedback network that can provide a very large effective R using resistors normally found in the laboratory, thus making possible decay times of the order of 10 sec, is shown in Fig. C.4. Note that if the integrating capacitors in Figs. C.2, C.3, and C.4 are replaced by resistors, the current through the resistors is $\varepsilon_0 A(dE_n/dt)$, and hence the voltage across the resistors is proportional to the derivative of the field.

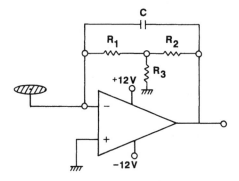

Fig. C.4 An electronic integrator having a resistive feedback network of effective resistance $R = (R_1 R_2 + R_1 R_3 + R_2 R_3)/R_3$. If $R_3 \ll R_1$, R_2, $R \cong R_1 R_2/R_3$, and a decay time constant $RC = 10\,\text{sec}$ can be obtained with, for example, $R_1 = R_2 = 10^6\,\Omega$, $R_3 = 100\,\Omega$, yielding $R \cong 10^{10}\,\Omega$, and $C = 10^{-9}\,\text{F}$.

Electric field measuring systems employing an elevated flat plate antenna with the electronics similar to that of Fig. C.2a have been described by Kitagawa and Brook (1960) and by Fisher and Uman (1972). The upper frequency cutoff in both of these systems was above 1 MHz. An electric field system using electronics similar to those in Fig. C.2b with a lower cutoff frequency near 0.1 Hz is described by Krehbiel *et al.* (1979). Krider *et al.* (1977) used a flat plate antenna surrounded by a grounded guard ring that was either mounted flush in the ground or on the roof of a grounded research vehicle. The sensor and electronics were used with two different types of recording systems to display the electric field waveforms, as illustrated in Fig. C.3. Baum *et al.* (1982) describe electric field sensors that serve the same purpose as a plate, sphere, or whip but are of more sophisticated geometry so as to extend the upper frequency response of the system to 100 MHz and above, frequencies at which the wavelengths being detected become of comparable size to the sensor.

The electric field measuring systems referenced above are designed to measure the vertical component of the electric field at ground. Spherical antennas designed to measure simultaneously the three perpendicular components of the electric field have been used by Weber *et al.* (1982) and Thomson *et al.* (1985). In the system used by Thomson *et al.* (1985) a metallic sphere having three small antenna patches cut out and isolated from the sphere on three perpendicular axes was suspended about 5 sphere diameters above the ground. The isolated metal sphere enhances each of the components of the ambient field by a factor of 3. The measuring electronics were within the sphere, and signals were sent from the sphere by fiber optic links.

The previous discussion has referred to wideband systems, those capable of having a flat frequency response from near dc to over 10 MHz. Narrow-band systems, those which detect only a narrow range of frequencies around a central frequency, by virtue of electronic filters, have been used to study lightning by LeVine and Krider (1977), Krider *et al.* (1979), Brook and Kitagawa (1964), and others (see review by Pierce, 1977). A system that has been used to measure simultaneously both narrow-band and wideband signals is shown in Fig. C.5.

Calibration of an electric field measuring system for field magnitude and for bandwidth or risetime is obviously crucial. Flush-mounted flat plate antennas or spherical antennas with isolated cutouts can be calibrated theoretically for field magnitude knowing the antenna area and electronic circuit parameters since the field enhancement factors are known. For the flush-mounted plate the ambient field is unenhanced, and for the sphere it is enhanced a factor of three from the ambient value. Antennas can be experimentally calibrated (1) by placing them in a large parallel plate capacitor across which is applied a known voltage and (2) by applying a known charge to the antenna via a series combination of an external voltage source and a known capacitor. If elevated antennas are calibrated in the latter fashion the field enhancement factor due to the antenna geometry must be calculated or determined by comparing the antenna output with a flush-mounted reference antenna. Fisher and Uman (1972) found experimentally that the parallel plate capacitor having capacitor plates 5 by 5 m in which they calibrated a 93-cm-high, flat plate antenna of 43 cm diameter had to be

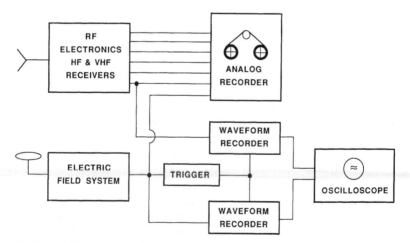

Fig. C.5 A block diagram of a data recording system in which narrow-band and wideband signals are recorded simultaneously. Adapted from LeVine and Krider (1977).

3–5 m in vertical gap spacing in order that the capacitor properly simulated a uniform field.

Care must be taken if horizontal coaxial cables are employed between the antenna and the measuring electronics since the horizontal component of the electric field present due to the finite ground conductivity (Section 7.3) can induce unwanted voltages in these cables. The general symptom of this effect is the measurement of sharper peaks and valleys than actually exist in the vertical field, as if a differentiated signal were being added to the field. This is the case because the horizontal electric field is similar to the derivative of the vertical field (see Section 7.3). The problem can be avoided by using a optical fiber system for horizontal signal transmission.

Perhaps the best confirmation that an electric field system is working properly is obtained by comparing individual electric radiation fields from distant lightning with the corresponding magnetic radiation fields recorded with loop antennas and associated electronics as discussed in the next section (see, for example, Krider and Noggle, 1975). These distant electric and magnetic radiation fields should have identical waveshapes (Section A.3).

C.2   MAGNETIC FIELD MEASUREMENTS

The simplest sensor for a magnetic field measurement is an open-circuited loop of wire. The voltage induced at the open-circuit terminals of such a loop is equal to the loop area multiplied by the derivative of the magnetic flux density perpendicular to the loop surface. Thus the signal in a vertical loop antenna is proportional to the cosine of the angle between the lightning direction as viewed from the antenna and the plane of the loop antenna. It follows that two perpendicular loop antennas can be used for magnetic direction finding, as discussed in Section C.7.2. A drawing of a magnetic field antenna and associated electronics is shown in Fig. C.6. Since the signal out of a loop antenna is proportional to the derivative of the magnetic flux density, that signal must be integrated to obtain a signal that represents the field. The loop antenna shown in Fig. C.6 is operated differentially and is shielded from electric fields with a grounded shield symmetrically broken at the top of the antenna so that no circulating currents can flow in the shield. Although many characteristics of the lightning magnetic field were studied by Norinder and co-workers in Sweden (e.g., Norinder and Dahle, 1945; Norinder, 1956), Krider and Noggle (1975) and Krider *et al.* (1976) developed the first wideband magnetic field measuring system capable of resolving the initial radiation field peaks. The magnetic field systems devised by the University of Arizona group can accurately measure risetimes considerably faster than 0.1 μsec.

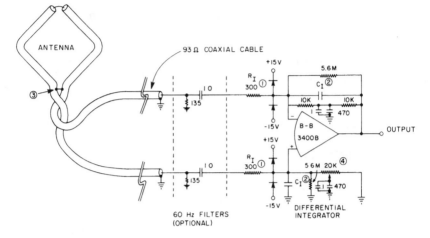

Fig. C.6 A magnetic field antenna formed from a single loop of 93-Ω coaxial cable and associated electronic differential integrator used to obtain an output voltage proportional to the field. (1) 1% noninductive resistors; (2) low-loss 1% capacitors: 100 to 10,000 pF; (3) outside coaxial shields connected here; (4) adjust for optimum common mode rejection. All diodes IN4447; all capacitor values in microfarads. Adapted from Krider and Noggle (1975).

Baum *et al.* (1982) describe several sophisticated geometries for looplike antennas that make possible the extension of the upper frequency response above that of the conventional loop antenna, which are limited to the detection of wavelengths larger than the antenna size.

Besides magnetic field loops, and more sophisticated antennas that operate on the same principle, several other types of sensors have been used to detect lightning magnetic fields. Williams and Brook (1963) have described magnetic field measurements made with a fluxgate magnetometer (Marshall, 1967). For relative low-frequency measurements, ballistic magnetometers that basically operate like compass needles have been used (Meese and Evans, 1962; Nelson, 1968; Pierce, 1968). Hall effect or other solid-state magnetometers could be used to measure lightning magnetic fields with very fast time response, but apparently have not been used for this purpose to date.

C.3   PHOTOELECTRIC MEASUREMENTS

A variety of photoelectric sensors have been used to observe lightning from Earth-orbiting satellites (Sparrow and Ney, 1968; Vorphal *et al.*, 1970; Turman, 1977, 1978, 1979; Edgar, 1978, 1982; Orville and Spencer, 1979; Turman and Tettelbach, 1980; Orville, 1981; Turman and Edgar, 1982;

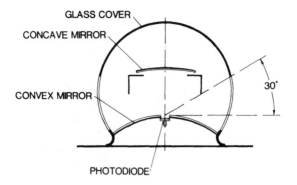

Fig. C.7 Sketch of a photoelectric detector which views 360° in azimuth and 30° in elevation. Adapted from Guo and Krider (1982).

see also Section 2.4), from aircraft (Vonnegut and Passarelli, 1978; Brook *et al.*, 1980), and from ground-based stations (Kitagawa and Kobayashi, 1959; Brook and Kitagawa, 1960; Clegg, 1971; Mackerras, 1973; Kidder, 1973; Griffiths and Vonnegut, 1975; Thomson, 1978; Hubert and Mouget, 1981; Guo and Krider, 1982, 1983, 1985; Ganesh *et al.*, 1984; Beasley *et al.*, 1983a,b).

With the exception of Hubert and Mouget (1981), Ganesh *et al.* (1984), and Beasley *et al.* (1983a,b) all of the photoelectric detectors have been essentially all-sky devices, that is, they view essentially a hemisphere or a significant portion of a hemisphere. The three papers referenced above describe systems that view a relatively narrow vertical angle in order to detect the light from only a limited section of lightning channel. Figure C.7 shows an essentially all-sky photoelectric detector system that detects light from all azimuths and up to 30° in elevation. This system was used by Guo and Krider (1982, 1983, 1985) to study return-stroke and dart-leader optical signals (Section 7.2.4). The system comprised a 1-cm^2 silicon photodiode (EG&G type SGD-444) that viewed the all-sky mirror system shown. The photodiode response and total system response are given by Guo and Krider (1982). The photodiode is sensitive to the visible and near infrared.

C.4   BOYS AND STREAK-CAMERA MEASUREMENTS

A streak camera is a device in which there is a relative, continuous motion between the lens and the film. A two-lens streak camera, called a Boys camera after its inventor (Boys, 1926), that was used by McEachron (1939) is illustrated in Fig. C.8. Earlier versions of the Boys camera are described in Schonland and Collens (1934) and Schonland *et al.* (1935). Luminosity

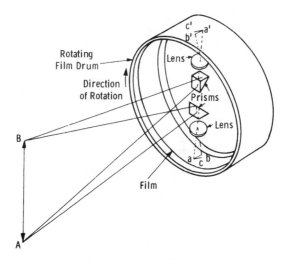

Fig. C.8   Diagram of a Boys camera with moving film and stationary optics. Luminosity progressing from A to B leaves the image ab (a'b') when the drum is stationary and ac (a'c') when the drum is rotating. Adapted from McEachron (1939).

moving, for example, vertically upward, such as in a return stroke, is streaked in different directions through each of the two lenses, thus making possible a determination of return stroke speed by comparison of the two streak images and knowledge of the speed of motion of the film. Individual stepped-leader steps and individual return strokes can be clearly separated on such film as illustrated by Figs. 1.6, 1.7, and 1.8, and the time between individual events measured. A modern version of the Boys camera, employing a modified version of the Beckman and Whitley model 318 streak camera, has been described by Idone and Orville (1982), Jordan and Uman (1983), and Idone *et al.* (1984). The maximum time resolution of the streak or Boys cameras used to date is of the order of 1 μsec. A discussion of the time resolution obtained in measuring a variety of ground flash processes is given by Malan and Collens (1937). Electronic versions of streak camera, called image converters or image amplifiers, have been used to study long laboratory sparks (e.g., Uman *et al.*, 1968). They have the advantage of faster time resolution than cameras employing film but the disadvantage of poorer spatial resolution.

C.5   SPECTROMETERS

A schematic drawing of the spectrometer used by Orville (1968) to obtain time-resolved spectra of 10-m sections of return stroke channels is shown

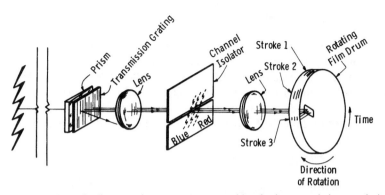

Fig. C.9 Schematic diagram of a spectrometer capable of microsecond time resolution of the spectrum of a 10-m or smaller section of the lightning channel. Adapted from Orville (1968).

in Fig. C.9. These spectra were obtained by passing the light from the return stroke through a prism and transmission grating and blocking all but a portion of the spectrum from a short vertical section of channel. The spectrum was then recorded on a streak camera, a Beckman and Whitley model 318. The time resolution obtained by Orville (1968) was 2–5 μsec and the wavelength resolution about 10 Å (Section 7.2.4; Fig. 7.10).

Instead of using film to record the spectral data, various photoelectric sensors can be employed. Orville and Henderson (1984) describe such a system which time-integrates the light from a stroke but can, under computer control, measure the background light before and after the stroke and subtract it from the total as well as provide absolute spectral intensities (Section 7.2.4). Reviews of lightning spectroscopic techniques are found in Uman (1969) and in Orville (1977).

C.6   THUNDER MEASUREMENTS

Thunder is measured using microphones whose output signals are recorded on tape recorders for later analysis. Bohannon *et al.* (1977) used Globe 100-D capacitor microphones to record the infrasonic component of thunder with a 3-db frequency response from 0.1 to 450 Hz. Balachandran (1983) used similar low-frequency condensor microphones with a frequency response from 0.1 to 300 Hz. Holmes *et al.* (1971a) used modified Brevil and Kjaer type 4140 capacitor microphones having a flat frequency response from 0.3 Hz to 20 kHz to study both audible and infrasonic thunder.

C.7   LIGHTNING LOCATION TECHNIQUES

A variety of techniques can be used to locate lightning. Some of these techniques can also delineate the details of channel formation and propagation. The location techniques can be conveniently divided into six categories based on the principles of operation: (1) electric field amplitude, (2) magnetic field direction, (3) radiation field time-of-arrival and interferometry, (4) radar, (5) visible light direction, and (6) thunder time-of-arrival.

C.7.1   ELECTRIC FIELD AMPLITUDE

Location techniques using electric field amplitude can be divided into two general types: (1) single station and (2) multiple station. (1) Single-station devices are generally called flash counters and are mainly useful in providing an estimate of the number of lightning events within a range of some tens of kilometers over a period of months or years. Location of individual events or short-term flash density measurements by single-station devices is, in general, quite inaccurate due to the large variation in the amplitude of the electric field from lightning to lightning at a given range. The applications and limitations of flash counters are considered in Section 2.2. A single-station device that apparently produces relatively accurate storm locations by averaging, for individual azimuthal octants, the initial radiation field peaks from many ground strokes and comparing this average with the averages found in previous research for storms at known ranges (Table 7.1) is described in the weekly magazine *Aviation Week and Space Technology*, Vol. 123, No. 14, pp. 83–85, 1985. (2) The use of multiple-station electric field measurements to determine the location of the lightning-caused changes in the cloud charge distribution has been discussed by Jacobson and Krider (1976) and by Krehbiel et al. (1979). Jacobson and Krider (1976) have presented a least-squares optimization method for fitting parameters of assumed models of cloud charges to electric field changes measured at multiple stations (Section B.5). They have illustrated the method using flash field change data from the Kennedy Space Center field mill network. This method has been compared to an analytical inversion technique by Krehbiel et al. (1979) who found that the least-squares technique was best for finding charge magnitude, location, and movement for various individual lightning processes (e g , Section 3.2).

C.7.2   MAGNETIC FIELD DIRECTION

Vertical and orthogonal magnetic field loops can be used to obtain lightning direction because the ratio of the signals in the two detectors is proportional to the tangent of the angle to the source. Crossed-loop magnetic direction finders (DFs) can be divided into two general types: (1) narrow band and (2) wideband. (1) Narrow-band DFs have been used to detect distant lightning since the 1920s (Horner, 1954, 1957). They generally operate in a narrow frequency band near 5 kHz where attenuation in the earth-ionosphere waveguide is minimum and where the lightning signal is maximum. Two or more such DFs operated in concert can locate lightning activity at distances of the order of 1000 km, but no discrimination is provided between cloud and ground discharges. One version of the tuned DF provides range information from a single-station measurements by measuring the difference in the time-of-arrival of signals at 6 and 8 kHz (Heydt and Volland, 1964; Harth et al., 1978). At ranges less than about 200 km, tuned DFs exhibit inherent azimuth errors of the order of 10° (Nishino et al., 1973; Kidder, 1973). These so-called polarization errors are due to the magnetic fields from nonvertical channel sections, ionospheric reflections, and other effects (Yamamhshita et al., 1974a,b; Uman et al., 1980). (2) Wideband direction finders have been used to obtain an angular accuracy of the order of 1° for lightning at ranges from a few kilometers to a few hundred kilometers (Krider et al., 1976). In this type of system, direction finding is accomplished using the NS and the EW components of the initial peak of the return-stroke magnetic field which is radiated early in the return stroke and hence is from the roughly vertical channel section near ground. Thus the polarization errors associated with tuned DFs, which view the magnetic field radiated from the whole channel including branches and other horizontal channel sections, are minimized, and the DF locates the ground strike point. Further, Krider et al. (1980) have designed a DF of this type that responds only to ground flashes. A network of these wideband DFs presently monitors about 40% of the land area of the United States for the purpose of forest fire detection (Krider et al., 1980; Reap, 1986) and another network covers the eastern part of the United States for meteorological studies (Orville et al., 1983, 1986). An evaluation of the operating charac-teristics of a ground-flash locating system composed of four wideband DFs is given by Mach et al. (1986).

C.7.3 RADIATION FIELD TIME-OF-ARRIVAL AND INTERFEROMETRY

Radiation field time-of-arrival and interferometric techniques can be divided into three general types: (1) short baseline time-of-arrival, (2) long baseline time-of-arrival, and (3) interferometry. All systems to be discussed operate at VHF, that is, at frequencies from 30 to about 300 MHz, except the VLF long baseline time-of-arrival system of Lee (1986) and the VLF short baseline time-of-arrival system of Lewis et al. (1960).

1. Oetzel and Pierce (1969) first suggested that a short baseline time-of-arrival technique could be used for locating lightning VHF sources. A criticism of short baseline systems and a discussion of the advantages and disadvantages of long and short baseline systems are given by Proctor (1973). A reply to Proctor (1973) is given by Cianos *et al.* (1973). Basically, with a short baseline system, pulse identification is no problem since the same pulse arrives at each of the closely spaced receivers in a time that is short compared to the time between pulses, and thus sequences of pulses arrive at each receiver in the same order. On the other hand, a short baseline system produces at best two-dimensional direction: with two receivers one can obtain azimuth; with three receivers one can obtain both azimuth and elevation. Cianos *et al.* (1972), Murty and MacClement (1973), and MacClement and Murty (1978) have tested two-receiver systems. Taylor (1978) used three closely spaced receivers to determine elevation and bearing at each of two sites separated by about 10 km, but his system did not have the capability of matching individual pulses over the 10-km path. The original short baseline system is due to Lewis *et al.* (1960). It consisted of four VLF (4–45 kHz) stations in New England, each separated by over 100 km, and was used to determine the direction to transatlantic lightning.

2. Proctor (1971, 1976, 1981a) in South Africa pioneered the long baseline time-of-arrival technique using five ground stations each separated by between 10 and 30 km. The resultant lightning channel reconstructions have provided much valuable information on the development of lightning channels in the cloud (e.g., Section 13.4). The system operated at 253 MHz with a 5-MHz bandwidth. Spatial resolution in these studies was of the order of 100 m. Recorded VHF envelope pulse widths ranged from about $0.2 \, \mu sec$, the system limit, to about $2 \, \mu sec$. On the average, a VHF location was obtained each $70 \, \mu sec$. Rustan (1979) and Rustan *et al.* (1980) have used a similar long baseline system developed by Lennon (1975) at the Kennedy Space Center, Florida. The system operated in the 50-MHz range with about a 5-MHz bandwidth providing a spatial resolution of about 250 m and a location each 10–$100 \, \mu sec$. Long baseline VHF time-of-arrival systems have the advantage of providing three-dimensional locations, but suffer the disadvantage of difficulty in identifying VHF pulses from two or more simultaneous separated sources because the pulses can arrive in a different order at each receiver. Lee (1986) has discussed the modification of the British Meterological Office's narrow-band magnetic direction-finding network to a long baseline VLF time-of-arrival system for identifying the location of flashes. Seven receiving stations operating at 9 kHz with a 250-Hz bandwidth are located over the United Kingdom and the Mediterranean. With this time-of-arrival system, lightning groupings characteristic of the size of active thunderstorm cells were frequently observed at ranges of the order of a thousand kilometers.

3. VHF interferometry is a technique by which changes in phase between narrow-band signals received at each of two closely spaced receivers are compared electronically to determine the direction to the source. Hayenga (1979, 1984), Warwick *et al.* (1979), Hayenga and Warwick (1981), and Richard and Auffray (1985) have used interferometers to study lightning VHF sources. The interferometer of Hayenga and Warwick (1981) had two pairs of receivers on baselines perpendicular to one another, operated at a frequency of 34 MHz with a bandwidth of 3.4 MHz, and provided an average two-dimensional location each 2.5 μsec. Richard and Auffrey (1985) describe in great detail an interferometer that operated at 300 MHz with a 10-MHz bandwidth and consisted of a small system (distance between the two quarter-wavelength vertical monopole antennas about equal to a wavelength) nestled within a larger system. The time resolution was 1.6 μsec.

C.7.4   RADAR

Observations of transient radar returns from lightning have been reported by Ligda (1950, 1956), Browne (1951), Marshall (1953), Miles (1953), Pawsey (1957), Hewitt (1957), Atlas (1958), Cerni (1976), Holmes *et al.* (1980), Szymanski and Rust (1979), Szymanski *et al.* (1980), Proctor (1981b), Zrnic *et al.* (1982), Mazur and Rust (1983), Mazur *et al.* (1984a,b, 1985, 1986a,b), and Mazur (1986). Dawson (1972) has applied the scattering theory originally developed for meteor trails to cloud-to-ground return stroke channels. Additional theory is presented by Mazur and Walker (1982) and Mazur and Doviak (1983).

The most thorough work to date on the location of lightning and the measurement of its properties via radar is due to Holmes *et al.* (1980) in New Mexico and Proctor (1981b) in South Africa. Holmes *et al.* (1980) used a 10.9-cm radar with a 1-msec pulse-repetition period to detect echoes from 156 lightning flashes. The flashes were located at altitudes ranging from 5 to 14 km above sea level with extents along the fixed radar beam of from less than 300 m to over 2 km. Most echoes rose to peak intensity in less than the 1-msec resolution of the radar and had a duration between 10 and 600 msec.

Proctor (1981b) observed radar echoes from lightning at 5.5, 50, and 111 cm. From an analysis of his measurements and those of Pawsey (1957), he reaches the same conclusion as Holmes *et al.* (1980): that the echoes are due to many reflectors distributed throughout a volume of cloud. Proctor (1981b) gives values for measured effective radar cross-sections and compares those results with the theory of Dawson (1972). Proctor (1981b) reports that lightning echoes at 50 and 111 cm, unlike echoes from precipitation, do not fluctuate greatly in amplitude from one pulse to the next except for sudden rises in amplitude at the start and when subsequent strokes occur. After such an increase, the echoes decayed smoothly in tens of milliseconds, as previously reported by Hewitt (1957). It is interesting to note that Holmes *et al.* (1980) find that echoes at 10.9 cm exhibit short-term fluctuations in amplitude similar to those from precipitation.

Szymanski *et al.* (1980), as part of the study of Holmes *et al.* (1980), describe one lightning echo that was quickly followed by the development of precipitation in the same area of the cloud, and they reference previous literature on this so-called rain-gush phenomenon.

C.7.5   VISIBLE LIGHT DIRECTION

Two classes of optical detectors have been used to detect lightning: (1) television cameras and (2) photoelectric detectors. The latter have been discussed in Section C.3. Television cameras and associated videotape recorders can be used not only to locate lightning but to measure its properties on a 16.7-msec time scale, the standard TV frame rate. Winn *et al.* (1973), Brantley *et al.* (1975), Clifton and Hill (1980), and Thomson *et al.* (1984) have used television cameras to measure a variety of lightning properties such as the number of strokes per flash, the number of separate channels to ground per flash, interstroke interval, and flash duration. Further, Thomson *et al.* (1984) made correlated television and electric field measurements to investigate the errors made in determining strokes per flash, channels per flash, and interstroke interval using only television. About 80% of the strokes were detected by the video system. The television technique has the advantage over photography of not missing portions of the event due to shutter action and of having relatively high sensitivity (e.g., Clifton and Hill, 1980).

C.7.6   THUNDER RAY TRACING AND RANGING

The reconstruction of the sources of acoustic radiation from lightning has been discussed in Section 15.5. References are given there both to articles in which the techniques are described and to articles in which acoustically reconstructed channels are presented.

REFERENCES

Atlas, D.: Radar Lightning Echoes and Atmospherics in Vertical Cross-Section. *In* "Recent Advances in Atmospheric Electricity" (L. G. Smith, ed.), pp. 441–459. Pergamon, Oxford, 1958.

Austin, G. I., and E. J. Stansbury: The Location of Lightning and Its Relation to Precipitation Detected by Radar. *J. Atmos. Terr. Phys.*, **33**:841–844 (1971).

Balachandran, N. K.: Acoustic and Electric Signals from Lightning. *J. Geophys. Res.*, **88**:3879-3884 (1983).

Baum, C. E., E. L. Breen, F. L. Pitts, G. D. Sower, and M. E. Thomas: The Measurement of Lightning Environmental Parameters Related to Interaction with Electronic Systems. *IEEE Trans. Electromagn. Compat.*, **EMC-24**:123-137 (1982).

Beasley, W. H., M. A. Uman, D. M. Jordan, and C. Ganesh: Positive Cloud to Ground Lightning Return Strokes. *J. Geophys. Res.*, **88**:8475-8482 (1983a).

Beasley, W. H., M. A. Uman, D. M. Jordan, and C. Ganesh: Simultaneous Pulses in Light and Electric Field from Stepped Leaders near Ground Level. *J. Geophys. Res.*, **88**:8617-8619 (1983b).

Bohannon, J. L.: Infrasonic Thunder: Explained. Ph.D. thesis, Department of Space Physics and Astronomy, Rice University, Houston, Texas, 1980.

Bohannon, J. L., A. A. Few, and A. J. Dessler: Detection of Infrasonic Pulses from Thunderclouds. *Geophys. Res. Lett.*, **4**:49-52 (1977).

Boys, C. V.: Progressive Lightning. *Nature (London)*, **118**:749-750 (1926).

Brantley, R. D., J. A. Tiller, and M. A. Uman: Lightning Properties in Florida Thunderstorms from Video Tape Records. *J. Geophys. Res.*, **80**:3402-3406 (1975).

Brook, M., and N. Kitagawa: Electric Field Changes and the Design of Lightning-Flash Counters. *J. Geophys. Res.*, **65**:1927-1931 (1960).

Brook, M., and N. Kitagawa: Radiation from Lightning Discharges in the Frequency Range 400 to 1000 Mc/s. *J. Geophys. Res.*, **69**:2431-2434 (1964).

Brook, M., R. Tennis, C. Rhodes, P. Krehbiel, B. Vonnegut, and O. H. Vaughan, Jr.: Simultaneous Observations of Lightning Radiations from above and below Clouds. *Geophys. Res. Lett.*, **7**:267-270 (1980).

Browne, I. C.: A Radar Echo from Lightning. *Nature (London)*, **167**:438 (1951).

Carte, A. E., and R. E. Kidder: Lightning in Relation to Precipitation. *J. Atmos. Terr. Phys.*, **39**:139-148 (1977).

Cerni, T. A.: Experimental Investigation of the Radar Cross-Section of Cloud-to-Ground Lightning. *J. Appl. Meteorol.*, **15**:795-798 (1976).

Cianos, N., G. N. Oetzel, and E. T. Pierce: A Technique for Accurately Locating Lightning at Close Ranges. *J. Appl. Meteorol.*, **11**:1120-1127 (1972).

Cianos, N., G. N. Oetzel, and E. T. Pierce: Reply. *J. Appl. Meteorol.*, **12**:1421-1423 (1973).

Clegg, R. J.: A Photoelectric Detector of Lightning. *J. Atmos. Terr. Phys.*, **33**:1431-1439 (1971).

Clifton, K. S., and C. K. Hill: Low-Light-Level Television Measurement of Lightning. *Bull. Am. Meteorol. Soc.*, **61**:987-992 (1980).

Cooray, V.: Errors in Direction Finding Due to Nonvertical Channels: Effects of the Finite Ground Conductivity. *Radio Sci.*, **21**:857-862 (1986).

Dawson, G. A.: Radar as a Diagnostic Tool for Lightning. *J. Geophys. Res.*, **77**:4518-4527 (1972).

Edgar, B. C.: Global Lightning Distribution at Dawn and Dusk for August-December, 1977, as Observed by the DMSP Lightning Detector. Rep. SSL-78 (3639-02)-1, Aerospace Corporation Space Science Laboratory, Los Angeles, California, August 1978.

Few, A. A.: Lightning Channel Reconstruction from Thunder Measurements. *J. Geophys. Res.*, **75**:7517-7523 (1970).

Few, A. A., and T. L. Teer: The Accuracy of Acoustic Reconstructions of Lightning Channels. *J. Geophys. Res.*, **79**:5007-5011 (1974).

Fisher, R. J., and M. A. Uman: Measured Electric Field Risetimes for First and Subsequent Lightning Return Strokes. *J. Geophys. Res.*, **77**:399-406 (1972).

Ganesh, C., M. A. Uman, W. H. Beasley, and D. M. Jordan: Correlated Optical and Electric Field Signals Produced by Lightning Return Strokes. *J. Geophys. Res.*, **89**:4905-4909 (1984).

Griffiths, R. F., and B. Vonnegut: Tape Recorder Photocell Instrument for Detecting and Recording Lightning Strokes. *Weather*, 30:254-257 (1975).

Guo, C., and E. P. Krider: The Optical and Radiation Field Signatures Produced by Lightning Return Strokes. *J. Geophys. Res.*, 87:8913-8922 (1982).

Guo, C., and E. P. Krider: The Optical Power Radiated by Lightning Return Strokes. *J. Geophys. Res.*, 88:8621-8622 (1983).

Guo, C., and E. P. Krider: Anomalous Light Output from Lightning Dart Leaders. *J. Geophys. Res.*, 90:13,073-13,075 (1985).

Harth, W., C. A. Hofmann, H. Falcoz, and G. Heydt: Atmospherics Measurements in San Miguel, Argentina. *J. Geophys. Res.*, 83:6231-6237 (1978).

Hayenga, C. O.: Positions and Movement of VHF Lightning Sources Determined with Microsecond Resolution by Interferometry. Ph.D. thesis, University of Colorado, Boulder, Colorado, 1979.

Hayenga, C. O.: Characteristics of Lightning VHF Radiation near the Time of Return Strokes. *J. Geophys. Res.*, 89:1403-1410 (1984).

Hayenga, C. O., and J. W. Warwick: Two Dimensional Interferometric Positions of VHF Lightning Sources. *J. Geophys. Res.*, 87:7451-7462 (1981).

Herrman, B. D., M. A. Uman, R. D. Brantley, and E. P. Krider: Test of a Wideband Magnetic Direction Finder for Lightning Return Strokes. *J. Appl. Meteorol.*, 15:402-405 (1976).

Hewitt, F. J.: Radar Echoes from Inter-Stroke Processes in Lightning. *Proc. Phys. Soc. London, Ser. B*, 70:961-979 (1957).

Heydt, G., and H. Volland: A New Method for Locating Thunderstorms and Counting Their Lightning Discharges from a Single Observing Station. *J. Atmos. Terr. Phys.*, 26:780-782 (1964).

Hiser, H. W.: Sferics and Radar Studies of South Florida Thunderstorms. *J. Appl. Meteorol.*, 12:479-483 (1973).

Holmes, C. R., M. Brook, P. Krehbiel, and R. McCrory: On the Power Spectrum and Mechanism of Thunder. *J. Geophys. Res.*, 76:2106-2115 (1971a).

Holmes, C. R., M. Brook, P. Krehbiel, and R. McCrory: Reply. *J. Geophys. Res.*, 76:7443 (1971b).

Holmes, C. R., E. W. Szymanski, S. J. Szymanski, and C. B. Moore: Radar and Acoustic Study of Lightning. *J. Geophys. Res.*, 85:7517-7532 (1980).

Horner, F.: The Accuracy of the Location Sources of Atmospherics by Radio Direction Finding. *Proc. IEEE*, 101:383-390 (1954).

Horner, F.: Very-Low-Frequency Propagation and Direction Finding. *Proc. IEEE*, 101B:73-80 (1957).

Hubert, P., and G. Mouget: Return Stroke Velocity Measurements in Two Triggered Lightning Flashes. *J. Geophys. Res.*, 86:5253-5261 (1981).

Idone, V. P., and R. E. Orville: Lightning Return Stroke Velocities in the Thunderstorm Research International Program (TRIP). *J. Geophys. Res.*, 87:4903-4915 (1982).

Idone, V. P., R. E. Orville, P. Hubert, L. Barret, and A. Eybert-Bernard: Correlated Observations of Three Triggered Lightning Flashes. *J. Geophys. Res.*, 89:1385-1394 (1984).

Iwai, A., M. Kashwagi, M. Nishino, Y. Katoh, and A. Kengpol: On the Accuracy of Direction Finding Methods for Atmospheric Sources in South-East Asia. *Proc. Res. Inst. Atmos., Nagoya Univ.*, 29:35-46 (1982).

Jacobson, E. A., and E. P. Krider: Electrostatic Field Changes Produced by Florida Lightning. *J. Atmos. Sci.*, 33:113-116 (1976).

Johnson, R. L., and D. E. Janota: An Operational Comparison of Lightning Warning Systems. *J. Appl. Meteorol.*, 21:703-707 (1982).

Jordan, D. J., and M. A. Uman: Variation in Light Intensity with Height and Time from Subsequent Lightning Return Strokes. *J. Geophys. Res.*, **88**:6555-6562 (1983).

Kashiwagi, M., A. Iwai, and M. Nishino: Fixing of the Sources of Atmospherics Using the Measurement of the Arrival Time Difference of Atmospherics between Toyokawa and Bangkok. *Res. Lett. Atmos. Electr.*, **1**:119-124 (1981).

Kidder, R. E.: The Location of Lightning Flashes at Ranges Less Than 100 km. *J. Atmos. Terr. Phys.*, **35**:283-290 (1973).

Kidder, R. E.: Location of Lightning Flashes to Ground with a Single Camera. *Weather*, **30**:72-77 (1975).

Kinzer, G. D.: Cloud-to-Ground Lightning versus Radar Reflectivity in Oklahoma Thunderstorm. *J. Atmos. Sci.*, **31**:787-799 (1974).

Kitagawa, N., and M. Brook: A Comparison of Intracloud and Cloud-to-Ground Lightning Discharges. *J. Geophys. Res.*, **65**:1189-1201 (1960).

Kitagawa, N., and M. Kobayashi: Field Changes and Variations of Luminosity Due to Lightning Flashes. *In* "Recent Advances in Atmospheric Electricity" (L. G. Smith, ed.), pp. 485-501. Pergamon, Oxford, 1959.

Kohl, D. A.: A 500 kHz Sferics Range Detector. *J. Appl. Meteorol.*, **8**:610-617 (1969).

Krehbiel, P. R., M. Brook, and R. McCrory: Analysis of the Charge Structure of Lightning Discharges to Ground. *J. Geophys. Res.*, **84**:2432-2456 (1979).

Krider, E. P., and R. C. Noggle: Broadband Antenna Systems for Lightning Magnetic Fields. *J. Appl. Meteorol.*, **14**:252-256 (1975).

Krider, E. P., R. C. Noggle, and M. A. Uman: A Gated Wideband Magnetic Direction Finder for Lightning Return Strokes. *J. Appl. Meteorol.*, **15**:301-306 (1976).

Krider, E. P., C. D. Weidman, and R. C. Noggle: The Electric Fields Produced by Lightning Stepped Leaders. *J. Geophys. Res.*, **82**:951-960 (1977).

Krider, E. P., C. D. Weidman, and D. M. LeVine: The Temporal Structure of the HF and VHF Radiation Produced by Intracloud Lightning Discharges. *J. Geophys. Res.*, **84**:5760-5762 (1979).

Krider, E. P., R. C. Noggle, A. E. Pifer, and D. L. Vance: Lightning Direction-Finding Systems for Forest Fire Detection. *Bull. Am. Meteorol. Soc.*, **61**:980-986 (1980).

Larsen, H. R., and E. J. Stansbury: Association of Lightning Flashes with Precipitation Cores Extending to Height 7 km. *J. Atmos. Terr. Phys.*, **36**:1547-1553 (1974).

Lee, A. C. L.: An Experimental Study of the Remote Location of Lightning Flashes Using a VLF Arrival Time Difference Technique. *Q. J. R. Meteorol. Soc.*, **112**:203-229 (1986).

Lennon, C. L.: LDAR—A New Lightning Detection and Ranging System. *EOS Trans. AGU*, **56**(12):991 (1975).

LeVine, D. M., and E. P. Krider: The Temporal Structure of HF and VHF Radiations during Florida Lightning Return Strokes. *Geophys. Res. Lett.*, **4**:13-16 (1977).

Lewis, E. A., R. B. Harvey, and J. E. Rasmussen: Hyperbolic Direction Finding with Sferics of Transatlantic Origin. *J. Geophys. Res.*, **65**:1879-1905 (1960).

Lhermitte, R., and P. R. Krehbiel: Doppler Radar and Radio Observations of Thunderstorms. *IEEE Trans Geosci. Electron.*, **GE-17**:162-171 (1979).

Ligda, M. G. H.: Lightning Detection by Radar *Bull. Am. Meteorol. Soc.*, **31**:279-283 (1950).

Ligda, M. G. H.: The Radar Observations of Lightning. *J. Atmos. Terr. Phys.*, **9**:329-346 (1956).

MacClement, W. D., and R. C. Murty: VHF Direction Finder Studies of Lightning. *J. Appl. Meteorol.*, **17**:786-795 (1978).

McDonald, T. B., M. A. Uman, J. A. Tiller, and W. H. Beasley: Lightning Location and Lower Inospheric Height Determination from Two Station Magnetic Field Measurements. *J. Geophys. Res.*, **84**:1727-1734 (1979).

McEachron, K. B.: Lightning to the Empire State Building. *J. Franklin Inst.*, **227**:149–217 (1939).

MacGorman, D. R.: Lightning Location in a Storm with Strong Wind Shear. Ph.D. thesis, Department of Space Physics and Astronomy, Rice University, Houston, Texas, 1978.

Mach, D. M., D. R. MacGorman, W. D. Rust, and R. T. Arnold: Site Errors and Detection Efficiency in a Magnetic Direction-Finder Network for Locating Lightning Strikes to Ground. *J. Atmos. Oceanic Tech.*, **3**:67–74 (1986).

Mackerras, D.: Photoelectric Observations of the Light Emitted by Lightning Flashes. *J. Atmos. Terr. Phys.*, **35**:521–535 (1973).

Mackerras, D.: Automatic Short-Range Measurement of the Cloud Flash to Ground Flash Ratio in Thunderstorms. *J. Geophys. Res.*, **90**:6195–6201 (1985).

Malan, D. J.: "Physics of Lightning." English Univ. Press, London, 1963.

Malan, D. J., and H. Collens: Progressive Lightning III—The Fine Structure of Return Lightning Strokes. *Proc. R. Soc. London Ser. A*, **162**:175–203 (1937).

Malan, D. J., and B. F. J. Schonland: An Electrostatic Fluxmeter of Short Response-Time for Use in Studies of Transient Field-Changes. *Proc. Phys. Soc. London Ser. B*, **63**:402–408 (1950).

Marshall, J. S.: Frontal Precipitation and Lightning Observed by Radar. *Can. J. Phys.*, **31**:194–203 (1953).

Marshall, J. S., and S. Radhakant: Radar Precipitation Maps as Lightning Indicators. *J. Appl. Meteorol.*, **17**:206–212 (1978).

Marshall, S. V.: An Analytical Model for the Fluxgate Magnetometer. *IEEE Trans. Magn.*, **MAG-3**:459–463 (1967).

Marshall, S. V.: Impulse Response of a Fluxgate Sensor-Application to Lightning Discharge Location and Measurement. *IEEE Trans. Magn.*, **MAC-9**:235–238 (1973).

Mazur, V.: Rapidly Occurring Short Duration Discharges in Thunderstorms as Indicators of a Lightning-Triggering Mechanism. *Geophys. Res. Lett.*, **13**:355–358 (1986).

Mazur, V., and R. Doviak: Radar Cross Section of a Lightning Element Modeled as a Plasma Cylinder. *Radio Sci.*, **18**:381–390 (1983).

Mazur, V., and W. D. Rust: Lightning Propagation and Flash Density in Squall Lines as Determined with Radar. *J. Geophys. Res.*, **88**:1495–1502 (1983).

Mazur, V., and G. B. Walker: The Effect of Polarization on Radar Detection of Lightning. *Geophys. Res. Lett.*, **9**:1231–1234 (1982).

Mazur, V., J. C. Gerlach, and W. D. Rust: Lightning Flash Density versus Altitude and Storm Structure from Observations with UHF- and S-band Radars. *Geophys. Res. Lett.*, **11**:61–64 (1984a).

Mazur, V., B. D. Fisher, and J. C. Gerlach: Lightning Strikes to an Airplane in a Thunderstorm. *J. Aircraft*, **21**:607–611 (1984b).

Mazur, V., D. S. Zrnic, and W. D. Rust: Lightning Channel Properties Determined with a Vertically Pointing Doppler Radar. *J. Geophys. Res.*, **90**:6165–6174 (1985).

Mazur, V., W. D. Rust, and J. C. Gerlach: Evolution of Lightning Flash Density and Reflectivity Structure in a Multicell Thunderstorm. *J. Geophys. Res.*, **91**:8690–8700 (1986a).

Mazur, V., B. D. Fisher, and J. C. Gerlach: Lightning Strikes to a NASA Airplane Penetrating Thunderstorms at Low Altitude. *J. Aircraft*, **23**:499–505 (1986b).

Meese, A. D., and W. H. Evans: Charge Transfer in the Lightning Stroke as Determined by the Magnetograph. *J. Franklin Inst.*, **273**:375–382 (1962).

Miles, V. H.: Radar Echoes Associated with Lightning. *J. Atmos. Terr. Phys.*, **3**:258–263 (1953).

Murty, R. C., and W. D. MacClement: VHF Direction Finder for Lightning Location. *J. Appl. Meteorol.*, **12**:1401–1405 (1973).

Nakano, M.: Lightning Channel Determined by Thunder. *Proc. Res. Inst. Atmos.*, *Nagoya Univ.*, *Japan*, **20**:1-9 (1973).

Nakano, M.: Characteristics of Lightning Channel in Thunderclouds Determined by Thunder. *J. Meteorol. Soc. Japan*, **54**:441-447 (1976).

Nelson, L. D.: Magnetographic Measurements of Charge Transfer in the Lightning Flash. *J. Geophys. Res.*, **73**:5967-5972 (1968).

Nishino, M., A. Iwai, and M. Kashiwagi: Location of the Sources of Atmospherics in and around Japan. *Proc. Res. Inst. Atmos. Nayoga Univ.*, *Japan*, **20**:9-18 (1973).

Nishizawa, Y., A. Iwai, and M. Satoh: VHF Direction Finding for Lightnings at Close Ranges. *Proc. Res. Inst. Atmos. Nagoya Univ.*, *Japan*, **27**:11-24 (1980).

Norinder, H.: Magnetic Field Variations from Lightning Strokes in Vicinity Thunderstorms. *Ark. Geofys.*, **2**:423-451 (1956).

Norinder, H., and O. Dahle: Measurements by Frame Aerials of Current Variations in Lightning Discharges. *Ark. Mat. Astron. Fys.*, **32A**:1-70 (1945).

Oetzel, G. N., and E. T. Pierce: VHF Technique for Locating Lightning. *Radio Sci.*, **4**:199-201 (1969).

Orville, R. E.: A High-Speed Time-Resolved Spectroscopic Study of the Lightning Return Stroke, 1, A Qualitative Analysis. *J. Atmos. Sci.*, **25**:827-838 (1968).

Orville, R. E.: Lightning Spectroscopy. *In* "Lightning, Vol. 1, Physics of Lightning" (R. H. Golde, ed.), pp. 281-306. Academic Press, New York, 1977.

Orville, R. E.: Global Distribution of Midnight Lightning—September to November 1977. *Mon. Weather Rev.*, **109**:391-395 (1981).

Orville, R. E., and R. W. Henderson: Absolute Spectral Irradiance Measurements of Lightning from 375 to 880 nm. *J. Atmos. Sci.*, **41**:3180-3187 (1984).

Orville, R. E., R. B. Pyle, and R. W. Henderson: The East Coast Lightning Detection Network. *IEEE Trans. Power Systems*, **PWRS-1**:243-246 (1986).

Orville, R. E., and D. W. Spencer: Global Lightning Flash Frequency. *Mon. Weather Rev.*, **107**:934-943 (1979).

Orville, R. E., R. W. Henderson, and L. F. Bosart: An East Coast Lightning Detection Network. *Bull. Am. Meteorol. Soc.*, **64**:1029-1037 (1983).

Pawsey, J. L.: Radar Observations of Lightning on 1.5 Meters. *J. Atmos. Terr. Phys.*, **11**:289-290 (1957).

Pierce, E. T.: The Charge Transferred to Earth by a Lightning Flash. *J. Franklin Inst.*, **286**:353-354 (1968).

Pierce, E. T.: Atmospherics and Radio Noise. *In* "Lightning, Vol. 1, Physics of Lightning" (R. H. Golde, ed.), pp. 351-384. Academic Press, New York, 1977.

Proctor, D. E.: A Hyperbolic System for Obtaining VHF Radio Pictures of Lightning. *J. Geophys. Res.*, **76**:1478-1489 (1971).

Proctor, D. E.: Comments on "A Technique for Accurately Locating Lightning at Close Range." *J. Appl. Meteorol.*, **12**:1419-1423 (1973).

Proctor, D. E.: A Radio Study of Lightning. Ph.D. thesis, University of Witwatersrand, Johannesburg, South Africa, 1976.

Proctor, D. E.: VHF Radio Pictures of Cloud Flashes. *J. Geophys. Res.*, **86**:4041-4171 (1981a).

Proctor, D. E.: Radar Observations of Lightning. *J. Geophys. Res.*, **86**:12,109-12,114 (1981b).

Reap, R. M.: Evaluation of Cloud-to-Ground Lightning Data from the Western United States for the 1983-1984 Summer Seasons. *J. Climate Appl. Meteorol.*, **25**:785-799 (1986).

Richard, P., and G. Auffray: VHF-UHF Interferometric Measurements, Applications to Lightning Discharge Mapping. *Radio Sci.*, **20**:171-192 (1985).

Rust, W. D., and R. J. Doviak: Radar Research on Thunderstorms and Lightning. *Nature (London)*, **297**:461-468 (1982).

Rustan, P. L., Jr.: Properties of Lightning Derived from Time Series Analysis of VHF Radiation Data. Ph.D. thesis, University of Florida, Gainesville, Florida, 1979.

Rustan, P. L., Jr.: The Lightning Threat to Aerospace Vehicles. *J. Aircraft*, **23**:62–67 (1986).

Rustan, P. L., M. A. Uman, D. G. Childers, W. H. Beasley, and C. L. Lennon: Lightning Source Locations from VHF Radiation Data for a Flash at Kennedy Space Center. *J. Geophys. Res.*, **85**:4893–4903 (1980).

Schonland, B. F. J., and H. Collens: Progressive Lightning—I. *Proc. R. Soc. London Ser. A*, **143**:654–674 (1934).

Schonland, B. F. J., D. J. Malan, and H. Collens: Progressive Lightning—II. *Proc. R. Soc. London Ser. A*, **152**:595–625 (1935).

Sparrow, J. G., and E. P. Ney: Discrete Light Sources Observed by Satellite OSO-B. *Science*, **161**:459–460 (1968).

Sparrow, J. G., and F. E. Ney: Lightning Observations by Satellite. *Nature (London)*, **232**:540–541 (1971).

Stansbury, E. J., E. Cherna, and J. Percy: Lightning Flash Locations Related to the Precipitation Pattern of the Storm. *Atmosphere-Ocean*, **17**:291–305 (1979).

Szymanski, E. W., and W. D. Rust: Preliminary Observations of Lightning Radar Echoes and Simultaneous Electric Field Changes. *Geophys. Res. Lett.*, **6**:527–530 (1979).

Szymanski, E. W., S. J. Szymanski, C. R. Holmes, and C. B. Moore: An Observation of a Precipitation Echo Intensification Associated with Lightning. *J. Geophys. Res.*, **85**:1591–1593 (1980).

Taylor, W. L.: Determining Lightning Stroke Height from Ionospheric Components of Atmospheric Waveforms. *J. Atmos. Terr. Phys.*, **32**:983–990 (1969).

Taylor, W. L.: An Electromagnetic Technique for Tornado Detection. *Weatherwise*, **26**:70–71 (1973).

Taylor, W. L.: A VHF Technique for Space-Time Mapping of Lightning Discharge Processes. *J. Geophys. Res.*, **83**:3575–3583 (1978).

Taylor, W. L., E. A. Brandes, W. D. Rust, and D. R. MacGorman: Lightning Activity and Severe Storm Structure. *Geophys. Res. Lett.*, **11**:545–548 (1984).

Teer, T. L.: Lightning Channel Structure Inside an Arizona Thunderstorm. Ph.D. dissertation, Rice University, Department of Space Physics and Astronomy, Houston, Texas, 1973.

Thomason, L. W., and E. P. Krider: The Effects of Clouds on the Light Produced by Lightning. *J. Atmos. Sci.*, **39**:2051–2065 (1982).

Thomson, E. M.: Photoelectric Detector for Daytime Lightning. *Electron. Lett.*, **14**:337–339 (1978).

Thomson, E. M., M. A. Galib, M. A. Uman, W. H. Beasley, and M. J. Master: Some Features of Stroke Occurrence in Florida Lightning Flashes. *J. Geophys. Res.*, **89**:4910–4916 (1984).

Thomson, E. M., M. A. Uman, and W. H. Beasley: Speed and Current for Lightning Stepped Leaders near Ground as Determined from Electric Field Records. *J. Geophys. Res.*, **90**:8136–8142 (1985).

Thomson, E. M., P. Medelius, M. Rubinstein, M. A. Uman, J. Johnson, and J. W. Stone: Horizontal Electric Fields from Lightning Return Strokes. *In* "10th International Aerospace and Ground Conference on Lightning, and Static Electricity, Paris, June 1985," pp. 167–173. Available from Les Éditions de Physique, BP 112, 91944 Les Ulis Cedex, France.

Tsuruda, K., and M. Ikeda: Comparison of Three Different Types of VLF Direction-Finding Techniques. *J. Geophys. Res.*, **84**:5325–5332 (1979).

Turman, B. N.: Detection of Lightning Superbolts. *J. Geophys. Res.*, **82**:2566–2568 (1977).

Turman, B. N.: Analysis of Lightning Data from the DMSP Satellite. *J. Geophys. Res.*, **83**:5019–5024 (1978).

Turman, B. N.: Lightning Detection from Space. *Am. Sci.*, **67**:321–329 (1979).

Turman, B. N., and B. C. Edgar: Global Lightning Distributions at Dawn and Dusk. *J. Geophys. Res.*, **87**:1191–1206 (1982).

Turman, B. N., and R. J. Tettelbach: Synoptic-Scale Satellite Observations in Conjunction with Tornadoes. *Mon. Weather Rev.*, **108**:1878–1882 (1980).

Uman, M. A.: "Lightning." McGraw-Hill, New York, 1969. See also, Dover, New York, 1984, Chap. 5.

Uman, M. A., R. E. Orville, A. M. Sletten, and E. P. Krider: Four-Meter Sparks in Air. *J. Appl. Phys.*, **39**:5162–5168 (1968).

Uman, M. A., Y. T. Lin, and E. P. Krider: Errors in Magnetic Direction Finding Due to Non-Vertical Lightning Channels. *Radio Sci.*, **15**:35–39 (1980).

Volland, H., J. Schafer, P. Ingmann, W. Harth, G. Heydt, A. J. Eriksson, and A. Manes: Registration of Thunderstorm Centers by Automatic Atmospherics Stations. *J. Geophys. Res.*, **88**:1503–1518 (1983).

Vonnegut, B., and R. E. Passarelli: Modified Cine Sound Camera for Photographing Thunderstorms and Recording Lightning. *J. Appl. Meteorol.*, **17**:1078–1081 (1978).

Vorphal, J. A., J. G. Sparrow, and E. P. Ney: Satellite Observations of Lightning. *Science*, **169**:860–862 (1970).

Warwick, J. W., C. O. Hayenga, and J. W. Brosnahan: Interferometric Position of Lightning Sources at 34 MHz. *J. Geophys. Res.*, **84**:2457–2468 (1979).

Weber, M. E., H. J. Christian, A. A. Few, and M. F. Stewart: A Thundercloud Electric Field Sounding: Charge Distribution and Lightning. *J. Geophys. Res.*, **87**:7158–7169 (1982).

Williams, D. P., and M. Brook: Magnetic Measurement of Thunderstorm Currents, 1. Continuing Currents in Lightning. *J. Geophys. Res.*, **68**:3243–3247 (1963).

Winn, W. P., T. V. Aldridge, and C. B. Moore: Video-Tape Recordings of Lightning Flashes. *J. Geophys. Res.*, **78**:4515–4519 (1973).

Yamashita, H., A. Iwai, M. Satoh, and T. Katoh: VHF Direction Finder for Locating Lightnings at Close Ranges. *Proc. Res. Inst. Atmos. Nagoya Univ.*, *Japan*, **30**:15–24 (1983).

Yamashita, M., and K. Sao: Some Considerations of the Polarization Error in Direction Finding of Atmospherics, I, Effects of the Earth's Magnetic Field. *J. Atmos. Terr. Phys.*, **36**:1623–1632 (1974a).

Yamashita, M., and K. Sao: Some Considerations of the Polarization Error in Direction Finding of Atmospherics, II, Effects of the Inclined Electric Dipole. *J. Atmos. Terr. Phys.*, **36**:1633–1641 (1974b).

Zrnic, D. S., W. D. Rust, and W. L. Taylor: Doppler Radar Echoes of Lightning and Precipitation at Vertical Incidence. *J. Geophys. Res.*, **87**:7179–7191 (1982).

Appendix D | Books Containing Information on Lightning

Barry, J. D.: "Ball Lightning and Bead Lightning." Plenum, New York, 1980.

Battan, L. J.: "Radar Meteorology." Univ. of Chicago Press, Chicago, Illinois, 1959.

Battan, L. J.: "The Thunderstorm." New American Library, Signet, New York, 1964.

Battan, L. J.: "Radar Observation of the Atmosphere." Univ. of Chicago Press, Chicago, Illinois, 1973.

Bell, T. H.: "Thunderstorms." Dobson, London, 1962.

Bewley, L. V.: "Travelling Waves on Transmission Systems," 2nd Ed. Dover, New York, 1951.

Brand, W.: "Der Kugelblitz." Grand, Hamburg, 1923.

*Byers, H. R. (ed.): "Thunderstorm Electricity." Univ. of Chicago Press, Chicago, Illinois, 1953.

Byers, H. R., and R. R. Braham: "The Thunderstorm." U.S. Weather Bureau, Washington, D.C., 1949.

Cade, C. M., and D. Davis: "The Taming of the Thunderbolts." Abelard-Schuman, New York, 1969.

Chalmers, J. A.: "Atmospheric Electricity," 2nd Ed. Pergamon, Oxford, 1967.

†Cianos, N., and E. T. Pierce: A Ground-Lightning Environment for Engineering Usage. Stanford Research Institute Technical Report, Project 1834, 1972.

*Coroniti, S. C. (ed.): "Problems of Atmospheric and Space Electricity." American Elsevier, New York, 1965.

*Coroniti, S. C., and J. Hughes (eds.): "Planetary Electrodynamics," Vols. I and II. Gordon & Breach, New York, 1969.

Davies, K.: "Ionospheric Radio Propagation." Dover, New York, 1966.

*Dolezalek, H., and R. Reiter (eds.): "Electrical Processes in Atmospheres." Steinkopff, Darmstadt, 1977.

Fisher, F. A., and J. A. Plumer: Lightning Protection of Aircraft. NASA Reference Publication 1008, NASA Lewis Reseach Center, October, 1977.

*Forrest, J. S., P. R. Howard, and D. J. Littler (eds.): "Gas Discharges and the Electricity Supply Industry." Butterworths, London, 1962.

†Golde, R. H.: "Lightning Protection." Arnold, London, 1973.

†Golde, R. H. (ed.): "Lightning, Vol. 1, Physics of Lightning" and "Vol. 2, Lightning Protection." Academic Press, New York, 1977.

Hart, W. C., and E. W. Malone: "Lightning and Lightning Protection." Don White Consultants, P.O. Box D, Gainesville, Virginia 22065, 1979.

* Indicates "Proceedings of International Conferences on Atmospheric Electricity."
† Indicates books of particular interest.

Hasse, P., and J. Weisinger: "Handbuch für Blitzschute und Erdung." Richard Pflaum Verlag KG, München, 1982.

Israel, H.: Atmospheric Electricity, 1. Authorized and Revised Translation from the 2nd German Ed. Israel Program for Scientific Publications, Jerusalem, 1973, TT68-5194/2, Clearinghouse for Federal Scientific and Technical Information.

Israel, H.: Atmospheric Electricity, 2. Authorized and Revised Translations from the 2nd German Ed. Israel Program for Scientific Publications, Jerusalem, 1973, TT68-5194/2, Clearinghouse for Federal Scientific and Technical Information.

Kessler, E. (ed.): "Thunderstorms: A Social, Scientific, and Technological Documentary," 3 Vols. Univ. of Oklahoma Press, Norman, Oklahoma, 1983–1986.

Krider, E. P., and R. G. Roble (panel co-chairman): "The Earth's Electrical Environment, Studies in Geophysics." National Academy Press, Washington, D.C., 1986.

Lewis, W. W.: "The Protection of Transmission Systems against Lightning." Dover, New York, 1965.

†Malan, D. J.: "Physics of Lightning." English Univ. Press, London, 1964.

Marshall, J. L.: "Lightning Protection." Wiley, New York, 1973.

Mason, B. J.: "The Physics of Clouds." Oxford Univ. Press (Clarendon), London and New York, 1957.

Mogono, C.: "Thunderstorms." Elsevier, Amsterdam, 1980.

*Orville, R. E. (ed.): Preprints from 7th International Conference on Atmospheric Electricity, June 3–8, 1984, Albany, New York. Am. Meteorol. Soc., Boston, Massachusetts, 1984.

*Orville, R. E. (ed.): Selected Papers from 7th International Conference on Atmospheric Electricity, June 3–8, 1984, Albany, New York. *J. Geophys. Res.*, **90** (D4): June 30 (1985).

Pierce, E. T.: The Thunderstorm as a Source of Atmospheric Noise at Frequencies between 1 and 100 kHz. Stanford Research Institute Technical Report, Project 7045, DASA 2299, June, 1969.

*Ruhnke, L. H., and J. Latham (eds.): "Proceedings in Atmospheric Electricity." Deepak, Hampton, Virginia, 1983.

Salanave, L. E.: "Lightning and Its Spectrum." Univ. of Arizona Press, Tucson, Arizona, 1980.

Schonland, B. F. J.: "Atmospheric Electricity," 2nd Ed. Methuen, London, 1953.

†Schonland, B. F. J.: "The Flight of Thunderbolts," 2nd Ed. Oxford Univ. Press (Clarendon), London and New York, 1964.

Singer, S.: "The Nature of Ball Lightning." Plenum, New York, 1971.

*Smith, L. G. (ed.): "Recent Advances in Atmospheric Electricity." Pergamon, Oxford, 1959.

Sunde, E. D.: "Earth Conduction Effects in Transmission Systems." Van Nostrand-Reinhold, Princeton, New Jersey, 1949. See also, Dover, New York, 1967.

†Uman, M. A.: "Lightning." McGraw-Hill, New York, 1969. See also, Dover, New York, 1984.

Uman, M. A.: "Understanding Lightning." BEK Technical Publications, Pittsburgh, Pennsylvania, 1971. See also, Rev. Ed. "All about Lightning." Dover, New York, 1986.

Viemeister, P. E.: "The Lightning Book." Doubleday, Garden City, New York, 1961.

Wesinger, J.: "Blitzforschung und Blitzschutz." Oldenbourg Verlag, München, 1972.

Index

T

A CATALOG OF SELECTED
DOVER BOOKS
IN ALL FIELDS OF INTEREST

A CATALOG OF SELECTED DOVER
BOOKS IN ALL FIELDS OF INTEREST

CONCERNING THE SPIRITUAL IN ART, Wassily Kandinsky. Pioneering work by father of abstract art. Thoughts on color theory, nature of art. Analysis of earlier masters. 12 illustrations. 80pp. of text. 5⅜ x 8½. 23411-8 Pa. $4.95

ANIMALS: 1,419 Copyright-Free Illustrations of Mammals, Birds, Fish, Insects, etc., Jim Harter (ed.). Clear wood engravings present, in extremely lifelike poses, over 1,000 species of animals. One of the most extensive pictorial sourcebooks of its kind. Captions. Index. 284pp. 9 x 12. 23766-4 Pa. $14.95

CELTIC ART: The Methods of Construction, George Bain. Simple geometric techniques for making Celtic interlacements, spirals, Kells-type initials, animals, humans, etc. Over 500 illustrations. 160pp. 9 x 12. (Available in U.S. only.) 22923-8 Pa. $9.95

AN ATLAS OF ANATOMY FOR ARTISTS, Fritz Schider. Most thorough reference work on art anatomy in the world. Hundreds of illustrations, including selections from works by Vesalius, Leonardo, Goya, Ingres, Michelangelo, others. 593 illustrations. 192pp. 7⅛ x 10¼. 20241-0 Pa. $9.95

CELTIC HAND STROKE-BY-STROKE (Irish Half-Uncial from "The Book of Kells"): An Arthur Baker Calligraphy Manual, Arthur Baker. Complete guide to creating each letter of the alphabet in distinctive Celtic manner. Covers hand position, strokes, pens, inks, paper, more. Illustrated. 48pp. 8¼ x 11. 24336-2 Pa. $3.95

EASY ORIGAMI, John Montroll. Charming collection of 32 projects (hat, cup, pelican, piano, swan, many more) specially designed for the novice origami hobbyist. Clearly illustrated easy-to-follow instructions insure that even beginning papercrafters will achieve successful results. 48pp. 8¼ x 11. 27298-2 Pa. $3.50

THE COMPLETE BOOK OF BIRDHOUSE CONSTRUCTION FOR WOOD-WORKERS, Scott D. Campbell. Detailed instructions, illustrations, tables. Also data on bird habitat and instinct patterns. Bibliography. 3 tables. 63 illustrations in 15 figures. 48pp. 5¼ x 8½. 24407-5 Pa. $2.50

BLOOMINGDALE'S ILLUSTRATED 1886 CATALOG: Fashions, Dry Goods and Housewares, Bloomingdale Brothers. Famed merchants' extremely rare catalog depicting about 1,700 products: clothing, housewares, firearms, dry goods, jewelry, more. Invaluable for dating, identifying vintage items. Also, copyright-free graphics for artists, designers. Co-published with Henry Ford Museum & Greenfield Village. 160pp. 8¼ x 11. 25780-0 Pa. $10.95

HISTORIC COSTUME IN PICTURES, Braun & Schneider. Over 1,450 costumed figures in clearly detailed engravings–from dawn of civilization to end of 19th century. Captions. Many folk costumes. 256pp. 8⅜ x 11¼. 23150-X Pa. $12.95

CATALOG OF DOVER BOOKS

ANATOMY: A Complete Guide for Artists, Joseph Sheppard. A master of figure drawing shows artists how to render human anatomy convincingly. Over 460 illustrations. 224pp. 8⅜ x 11¼. 27279-6 Pa. $11.95

MEDIEVAL CALLIGRAPHY: Its History and Technique, Marc Drogin. Spirited history, comprehensive instruction manual covers 13 styles (ca. 4th century through 15th). Excellent photographs; directions for duplicating medieval techniques with modern tools. 224pp. 8⅜ x 11¼. 26142-5 Pa. $12.95

DRIED FLOWERS: How to Prepare Them, Sarah Whitlock and Martha Rankin. Complete instructions on how to use silica gel, meal and borax, perlite aggregate, sand and borax, glycerine and water to create attractive permanent flower arrangements. 12 illustrations. 32pp. 5⅜ x 8½. 21802-3 Pa. $1.00

EASY-TO-MAKE BIRD FEEDERS FOR WOODWORKERS, Scott D. Campbell. Detailed, simple-to-use guide for designing, constructing, caring for and using feeders. Text, illustrations for 12 classic and contemporary designs. 96pp. 5⅜ x 8½. 25847-5 Pa. $3.95

SCOTTISH WONDER TALES FROM MYTH AND LEGEND, Donald A. Mackenzie. 16 lively tales tell of giants rumbling down mountainsides, of a magic wand that turns stone pillars into warriors, of gods and goddesses, evil hags, powerful forces and more. 240pp. 5⅜ x 8½. 29677-6 Pa. $6.95

THE HISTORY OF UNDERCLOTHES, C. Willett Cunnington and Phyllis Cunnington. Fascinating, well-documented survey covering six centuries of English undergarments, enhanced with over 100 illustrations: 12th-century laced-up bodice, footed long drawers (1795), 19th-century bustles, 19th-century corsets for men, Victorian "bust improvers," much more. 272pp. 5⅜ x 8¼. 27124-2 Pa. $9.95

ARTS AND CRAFTS FURNITURE: The Complete Brooks Catalog of 1912, Brooks Manufacturing Co. Photos and detailed descriptions of more than 150 now very collectible furniture designs from the Arts and Crafts movement depict davenports, settees, buffets, desks, tables, chairs, bedsteads, dressers and more, all built of solid, quarter-sawed oak. Invaluable for students and enthusiasts of antiques, Americana and the decorative arts. 80pp. 6½ x 9¼. 27471-3 Pa. $8.95

WILBUR AND ORVILLE: A Biography of the Wright Brothers, Fred Howard. Definitive, crisply written study tells the full story of the brothers' lives and work. A vividly written biography, unparalleled in scope and color, that also captures the spirit of an extraordinary era. 560pp. 6⅛ x 9¼. 40297-5 Pa. $17.95

THE ARTS OF THE SAILOR: Knotting, Splicing and Ropework, Hervey Garrett Smith. Indispensable shipboard reference covers tools, basic knots and useful hitches; handsewing and canvas work, more. Over 100 illustrations. Delightful reading for sea lovers. 256pp. 5⅜ x 8½. 26440-8 Pa. $8.95

FRANK LLOYD WRIGHT'S FALLINGWATER: The House and Its History, Second, Revised Edition, Donald Hoffmann. A total revision—both in text and illustrations—of the standard document on Fallingwater, the boldest, most personal architectural statement of Wright's mature years, updated with valuable new material from the recently opened Frank Lloyd Wright Archives. "Fascinating"—*The New York Times*. 116 illustrations. 128pp. 9¼ x 10¾. 27430-6 Pa. $12.95

PHOTOGRAPHIC SKETCHBOOK OF THE CIVIL WAR, Alexander Gardner. 100 photos taken on field during the Civil War. Famous shots of Manassas Harper's Ferry, Lincoln, Richmond, slave pens, etc. 244pp. 10⅝ x 8¼. 22731-6 Pa. $10.95

FIVE ACRES AND INDEPENDENCE, Maurice G. Kains. Great back-to-the-land classic explains basics of self-sufficient farming. The one book to get. 95 illustrations. 397pp. 5⅜ x 8½. 20974-1 Pa. $7.95

SONGS OF EASTERN BIRDS, Dr. Donald J. Borror. Songs and calls of 60 species most common to eastern U.S.: warblers, woodpeckers, flycatchers, thrushes, larks, many more in high-quality recording. Cassette and manual 99912-2 $9.95

A MODERN HERBAL, Margaret Grieve. Much the fullest, most exact, most useful compilation of herbal material. Gigantic alphabetical encyclopedia, from aconite to zedoary, gives botanical information, medical properties, folklore, economic uses, much else. Indispensable to serious reader. 161 illustrations. 888pp. 6½ x 9¼. 2-vol. set. (Available in U.S. only.) Vol. I: 22798-7 Pa. $9.95
Vol. II: 22799-5 Pa. $9.95

HIDDEN TREASURE MAZE BOOK, Dave Phillips. Solve 34 challenging mazes accompanied by heroic tales of adventure. Evil dragons, people-eating plants, blood-thirsty giants, many more dangerous adversaries lurk at every twist and turn. 34 mazes, stories, solutions. 48pp. 8¼ x 11. 24566-7 Pa. $2.95

LETTERS OF W. A. MOZART, Wolfgang A. Mozart. Remarkable letters show bawdy wit, humor, imagination, musical insights, contemporary musical world; includes some letters from Leopold Mozart. 276pp. 5⅜ x 8½. 22859-2 Pa. $7.95

BASIC PRINCIPLES OF CLASSICAL BALLET, Agrippina Vaganova. Great Russian theoretician, teacher explains methods for teaching classical ballet. 118 illustrations. 175pp. 5⅜ x 8½. 22036-2 Pa. $6.95

THE JUMPING FROG, Mark Twain. Revenge edition. The original story of The Celebrated Jumping Frog of Calaveras County, a hapless French translation, and Twain's hilarious "retranslation" from the French. 12 illustrations. 66pp. 5⅜ x 8½. 22686-7 Pa. $3.95

BEST REMEMBERED POEMS, Martin Gardner (ed.). The 126 poems in this superb collection of 19th- and 20th-century British and American verse range from Shelley's "To a Skylark" to the impassioned "Renascence" of Edna St. Vincent Millay and to Edward Lear's whimsical "The Owl and the Pussycat." 224pp. 5⅜ x 8½. 27165-X Pa. $5.95

COMPLETE SONNETS, William Shakespeare. Over 150 exquisite poems deal with love, friendship, the tyranny of time, beauty's evanescence, death and other themes in language of remarkable power, precision and beauty. Glossary of archaic terms. 80pp. 5 3/16 x 8¼. 26686-9 Pa. $1.00

BODIES IN A BOOKSHOP, R. T. Campbell. Challenging mystery of blackmail and murder with ingenious plot and superbly drawn characters. In the best tradition of British suspense fiction. 192pp. 5⅜ x 8½. 24720-1 Pa. $6.95

THE WIT AND HUMOR OF OSCAR WILDE, Alvin Redman (ed.). More than 1,000 ripostes, paradoxes, wisecracks: Work is the curse of the drinking classes; I can resist everything except temptation; etc. 258pp. 5⅜ x 8½. 20602-5 Pa. $6.95

SHAKESPEARE LEXICON AND QUOTATION DICTIONARY, Alexander Schmidt. Full definitions, locations, shades of meaning in every word in plays and poems. More than 50,000 exact quotations. 1,485pp. 6½ x 9¼. 2-vol. set.
Vol. 1: 22726-X Pa. $17.95
Vol. 2: 22727-8 Pa. $17.95

SELECTED POEMS, Emily Dickinson. Over 100 best-known, best-loved poems by one of America's foremost poets, reprinted from authoritative early editions. No comparable edition at this price. Index of first lines. 64pp. $5\frac{3}{16}$ x 8¼.
26466-1 Pa. $1.00

THE INSIDIOUS DR. FU-MANCHU, Sax Rohmer. The first of the popular mystery series introduces a pair of English detectives to their archnemesis, the diabolical Dr. Fu-Manchu. Flavorful atmosphere, fast-paced action, and colorful characters enliven this classic of the genre. 208pp. $5\frac{3}{16}$ x 8¼. 29898-1 Pa. $2.00

THE MALLEUS MALEFICARUM OF KRAMER AND SPRENGER, translated by Montague Summers. Full text of most important witchhunter's "bible," used by both Catholics and Protestants. 278pp. 6⅝ x 10. 22802-9 Pa. $12.95

SPANISH STORIES/CUENTOS ESPAÑOLES: A Dual-Language Book, Angel Flores (ed.). Unique format offers 13 great stories in Spanish by Cervantes, Borges, others. Faithful English translations on facing pages. 352pp. 5⅜ x 8½.
25399-6 Pa. $8.95

GARDEN CITY, LONG ISLAND, IN EARLY PHOTOGRAPHS, 1869–1919, Mildred H. Smith. Handsome treasury of 118 vintage pictures, accompanied by carefully researched captions, document the Garden City Hotel fire (1899), the Vanderbilt Cup Race (1908), the first airmail flight departing from the Nassau Boulevard Aerodrome (1911), and much more. 96pp. 8⅞ x 11¾. 40669-5 Pa. $12.95

OLD QUEENS, N.Y., IN EARLY PHOTOGRAPHS, Vincent F. Seyfried and William Asadorian. Over 160 rare photographs of Maspeth, Jamaica, Jackson Heights, and other areas. Vintage views of DeWitt Clinton mansion, 1939 World's Fair and more. Captions. 192pp. 8⅞ x 11. 26358-4 Pa. $12.95

CAPTURED BY THE INDIANS: 15 Firsthand Accounts, 1750-1870, Frederick Drimmer. Astounding true historical accounts of grisly torture, bloody conflicts, relentless pursuits, miraculous escapes and more, by people who lived to tell the tale. 384pp. 5⅜ x 8½. 24901-8 Pa. $8.95

THE WORLD'S GREAT SPEECHES (Fourth Enlarged Edition), Lewis Copeland, Lawrence W. Lamm, and Stephen J. McKenna. Nearly 300 speeches provide public speakers with a wealth of updated quotes and inspiration from Pericles' funeral oration and William Jennings Bryan's "Cross of Gold Speech" to Malcolm X's powerful words on the Black Revolution and Earl of Spenser's tribute to his sister, Diana, Princess of Wales. 944pp. 5⅜ x 8⅜. 40903-1 Pa. $15.95

THE BOOK OF THE SWORD, Sir Richard F. Burton. Great Victorian scholar/adventurer's eloquent, erudite history of the "queen of weapons"–from prehistory to early Roman Empire. Evolution and development of early swords, variations (sabre, broadsword, cutlass, scimitar, etc.), much more. 336pp. 6⅛ x 9¼.
25434-8 Pa. $9.95

AUTOBIOGRAPHY: The Story of My Experiments with Truth, Mohandas K. Gandhi. Boyhood, legal studies, purification, the growth of the Satyagraha (nonviolent protest) movement. Critical, inspiring work of the man responsible for the freedom of India. 480pp. 5⅜ x 8½. (Available in U.S. only.) 24593-4 Pa. $8.95

CELTIC MYTHS AND LEGENDS, T. W. Rolleston. Masterful retelling of Irish and Welsh stories and tales. Cuchulain, King Arthur, Deirdre, the Grail, many more. First paperback edition. 58 full-page illustrations. 512pp. 5⅜ x 8½. 26507-2 Pa. $9.95

THE PRINCIPLES OF PSYCHOLOGY, William James. Famous long course complete, unabridged. Stream of thought, time perception, memory, experimental methods; great work decades ahead of its time. 94 figures. 1,391pp. 5⅜ x 8½. 2-vol. set.
Vol. I: 20381-6 Pa. $14.95
Vol. II: 20382-4 Pa. $14.95

THE WORLD AS WILL AND REPRESENTATION, Arthur Schopenhauer. Definitive English translation of Schopenhauer's life work, correcting more than 1,000 errors, omissions in earlier translations. Translated by E. F. J. Payne. Total of 1,269pp. 5⅜ x 8½. 2-vol. set. Vol. 1: 21761-2 Pa. $12.95
Vol. 2: 21762-0 Pa. $12.95

MAGIC AND MYSTERY IN TIBET, Madame Alexandra David-Neel. Experiences among lamas, magicians, sages, sorcerers, Bonpa wizards. A true psychic discovery. 32 illustrations. 321pp. 5⅜ x 8½. (Available in U.S. only.) 22682-4 Pa. $9.95

THE EGYPTIAN BOOK OF THE DEAD, E. A. Wallis Budge. Complete reproduction of Ani's papyrus, finest ever found. Full hieroglyphic text, interlinear transliteration, word-for-word translation, smooth translation. 533pp. 6½ x 9¼.
21866-X Pa. $12.95

MATHEMATICS FOR THE NONMATHEMATICIAN, Morris Kline. Detailed, college-level treatment of mathematics in cultural and historical context, with numerous exercises. Recommended Reading Lists. Tables. Numerous figures. 641pp. 5⅜ x 8½.
24823-2 Pa. $11.95

PROBABILISTIC METHODS IN THE THEORY OF STRUCTURES, Isaac Elishakoff. Well-written introduction covers the elements of the theory of probability from two or more random variables, the reliability of such multivariable structures, the theory of random function, Monte Carlo methods of treating problems incapable of exact solution, and more. Examples. 502pp. 5³/₈ x 8¹/₂. 40691-1 Pa. $16.95

THE RIME OF THE ANCIENT MARINER, Gustave Doré, S. T. Coleridge. Doré's finest work; 34 plates capture moods, subtleties of poem. Flawless full-size reproductions printed on facing pages with authoritative text of poem. "Beautiful. Simply beautiful."–*Publisher's Weekly*. 77pp. 9¼ x 12. 22305-1 Pa. $7.95

NORTH AMERICAN INDIAN DESIGNS FOR ARTISTS AND CRAFTSPEOPLE, Eva Wilson. Over 360 authentic copyright-free designs adapted from Navajo blankets, Hopi pottery, Sioux buffalo hides, more. Geometrics, symbolic figures, plant and animal motifs, etc. 128pp. 8⅜ x 11. (Not for sale in the United Kingdom.) 25341-4 Pa. $9.95

SCULPTURE: Principles and Practice, Louis Slobodkin. Step-by-step approach to clay, plaster, metals, stone; classical and modern. 253 drawings, photos. 255pp. 8⅛ x 11.
22960-2 Pa. $11.95

THE INFLUENCE OF SEA POWER UPON HISTORY, 1660–1783, A. T. Mahan. Influential classic of naval history and tactics still used as text in war colleges. First paperback edition. 4 maps. 24 battle plans. 640pp. 5⅜ x 8½.   25509-3 Pa. $14.95

THE STORY OF THE TITANIC AS TOLD BY ITS SURVIVORS, Jack Winocour (ed.). What it was really like. Panic, despair, shocking inefficiency, and a little heroism. More thrilling than any fictional account. 26 illustrations. 320pp. 5⅜ x 8½.
20610-6 Pa. $8.95

FAIRY AND FOLK TALES OF THE IRISH PEASANTRY, William Butler Yeats (ed.). Treasury of 64 tales from the twilight world of Celtic myth and legend: "The Soul Cages," "The Kildare Pooka," "King O'Toole and his Goose," many more. Introduction and Notes by W. B. Yeats. 352pp. 5⅜ x 8½.   26941-8 Pa. $8.95

BUDDHIST MAHAYANA TEXTS, E. B. Cowell and others (eds.). Superb, accurate translations of basic documents in Mahayana Buddhism, highly important in history of religions. The Buddha-karita of Asvaghosha, Larger Sukhavativyuha, more. 448pp. 5⅜ x 8½.   25552-2 Pa. $12.95

ONE TWO THREE . . . INFINITY: Facts and Speculations of Science, George Gamow. Great physicist's fascinating, readable overview of contemporary science: number theory, relativity, fourth dimension, entropy, genes, atomic structure, much more. 128 illustrations. Index. 352pp. 5⅜ x 8½.   25664-2 Pa. $9.95

EXPERIMENTATION AND MEASUREMENT, W. J. Youden. Introductory manual explains laws of measurement in simple terms and offers tips for achieving accuracy and minimizing errors. Mathematics of measurement, use of instruments, experimenting with machines. 1994 edition. Foreword. Preface. Introduction. Epilogue. Selected Readings. Glossary. Index. Tables and figures. 128pp. 5³/₈ x 8¹/₂.
40451-X Pa. $6.95

DALÍ ON MODERN ART: The Cuckolds of Antiquated Modern Art, Salvador Dalí. Influential painter skewers modern art and its practitioners. Outrageous evaluations of Picasso, Cézanne, Turner, more. 15 renderings of paintings discussed. 44 calligraphic decorations by Dalí. 96pp. 5⅜ x 8½. (Available in U.S. only.)   29220-7 Pa. $5.95

ANTIQUE PLAYING CARDS: A Pictorial History, Henry René D'Allemagne. Over 900 elaborate, decorative images from rare playing cards (14th–20th centuries): Bacchus, death, dancing dogs, hunting scenes, royal coats of arms, players cheating, much more. 96pp. 9¼ x 12¼.   29265-7 Pa. $12.95

MAKING FURNITURE MASTERPIECES: 30 Projects with Measured Drawings, Franklin H. Gottshall. Step-by-step instructions, illustrations for constructing handsome, useful pieces, among them a Sheraton desk, Chippendale chair, Spanish desk, Queen Anne table and a William and Mary dressing mirror. 224pp. 8⅛ x 11¼.
29338-6 Pa. $13.95

THE FOSSIL BOOK: A Record of Prehistoric Life, Patricia V. Rich et al. Profusely illustrated definitive guide covers everything from single-celled organisms and dinosaurs to birds and mammals and the interplay between climate and man. Over 1,500 illustrations. 760pp. 7½ x 10⅛.   29371-8 Pa. $29.95

Prices subject to change without notice.

Available at your book dealer or write for free catalog to Dept. GI, Dover Publications, Inc., 31 East 2nd St., Mineola, N.Y. 11501. Dover publishes more than 500 books each year on science, elementary and advanced mathematics, biology, music, art, literary history, social sciences and other areas.